A Practical Guide to Construction of Hydropower Facilities

T0186624

A Practical Guide to Construction of Hydropower Facilities

Suchintya Kumar Sur

CRC Press
Taylor & Francis Group
Boca Raton London New York

CRC Press is an imprint of the
Taylor & Francis Group, an **informa** business

CRC Press
Taylor & Francis Group
6000 Broken Sound Parkway NW, Suite 300
Boca Raton, FL 33487-2742

First issued in paperback 2020

ISBN 13: 978-0-367-67054-2 (pbk)
ISBN 13: 978-0-8153-7805-1 (hbk)

Library of Congress Cataloging-in-Publication Data

Names: Sur, Suchintya Kumar, author.
Title: A practical guide to construction of hydropower facilities / Suchintya Kumar Sur.
Description: First edition. | New York, NY : CRC Press/Taylor & Francis Group, 2019. | Includes bibliographical references and index.
Identifiers: LCCN 2018047800| ISBN 9780815378051 (hardback : acid-free paper)| ISBN 9781351233279 (ebook)
Subjects: LCSH: Hydroelectric power plants--Design and construction. | Dams--Design and construction.
Classification: LCC TK1081 .S858 2019 | DDC 621.31/2134--dc23
LC record available at https://lccn.loc.gov/2018047800

Visit the Taylor & Francis Web site at
http://www.taylorandfrancis.com

and the CRC Press Web site at
http://www.crcpress.com

This book is dedicated to

my little grandson Shivansh Bhardwaj

and to my wife Ms. Bulbul Sur.

Contents

Preface...xxv
Author...xxvii

1 Concepts of a Hydropower Project 1
   1.1 Introduction to Objectives, Scope, and Outcomes 1
   1.2 Outcomes...4
   1.3 Theme and Philosophy 4
   1.4 Concept of Hydropower Project 4
        1.4.1 Role of Hydropower in Energy Mix.....................................5
        1.4.2 Energy Storage and Its Function 6
   1.5 Global Status of Hydropower...7
        1.5.1 Hydropower Installed Capacity by Region 7
        1.5.2 Top 10 Hydropower Producing Countries, 2016 7
        1.5.3 Ten Largest Hydropower Project in the World 7
   1.6 Definition of Hydropower 8
        1.6.1 Merits and Demerits..9
               1.6.1.1 Merits 9
               1.6.1.2 Demerits 10
        1.6.2 Type of Hydropower Projects 10
               1.6.2.1 Large Hydro Project 10
               1.6.2.2 Medium Hydro Project 10
               1.6.2.3 Small Hydro Project 11
               1.6.2.4 Mini Hydro Project...11
               1.6.2.5 Micro Hydro Project...11
               1.6.2.6 Pico Hydro Project 11
   Bibliography 11

2 Planning, Project Cost Estimation, and the Future of Small
  Hydropower (SHP): Large Hydro and Its Various Schemes and
  Components 13
   2.1 Global Definition and Various Rated Capacity of SHP 13
   2.2 Type of Facilities and Its Applicability 13
        2.2.1 Classification of Small Hydro by Head 13
   2.3 How Small Scale Hydro Is Utilized World Wide 14
   2.4 Planning of Small Hydropower (SHP)...16
        2.4.1 Feasibility/Detailed Project Report....................................17
        2.4.2 Scheme of SHP 18
               2.4.2.1 Selection of Type of Project 18
               2.4.2.2 Desilting Prevention..18

		2.4.2.3	Selection of Type of Structure	18
		2.4.2.4	Economic Layout	19
	2.4.3	Type of Scheme		19
	2.4.4	ROR Scheme of SHP and Its Components		19
	2.4.5	Canal Falls Scheme		21
		2.4.5.1	Layout of Canal Falls	21
		2.4.5.2	Component of Canal Falls Scheme	22
	2.4.6	Dam/Weir Based Scheme		22
		2.4.6.1	Component of Dam Toe Scheme	23
		2.4.6.2	Layout of Dam Toe Scheme	23
	2.4.7	Pumped Storage Scheme		23
2.5	Benefits of Small Hydropower (SHP)			24
2.6	Project Cost Estimation and the Future of SHP			24
	2.6.1	Break Up of Activities of SHP		25
		2.6.1.1	Enabling Works	25
		2.6.1.2	Civil Works	25
		2.6.1.3	Electro-Mechanical Work	26
	2.6.2	Preparation of Estimate		26
		2.6.2.1	Preliminary Estimate	26
		2.6.2.2	Detailed Estimate	27
		2.6.2.3	Formula Based Estimate	27
	2.6.3	Detailed Estimate for Evaluating Budget of the Work		27
		2.6.3.1	Direct Costs	27
		2.6.3.2	Indirect Costs	27
		2.6.3.3	Other Costs	27
		2.6.3.4	Preliminary Cost	28
		2.6.3.5	Land Cost	28
		2.6.3.6	Infrastructure Cost	28
		2.6.3.7	Cost of Civil Works	29
		2.6.3.8	Electro Mechanical Work	30
		2.6.3.9	Transmission Line	30
		2.6.3.10	Plantation Cost	31
		2.6.3.11	Maintenance Cost	31
		2.6.3.12	Plants and Erection Equipment	31
		2.6.3.13	Communication	31
		2.6.3.14	Miscellaneous	31
		2.6.3.15	Receipts and Recoveries	32
		2.6.3.16	Indirect Cost	32
		2.6.3.17	Other Cost	32
	2.6.4	Format for Detailed Estimate		33
	2.6.5	Type of Estimates		33
		2.6.5.1	Recasting of Estimate	33
		2.6.5.2	Supplementary Estimate	33
		2.6.5.3	Revised Estimate	33

2.6.6 Economic and Financial Cost Analysis 34
  2.6.6.1 Requisite Tools for Financial Analysis............34
  2.6.6.2 Capital Cost..35
  2.6.6.3 Funding and Its Means 35
2.6.7 Phase-Wise Distribution of Budget and Expenditures 36
  2.6.7.1 Capacity Factor..36
  2.6.7.2 Interest During Construction (IDC)................36
  2.6.7.3 Fiscal Incentives for Small Hydropower 37
  2.6.7.4 Debt to Equity Ratio 37
  2.6.7.5 Working Capital 37
  2.6.7.6 Discount Factors...38
  2.6.7.7 Internal Rate of Return 39
  2.6.7.8 Installation Cost 39
  2.6.7.9 Unit Generation Cost......................................39
  2.6.7.10 Energy Available for Sale................................39
  2.6.7.11 Sale Price of Electricity...................................40
  2.6.7.12 Payback Method..40
  2.6.7.13 Benefit Cost Ratio...40
  2.6.7.14 Debt Service Coverage Ratio 40
  2.6.7.15 Financial Analysis 41
2.6.8 Due Diligence..41
  2.6.8.1 Due Diligence of Financial Feasibility of
    Hydropower Project 42
2.7 Major Hydropower Project 47
 2.7.1 Run-of-River Project...47
  2.7.1.1 Advantage and Disadvantage of
    This Project 47
  2.7.1.2 Ten ROR of the World....................................48
 2.7.2 Storage Project...49
  2.7.2.1 Advantages and Disadvantages 49
 2.7.3 Pumped Storage...50
  2.7.3.1 Advantages 51
  2.7.3.2 World's 10 Pumped Storage Project................51
 2.7.4 Multipurpose Project 51
 2.7.5 Component of Hydel Project...54
 2.7.6 Difference between Dam and Weir................................55
 2.7.7 Classification of Dam, Other Components and Its
  Function 56
 2.7.8 Profile and Various Sections of Dam 58
  2.7.8.1 Non-Overflow Section of Dam 58
  2.7.8.2 Overflow Section of Dam................................58
 2.7.9 Energy Dissipation, Hydraulic Jump, Stilling Basin 58
  2.7.9.1 Hydraulic Jump...61
  2.7.9.2 Energy Dissipation 62
Bibliography 63

3 Concept of Forces Acting on Gravity Dam and Its Stability 65
   3.1   Terminology and Some Definitions Related to Gravity Dam 65
   3.2   Concept of Forces Acting on Gravity Dam 68
   3.3   Requirement of Stability 69
   3.4   Forces Acting on Dam 69
   3.5   Category of Forces to Be Resisted by Dam 71
   3.6   Assumptions for Consideration of Stability 71
   3.7   Load Combination 71
           3.7.1   Dead Load 72
           3.7.2   Reservoir and Tail Water Load 73
           3.7.3   Uplift Pressure 74
           3.7.4   Earthquake Force 76
           3.7.5   Ice Pressure 77
           3.7.6   Wave Pressure 77
           3.7.7   Thermal Load 77
   3.8   Reaction of Foundation 77
   3.9   Resistance against Over Turning 78
   3.10   Sliding Resistance 78
   3.11   Factor of Safety 78
   3.12   Discussion on Earthquake Engineering of Dam 78
           3.12.1   Integrity Check of Concrete Dam 81
   Bibliography 82

4 Contract and Administration Management 83
   4.1   Introduction 83
   4.2   Component of Contract Management 83
           4.2.1   Significance of Contract Management 84
           4.2.2   Challenges in Contract Management 84
           4.2.3   Contract Administration 84
           4.2.4   Service Delivery 85
           4.2.5   Relationship Domain 85
           4.2.6   Financial Analysis 85
   4.3   Function of Contract Management 85
   4.4   Type of Contract 86
           4.4.1   Lump Sum Contract 86
           4.4.2   Drawing and Design by Department 87
           4.4.3   Drawing and Design by Vendor 87
   4.5   Item Rate Contract 87
   4.6   Cost Plus Contract 88
   4.7   Turnkey Contract 88
   4.8   Estimation and Tendering Process of Concrete Gravity Dam 89
           4.8.1   Preparation of Cost Estimate 90
           4.8.2   Bill of Quantities 90
           4.8.3   Schedule of Rates (SOR) 90

4.9   Type of Estimates 91
       4.9.1   Preparation of Report of Detailed Estimate 91
       4.9.2   Technical Sanction of Estimate and Recasting
                of Estimate 91
       4.9.3   Preparation of Notice Inviting Tender (NIT) 91
       4.9.4   Preparation of Tender Document 92
       4.9.5   Publicity of Tender 92
       4.9.6   Tender for Sale 92
       4.9.7   Receipt of Tender 92
       4.9.8   Opening of Tender and Procedure of 2 Bid and
                3 Bid System Tender 92
       4.9.9   Comparative Statement 93
       4.9.10   Acceptance of the Contract 93
                4.9.10.1   Issuance of Letter of Intent 93
                4.9.10.2   Issuance of Work Order 94
                4.9.10.3   Glimpses on General Condition of
                            Contract (GCC) 94
       4.9.11   E-Tendering 97
                4.9.11.1   Different Phases of E-Tendering 97
                4.9.11.2   Work Breakdown Structure in
                            Hierarchical Structure 99
                4.9.11.3   How to Prepare WBS 99
                4.9.11.4   Type of WBS 100
                4.9.11.5   Work Breakdown Structure Diagram
                            and Outline with Work Packages 100
                4.9.11.6   WBS Dictionary 101
       4.9.12   Network Planning 101
                4.9.12.1   Concept of Planning and Scheduling
                            of a Project 102
                4.9.12.2   Components of Network 105
                4.9.12.3   Construction of Network 105
                4.9.12.4   Explanation of Dummy Activity 108
                4.9.12.5   Estimation of Time 109
                4.9.12.6   Probable Completion of the Project 110
                4.9.12.7   Time-Cost Trade Off 110
       4.9.13   Preparation of S-Curve 111
                4.9.13.1   Monitoring Mechanism 112
4.10   Start-Up, Testing, Pre-Commissioning, and Commissioning 113
       4.10.1   Planning for Commissioning 113
       4.10.2   Engineering Diligence 117
       4.10.3   Punch List Categorization 117
       4.10.4   Formation of Commissioning Group 118
       4.10.5   Documents Required 119
       4.10.6   Planning for Start-Up and Commissioning 119

4.11   Start-Up... 119
        4.11.1   Pre-Commissioning 119
        4.11.2   Commissioning.. 119
        4.11.3   Responsibility of Each Group 120
        4.11.4   Start-Up Test 120
                 4.11.4.1   Pre-Commissioning Test................. 121
                 4.11.4.2   Commissioning Test........................ 121
4.12   Liquidation of Punch List 123
4.13   Performance Test... 123
4.14   Project Close Out Report.. 123
Bibliography 124

5 **Site Mobilization and Kick-Off**.. 125
   5.1   Introduction 125
   5.2   Infrastructure Development by Project Authority....................... 126
          5.2.1   Construction of Road Network, Bridges, Culverts,
                   Cause Ways, and Others, in Project Area......................... 126
          5.2.2   Construction of Haul Roads................................ 126
          5.2.3   Service Road 127
          5.2.4   Project Townships and Utility 127
          5.2.5   Other Utility Accommodation and Facility
                   Buildings... 127
          5.2.6   Water Supply 127
          5.2.7   Construction Power.. 127
   5.3   Role of Construction Manager 127
          5.3.1   Phases of Construction Management............................ 127
          5.3.2   Assumptions of Construction Management 128
          5.3.3   Constraints of Construction Management 128
   5.4   Planning for Mobilization and Kick-Off Meeting....................... 128
          5.4.1   Kick Off Meeting 128
          5.4.2   Convener of the Meeting..................................... 128
          5.4.3   Documents to Be Furnished before Site
                   Mobilization 129
          5.4.4   Documents Required by Contractor before Site
                   Mobilization 129
                   5.4.4.1   By Contract Department.................... 129
                   5.4.4.2   By Engineering Department 130
                   5.4.4.3   By QA/QC Department 130
                   5.4.4.4   By HSE Department 130
                   5.4.4.5   By Planning Department..................... 130
                   5.4.4.6   By HR Department.............................. 130
                   5.4.4.7   By Account and Finance 131
                   5.4.4.8   By Construction Department............... 131
   5.5   Site Mobilization 131
          5.5.1   Site Infrastructures Development 132

5.5.2   Mobilization of Machinery and Construction
         Equipment.. 133
5.5.3   Manpower Mobilization 134
5.6   Initial Site Activities.. 134
5.6.1   Surveying of Dam Site 135
5.6.2   Stone Quarry 135
         5.6.2.1   Selection of Stone Quarry............................. 136
         5.6.2.2   Operational Requirement............................... 136
         5.6.2.3   Technical Requirement 136
         5.6.2.4   Alternative Proposal... 137
5.6.3   Batching Plant 137
         5.6.3.1   Location of Batching Plant............................ 137
         5.6.3.2   Selection of Batching Plant........................... 138
         5.6.3.3   Installation System of Batching Plant........... 138
         5.6.3.4   Calibration 139
         5.6.3.5   Routine Maintenance Check 140
5.7   Blasting Operations in Stone Quarry.................................... 141
5.7.1   Blasting Activity 141
5.7.2   Explosives 141
         5.7.2.1   How It Works... 142
         5.7.2.2   Efficiency of Blasting....................................... 142
         5.7.2.3   Creation of Face of Rock................................. 142
         5.7.2.4   Pattern and Spacing of Blast Holes............... 142
         5.7.2.5   Diameter of Blast Hole 143
         5.7.2.6   Bench Height 145
         5.7.2.7   Depth of Hole 145
         5.7.2.8   Burden 145
5.7.3   Type of Explosive.. 145
5.7.4   Design of Explosive Charge 146
5.7.5   Equipment Required for Drilling Blast Holes................ 146
5.7.6   Controlled Blasting... 148
         5.7.6.1   Method of Blasting in Dam Foundation....... 148
5.7.7   Storage of Explosive... 148
5.7.8   Transportation of Explosive 149
5.7.9   Precaution 149
5.7.10 Blast Log.. 150

6 **River Diversion** 151
6.1   Introduction 151
6.2   Diversion Requirement 151
6.3   Method of Diversion 152
6.4   Various River Diversion 152
6.4.1   River Diversion: Open Channel on Edge of the River 152
6.4.2   River Diversion: Diversion Channel and Coffer Dam 153
         6.4.2.1   Construction of Upstream Coffer Dam 154

6.4.2.2 Construction of Downstream Coffer Dam..... 154
6.4.2.3 Construction of Diversion Channel 155
6.4.3 River Diversion: Diversion Tunnel and Coffer Dams 156
6.4.3.1 Construction of Diversion Tunnel 156
6.4.3.2 Construction Methodology 157
6.4.3.3 Planning and Scheduling of Material,
Equipment, Machinery, and Others.............. 158
6.4.3.4 Construction of Upstream Coffer Dam 158
6.4.3.5 Construction of Downstream Coffer Dam 161
6.4.4 River Diversion: Diversion Tunnel/Channel and
Overtopped Coffer Dam....................................... 161
6.4.5 Discussion... 162
Bibliography 163

7 **Reinforcement Cutting and Bending Yard**.......................... 165
7.1 Purpose... 165
7.2 Type of Steel Used in Civil Engineering Work........................... 165
7.2.1 Chemical Properties 166
7.2.2 Mechanical Properties 166
7.2.3 Nominal Size of Rebar 166
7.2.4 Inspection and Various Checking on Rebar 167
7.3 Planning and Setting Up Bar Bending Yard............................... 169
7.3.1 Requirement 169
7.3.2 Preparation of Bar Bending Schedule............................. 169
7.3.3 Format of Bar Bending Schedule................................. 169
7.3.4 Guidelines for Preparation of BBS................................ 170
7.3.4.1 Calculation.. 171
7.3.4.2 Recommended End Hook for All Grades..... 171
7.4 Splicing and Its Types... 174
7.4.1 Conventional Lapping 174
7.4.2 Mechanical Splicing/Bar Coupler................................. 175
7.4.2.1 Standard Tapered Threaded Coupler 175
7.4.2.2 Parallel Threaded Coupler....................... 175
7.4.3 Installation of Coupler 176
7.4.4 Test for Coupler.. 176
7.4.5 Splicing by Welding 176
7.4.5.1 Electrodes.. 176
7.4.6 Welding Method 176
7.4.6.1 Butt Welding... 176
7.4.6.2 Sequence of Welding 177
7.4.6.3 Lap Weld 178
7.5 Setting up Bar Bending Yard... 178
7.5.1 Equipment/Machines Required................................... 179
7.5.2 Guide Lines for Selection of Machines........................... 179

7.5.3 Work as per BBS 180
7.5.4 Advantage of BBS 181
7.5.5 Safety Norms to Be Adopted at Bar Bending Yard 181
7.5.6 Safety Hazard Identification in Bar
Bending Yard 181
7.5.7 Advantage of BBS 182
Bibliography 182

8 **Investigation and Exploration of Dam and Reservoir Site** 183
8.1 Fundamentals 183
8.2 Dam Site Investigation 184
8.3 Geological Exploration 185
8.4 Stages of Geological Exploration 186
8.4.1 Reconnaissance or Pre Feasibility Stage....................... 187
8.4.2 Preliminary Investigation or Feasibility Stage 187
8.4.2.1 Method of Preliminary Investigation 187
8.4.2.2 Sub Surface Mapping 189
8.4.2.3 Feasibility Report 190
8.4.3 Detailed Investigation or DPR Stage............................. 190
8.4.3.1 Detailed Sub Surface Exploration................. 190
8.4.3.2 Geotechnical Property of Foundation
Material 192
8.4.3.3 Detailed Project Report................................... 192
8.4.4 Construction Stage 193
8.4.4.1 Exploration by Pit.. 193
8.4.4.2 Safety Precaution 193
8.4.4.3 Collection of Undisturbed Samples 194
8.4.4.4 Exploration by Trench 194
8.4.4.5 Exploration by Core Drilling........................... 194
8.4.4.6 Geologic Interpretation of Geological
Mapping 194
8.5 Permeability Assessment of Foundation 194
8.6 Definition of Grout and Determination Grout Holes 195
Bibliography 196

9 **Construction Methodology of Dam Foundation and
Technology of Its Foundation Treatment** 197
9.1 Foundation 197
9.2 Layout of Foundation 197
9.3 Identification and Removal of Over Burden 198
9.4 Geological Mapping 199
9.5 Foundation Treatment... 200
9.5.1 Drilling of Holes in Grid Pattern per Geologist............. 200
9.5.2 Safety Compliance during Drilling Operation 202

| | | 9.5.3 | Establish Connectivity between the Holes by Flushing for the Removal of Clay from the Clay Seam Existing within the Rock | 202 |

        9.5.3   Establish Connectivity between the Holes by
                Flushing for the Removal of Clay from the Clay
                Seam Existing within the Rock 202
        9.5.4   Washing Procedure 202
   9.6 Consolidation Grouting in Foundation Area 204
        9.6.1   Grouting Equipment 207
        9.6.2   Role of Ingredients of Grout Mix 208
   9.7 Discussion on Procedure of Grouting, Grouting Material,
        and Equipment 209
        9.7.1   Factors Affecting the Flow of Grout 209
        9.7.2   Grain Size Distribution of Cement 211
        9.7.3   Percolation Test 212
        9.7.4   Equipment Required 213
   9.8 Curtain Grouting 213
        9.8.1   Single Line Curtain 214
        9.8.2   Multiple Line Curtain 214
        9.8.3   Depth of Hole 215
        9.8.4   Drainage System 216
   9.9 Dental Treatment of Fault Zone, Shear Zone 216
        9.9.1   Protection against Piping 217
   9.10 Case Study 217
        9.10.1 Case Study Number 1 218
        9.10.2 Case Study Number 2 220
                9.10.2.1   Effect of This Particular Geologic
                          Condition on Construction 221
   9.11 FAQ on Grouting Grout Pressure, Multistage Grouting,
        Permeability Test, and Others 223
   9.12 Technical Discussion on Karst or Sinkholes 225
        9.12.1 How Is It Formed 225
   9.13 Sink Hole/Karst Topography Is a Matter of Concern for a
        Hydel Project 226
        9.13.1 Caves and Sinkholes and Their Characteristics 227
        9.13.2 Underground Inhabitants 227

10 Concrete and Its Application in Concrete Gravity Dam 229
   10.1 Introduction 229
   10.2 Group of Concrete 229
   10.3 Classification of Concrete with Type of Cement Used 230
   10.4 Classification of Concrete as per Weight 230
   10.5 Grades of Concrete 230
   10.6 Some Special Concrete 230
   10.7 Classification of Concrete Mix 231
   10.8 Parameters That Control the Character of Concrete 232
        10.8.1 Water-Cement Ratio 233

10.8.2 Water Content..233
10.8.3 Cement Content 234
10.8.4 Aggregates...235
    10.8.4.1 Classification of Aggregates...........................235
    10.8.4.2 Grading and Size of Aggregates.....................236
    10.8.4.3 Mechanical Properties of Aggregates...........237
    10.8.4.4 Physical Property 238
10.9 Mix Design and Its Requirement..238
    10.9.1 Advantages of Mix Design..239
    10.9.2 Information Required for Mix Design 239
    10.9.3 Flow Diagram of Concrete Mix 241
    10.9.4 Procedure of Design Mix..242
    10.9.5 Design Procedure of Mass Concrete per ACI Method 246
10.10 Example of Mix Design of Mass Concrete 252
10.11 Batching and Placement of Concrete.....................................255
    10.11.1 Definition of Batching and Mixing 255
    10.11.2 Concrete Production Flow Chart 256
    10.11.3 Batching Plant 257
    10.11.4 Component of Batching Plant and Its Function 257
    10.11.5 Placement, Compaction, and Curing of Dam
        Concrete 258
        10.11.5.1 Methodology of Placement of Dam
                Concrete...260
        10.11.5.2 Compaction of Dam Concrete.......................263
        10.11.5.3 Curing of Dam Concrete...............................265
Bibliography 265

11 **Design Requirement for Temperature Control in Mass Concrete ... 267**
11.1 Concept of Mass Concrete 267
11.2 Significance of Thermal Stresses in Mass Concrete....................267
11.3 Concept of Thermal Cracking..268
11.4 Parameters Initiates in Rise of Temperature...............................269
    11.4.1 Configuration of Dam Section 269
    11.4.2 Cement Composition..269
    11.4.3 Fineness of Cement..269
    11.4.4 Content of Cement...270
    11.4.5 Aggregate Content...270
11.5 Computation of Temperature of Concrete.................................270
    11.5.1 Background Information...271
    11.5.2 The Portland Association Method (PCA)........................273
        11.5.2.1 Schmidt Method...273
        11.5.2.2 Concrete Work Software Package..................274
    11.5.3 Experimental Procedure...274
    11.5.4 Practical Measure...274

11.6   Methodology of Controlling Temperature in Concrete 275
Bibliography 276

12 Joints in Concrete Structures 277
    12.1   Introduction 277
    12.2   Type of Joints 277
           12.2.1   Construction Joint 277
           12.2.2   Type of Construction Joint 278
           12.2.3   Location of Construction Joint in Beam and Slab 279
           12.2.4   Sequence of Pouring 279
           12.2.5   Preparation and Construction of Joint 279
           12.2.6   Construction Joint on Floor Slab on Grade 280
           12.2.7   Construction Method 280
    12.3   Construction Joint in Dam Construction 281
           12.3.1   Preparation of Joints 281
    12.4   Construction Joint in Tunnel Construction 281
           12.4.1   Longitudinal Joint 281
           12.4.2   Transverse Joints 282
    12.5   Construction Joint in Bridge 282
    12.6   Contraction Joints in Concrete Structures 282
           12.6.1   Contraction Joint in Building 283
           12.6.2   Contraction Joint in Slab on Grade 283
           12.6.3   Contraction Joints in Dams 283
           12.6.4   Contraction Joint in Water Retaining Structure 286
    12.7   Expansion Joints 287
           12.7.1   Expansion Joint in Building 287
           12.7.2   Expansion Joints in Bridges 287
    12.8   Cold Joints in Concrete 289
           12.8.1   Causes of Cold Joints 289
           12.8.2   Treatment of Cold Joints 289
           12.8.3   Effect of Cold Joints in Concrete 290
    12.9   Sealant Material 291
           12.9.1   Type of Sealant 291
           12.9.2   Selection of Sealant Material 292
           12.9.3   Various Type of Sealants 292
    12.10 Repair of Cracks and Grouting Procedures 294
           12.10.1 Causes of Cracks 294
           12.10.2 Cracks in Concrete Gravity Dams 295
                    12.10.2.1   Plastic Shrinkage Cracks 295
                    12.10.2.2   Drying Shrinkage Cracks 295
                    12.10.2.3   Crazing 296
                    12.10.2.4   Thermal Cracks 296
           12.10.3 Remedial Measures 297
           12.10.4 Crack due to Reactive Ingredients 297
                    12.10.4.1   Alkali-Silica Reaction 297

12.10.4.2   Precautions..297
12.10.5 Cracks and Deterioration of Concrete due to
            Corrosion 297
12.10.6 Precautionary and Remedial Measures 298
12.10.7 Grouting of Concrete Structures and Repair...................298
12.10.8 Classification of Grouting...................................298
12.10.9 Repairs of Cracks by Epoxy Grouting 298
12.10.10 Repairs of Defective Concrete by Epoxy Concrete 300
12.11 Case Study..300
12.11.1 Repair of Bauxite Secondary Crusher Foundation........300
Bibliography 302

13 Quality Assurance and Quality Control (QA/QC)...............................303
13.1   What Is Quality? 303
13.2   Quality Assurance 303
13.3   Quality Control 303
13.4   Quality Policy 304
13.5   Quality Manual 304
13.6   Job Procedure...304
13.7   Indicative Inspection and Test Plan 310
       13.7.1   ITP for Consolidation Grouting in Foundation
                Area of Gravity Dam..310
       13.7.2   Pour Card...312
13.8   Records Are to Be Maintained at Site 312

14 Health, Safety, and Environment...315
14.1   Introduction 315
14.2   Purpose..315
14.3   Safety Policy..316
14.4   Safety Oath..316
14.5   Action Plan..317
       14.5.1   HSE Meeting 317
       14.5.2   General Safety Meeting 317
       14.5.3   Agenda of the Meeting 318
       14.5.4   Monthly Safety Review Meetings.............................318
       14.5.5   Weekly Safety Meeting at Site Office.......................318
       14.5.6   Tool Box Talk 318
       14.5.7   Induction Training Program 319
       14.5.8   Displaying Safety Slogan in Key Locations..................319
       14.5.9   Safety Bulletins 320
       14.5.10 Safety Workshops 320
       14.5.11 Safety Slogan and Essay Competitions 320
       14.5.12 Competence and Awareness Training..........................320
14.6   Risk and Hazard Assessment 321
       14.6.1   Purpose of Risk Assessment 321

14.6.2   Procedure of Hazard Identification/Risk Assessment 321
            14.6.2.1   Identification of Hazard 321
            14.6.2.2   Assessment of Risk 321
14.7   Controlling Measures 322
        14.7.1   Details of PPE 322
14.8   Hazard Identification and Risk Assessment of Dam Work 323

15 Fabrication and Erection of Steel Structure and Penstock 329
15.1   Fabrication 329
15.2   Material 329
        15.2.1   Section 330
        15.2.2   Physical Property 330
        15.2.3   Classification 330
        15.2.4   Chemical, Mechanical, and Physical Property 331
        15.2.5   Designation 331
15.3   Plans and Drawings 332
        15.3.1   Plans 332
        15.3.2   Shop Drawings/Fabrication Drawings/Cut Sheet 332
        15.3.3   Tolerance 333
15.4   Fabrication Procedure 333
        15.4.1   Preparation 333
        15.4.2   Surface Cleaning 334
        15.4.3   Straightening 334
        15.4.4   Cutting 334
        15.4.5   Punching 335
        15.4.6   Fit Up and Clearance of Fit-Up 335
15.5   Welding Procedure 335
        15.5.1   Welding Procedure Specification (WPS) 336
        15.5.2   Welding Procedure Qualification (PQR) 336
        15.5.3   Welding Qualification Test Procedure (WQPT) 336
        15.5.4   Welder Qualification Test 337
15.6   Shop Fabrication 338
        15.6.1   Setting Out 338
        15.6.2   Surface Cleaning and Painting 338
        15.6.3   Erection 339
15.7   Glimpse on Penstock Fabrication 339
        15.7.1   Magnitude of Loss 340
        15.7.2   Water Hammer and Thickness of Penstock 341
        15.7.3   Pipe Shell 341
            15.7.3.1   Support 341
            15.7.3.2   Joints 341
            15.7.3.3   Material 341
            15.7.3.4   Penstock Accessories 342
            15.7.3.5   Equipments and Machinery Required 342
            15.7.3.6   Fabrication 342

15.7.3.7 Straightening and Levelling the Plates 342
15.7.3.8 Cutting the Plates to Required Dimensions 342
15.7.3.9 Preparation of Edge 343
15.7.3.10 Rolling and Bending 343
15.7.3.11 Welding Procedure 343
15.7.3.12 Inspection of Welding Job 343
15.7.3.13 Hydro Test 344
15.7.3.14 Quality Assurance Plan 344
Bibliography 344

16 **Environmental Assessment of Hydropower Projects** 345
16.1 Introduction 345
16.2 Environmental Impact Assessment (EIA) 346
16.2.1 Environmental Impact Assessment and Strategic Environmental Assessment 346
16.2.2 Aims and Objectives of EIA 347
16.2.3 Scope of EIA 347
16.2.4 General and Specific Principles of EIA Application 347
16.2.5 Some Key Factors of EIA Systems 348
16.3 EIA Process 349
16.4 Roles in EIA Process 350
16.4.1 Project Proponent 351
16.4.2 Project Management Consultant 351
16.4.3 Environmental Consultant 351
16.4.4 Environmental Scientist 351
16.5 Effectiveness of EIA 353
16.6 Benefits of ETA 353
16.7 Cost of ETA 353
16.8 Environmental Impacts due to Hydropower Project Construction 353
16.9 Terms of Reference of Hydropower Projects 355
16.10 Anticipated Impact, Mitigation Measures and Environmental Management Plan 358
16.10.1 Various Environmental Management Plan 358
16.10.1.1 Catchment Area Treatment (CAT) Plan 358
16.10.1.2 Management of Air and Noise Levels 359
16.10.1.3 Water Management 359
16.10.1.4 Public Health Management Plan 359
16.10.1.5 Waste Management Plan 360
16.10.1.6 Muck Disposal Plan 360
16.10.1.7 Fuel Wood Energy Management 360
16.10.1.8 Disaster Management Plan 361
16.10.1.9 Green Belt Development 361

16.10.1.10 Biodiversity Management Plan 361

16.10.1.11 Rehabilitation and Resettlement Plan 362

16.10.1.12 Fishery Development Plan 363

16.11 Application Prior to Environmental Clearance 363

16.11.1 Procedure for Environmental Clearance 363

16.11.2 Indicative Application Form (IAF NO 16.1) 364

16.12 Case Study: Tehri Hydropower Complex in India:
Problems Faced and Lessons Learned 368

Bibliography 369

17   **Estimation and Cost Analysis of Hydropower Project** 371

17.1   Introduction 371

17.2   Process of Capital Cost Estimate of Hydro Project 371

17.2.1   Initial Capital Cost (ICC) 372

17.2.2   Cost Break Down Structure 372

17.2.3   CAPEX Components of Project 372

17.2.4   Methods of Estimate 374

17.2.4.1   Method Number 1 374

17.2.4.2   Method Number 2 374

17.2.4.3   Example of Rate Analysis for Excavation
of Earthwork in All Kind of Soil by
Hydraulic Excavator with Necessary
Manpower Required for the Work 376

17.3   Break Down Cost Percentage 376

17.4   Levelized Cost of Electricity (LCOE) 378

17.5   Determination of Tariff of Hydropower Project 379

Bibliography 382

18   **Hydroelectric Power Houses** 383

18.1   Hydroelectric Power House 383

18.2   Classification of Power House 384

18.3   Function of Power House 384

18.4   What Does a Power House Need to Accomplish These
Activities? 385

18.5   Layout of Power House and Its Dimensions 385

18.5.1   Turbine and Generator Area 385

18.5.2   Erection Bay 386

18.5.3   Service Area 386

18.6   Hydraulic Turbine, Heads, and Its Types 386

18.6.1   Reaction Type of Turbine 387

18.6.1.1   Francis Turbines 388

18.6.1.2   Kaplan Turbine 390

18.6.2   Impulse Type of Turbine 391

18.6.2.1   Pelton Wheel 392

18.6.2.2   Component of Pelton Wheel 393

18.7 Selection of Turbine 394
18.8 Setting of Turbines 396
  18.8.1 Cavitations 397
    18.8.1.1 Control of Cavitations 398
18.9 Selection Procedure of Turbine 398
18.10 Example 401
18.11 Generator and Driving System 404
18.12 Rating and Various Characteristics 405
18.13 Components of Hydrogenerator 406
  18.13.1 Shaft, Couplings, and Alignment 407
  18.13.2 Rotor Assembles 407
    18.13.2.1 Stator Frame 408
  18.13.3 Stator Core 408
  18.13.4 Armature Windings 409
  18.13.5 Bearings 410
    18.13.5.1 Plane Bearing 410
    18.13.5.2 Rolling Element Bearing 411
18.14 Control System 411
  18.14.1 Turbine Controls 412
  18.14.2 Generator Controls 412
  18.14.3 Plant Automatic Control 412
Bibliography 413

Index 415

Preface

The global environmental change is a matter of great concern. The impact of these changes create an imbalance between basic elements of the universe and living creatures. The basic elements of the universe are earth, air, water, light, and ether. The constituents of the environment get disturbed if the percentage of any constituent goes beyond its limit, resulting in a disturbance in the natural balance. Any change in the natural balance causes huge problems to the living creatures of the universe.

Development is the order of the hour in this universe. Development indulges the utilization or overutilization of energy that transforms the environment. If energy required for development is produced through carbon-based technology like oil, gas, or coal, which create havoc in the environment, it will accelerate global warming beyond acceptable limits, and, who knows, one day temperatures may touch the temperature of Death Valley. Effective management of environmental change has become imperative in the last decade. Harnessing alternate energy sources to create a safe and balanced environment and to formulate other diverse plans of survival have gained top priority in high-level discussions on the international level. It has become a concern of all national governments, international organizations, PSUs, MNCs, private sectors, and individuals. Sustainable engineering is an emerging theme in the modern world, and there is need for environmentally friendly power generation systems in the present scenario for a better tomorrow. Renewable energy systems become the forerunner in this new race. Hydroenergy is the most effective tool in this scenario because it connects people, power, and prosperity and it is a cooler solution to a hotter earth.

There is a huge demand for electricity in developing countries. Approximately 12% of their population do not have access to electricity, and policymakers are ensuring their investment in hydropower. They have realized how the investment in hydropower is supporting national development priorities and the clean energy transition. Hydropower is a clean source of electricity as per the latest report from the IHA, which indicated that the world prevented approximately 4 billion tons of greenhouse gases by generating electricity from hydropower last year and avoided about a 10% rise in global emissions from fossil fuels while also successfully avoiding 150 million tons of air polluting particulates, 62 million tons of sulpher dioxide, and 8 million tons of nitrogen oxide from emission. Under the present trend, it is felt that a new category of book with improved course materials is essential to produce engineers suited to the newly metamorphosed power sector.

This book has been penned primarily as a reference book on conception of hydropower engineering with students and professionals in mind, and it is quite self-sufficient. It includes some unusual project activities generally not covered in other textbooks such as the theme and philosophy of hydropower, kick-off meetings and site mobilizations, philosophy of blasting operations per the Explosive Act, QA/QC, health, safety, and environment, environment impact assessments, contract management, and other topics related to the subject. The theoretical and practical aspects of engineering have been seamlessly integrated to ensure that the reader has a combination of engineering and managerial acumen to handle any contingency pertaining to a hydropower project.

This book provides readers with background information and guidance to help them to understand and deal more confidently with issues related to planning, construction, supervision, and analysis of hydropower projects. The author has attempted to address all issues both traditional and innovative in nature.

What motivated and rejuvenated my desire to write this book is my experience and knowledge acquired during construction of the hydropower project in Limestone Valley. This book has been written primarily as a reference book based on facts and figures obtained from that unique project. Design, planning, and construction of the project was a great challenge due to the existence of immense adverse geotechnical problems such as karst/sinkholes, fault/shear zones, and clay seams in the reservoir as well as in the dam foundation area, all of which made the project challenging and fascinating. This book will provide enriching technical and managerial insights to its readers, and will be a unique learning process. In addition to the aforementioned topics, prevalent issues such as corporate social responsibility measures, management of community risks, limiting the risks posed to the environment, and local/state level negotiations encompassing governments and local populations have also been covered. These aspects provide the reader a fundamental perspective on how a technical project can be engulfed in other issues that require astute management skills.

It is a pleasure to acknowledge the considerable help I received from my seniors and colleagues during my acquaintances with them. I convey my thanks to Mr. Sanjay Kumar Sharma and Mr. Kamal Kumar Sharma for their kind help during the process. I wish to express my thanks to the professionals of my publisher for their cooperation throughout the formative period of the book and particularly to Dr. Gagandeep Singh and his colleague Ms. Mouli Sharma for their editorial advice and guidance. I convey my thanks to Bipasha Bhardwaj, my daughter, for her necessary support during the process.

Author

Suchintya Kumar Sur was Visiting Professor at Jaipur National University and former Head of the Department of Civil Engineering at Jaipur Engineering College and Research Centre. He had also served as Vice President (Construction) at M/s Reliance Industries Ltd, Mumbai, and Construction Manager at Engineers India Limited, New Delhi, and SDO in North Eastern Electric Power Corporation Ltd, Shillong. He has expertise in the construction of a major hydel project in Limestone Valley and was instrumental in completion of the world's largest grassroots oil refinery and oil refining hub in Jamnagar, India in record time, which became a landmark of Indian success in the global market. He is equally adept in the handling of mechanical underground and aboveground piping networks with high temperature and pressure, along with erection and commissioning of dynamic, static, and rotary equipment. He has flair in earthquake engineering and design of RCC structures.

1

Concepts of a Hydropower Project

1.1 Introduction to Objectives, Scope, and Outcomes

Hydro is a word of Greek origin that means water. Water is the mother of all necessities and is a vital component of life. Without water, nothing is possible in our planet. It is synonymous with life. Water enabled evolution of life. It is a crucial factor for Earth's evolution and the survival of human grace. Without it, no life is possible on Earth and the ecological balance will be disturbed. Our Mother Earth would become barren and it would also have influence over the universe.

Water is going to be a critical and scarce commodity in the near future, which will be detrimental to development of mankind. Time has come to take serious steps to preserve rain water by creating ponds, lakes, reservoirs, dams, weirs, check dams, barrages and other surface water and subsequently recharge the ground water table so as to keep the Hydrologic Cycle alive in this planet. The Hydrologic Cycle (HC) is a phenomenon by which water moves from place to place above and below the earth and it changes states between liquid, vapor and ice. Water and air, both mediums essential to life, are the most critical source for potential energy. Hydropower engineering taps this huge quantity of energy available in the flowing water on the surface of Earth, which convert this potential energy to mechanical energy and subsequently to electrical energy for the generation of power.

Based on the forecast of scientists, it is indicated that our planet will be warmed up by between 1.0°C and 6.0°C in next 90–100 years, and 25% of World GDP would be spent to deal with it as suggested by economists of world repute. Emission models of the future predict about the amount of carbon emission which will be influenced by population growth, third world development, usage of fossil fuels and the rate at which we switch over to alternative energy, the rate of deforestation, and effectiveness of international agreement to cut down emission. Under the cloud of this global threat, it is essential to understand the history of the theory of global warming and evidence that supports it so as to welcome renewable energies for sustainable development of the universe.

Sustainable development is accepted and defined as "The development that meets the need of present without compromising the ability of future generations to meet their own needs," and it has three basic arms of a magic triangle such as Environment, Economy and Social aspects.

Renewable Energy (RE) is a golden thread which is a nucleus of development of the society. Renewable or Cleaner Energy (CE) is derived from water to meet the basic need of human being.

It retards global warming and carbon emissions which can be treated as a cool solution to a hotter earth.

A consensus concern all over the world is that of the increasing trend for utilization of renewable energy. This clean energy will retard the global warming process. Renewal or clean energy is derived from water, wind, solar and biogas to meet the basic need of human beings. Renewal energy in the present scenario is the nucleus of development of the country. India, a vast country, is having paramount potentiality of 1,45,000 MW of Hydropower that is still unexplored and only 27% of it remained explored. Hydropower came down from 45% to 18% as per reports, and growth has been reduced due to remoteness and inaccessibility to site, lack of infrastructure and land acquisition, environmental and deforestation, acute geotechnical constraint, etc.

Hydropower is synonymous with energy of water and the hydroelectric (hydel) project is a golden avenue to harness this energy for the benefit for mankind. River water is not subjected to market fluctuation. It is the only energy whose efficiency, flexibility and reliability are a matter of great consideration. Hydroelectricity means clean energy today and for tomorrow: an instrument for sustainable development. It is economically viable, environmentally sensible and socially responsible.

Obviously, access to Hydropower is one of the main factors for sustainability, but it has limitations due to geographic situation and geotechnical constraint. It is encouraging that the focus of India is on Hydropower in their 12th 5-year plan for 12.8 GIGA WATT for optimization, which is a morale booster and beneficial to every citizen.

Dam site geology has endless varieties of fascinating problems. It is treacherous as well as mysterious. Dam site geology has a tremendous influence on design and construction of dams.

The civil engineers who work in dam sites should be inquisitive and curious, careful and cautious, alert and astute. They should possess a high curiosity quotient. Their observations should be keen and investigative like detectives, and they should attempt to be more vigilant to formulate the problems in accurate terms to solve them under the light of relevant facts that can be collected by observations and experiments. Their insatiable minds should echo, now and then, some questions such as what and why? And these questions will challenge them to find a clue which may give them possible answers to the unforeseen problems on variable unpredictable sub surface ground conditions; otherwise, it might pose severe constraints in design and construction of safe hydel projects. Any indifference and lack of concern may cause havoc that may lead to a huge national disaster.

Dam and reservoir sites need extensive investigation before design and construction begin. Geological studies are most vital and essential in this process, especially for larger hydel projects. Site investigation involves exploration on and below the surface of the Earth. It is a mandatory and vital requirement for successful and economic design of dam structures. Questionnaires arise in the mind of sensible civil engineers before investigation regarding the perfect location of a dam and its geological features such as foundation condition, type of soil and rock with over burden, and weathering profile available at selected site. Minds of civil engineers suggest to assess carefully whether the type of rocks encountered will be able to support or sustain the weight of the dam. Are there any fault zones or shear zones, buried channels, sink holes or karsts, or clay seams? Are the bed rocks fractured and weathered, vulnerable and susceptible to hazardous seepage that may lead to instability of dam? The geologist sometimes needs to answer very difficult questions, whether a fault exists at site can become active or whether it can be deactivated by impounded water. All these data are to be obtained by investigative methods so as to supply adequate information to the designer in order to design a stable and safe hydraulic structure.

During construction, experienced geologists should be posted in areas like the dam site, low and high pressure tunnel, powerhouse, etc. Geologists work hand in hand with engineers who make geological logs of each working area to give overall ideas of geological features of the particular working site. The subject mapping comprises of the existence of fault, fold, clay seam, dip and strike, etc., along with detailed foundation treatment which will guide site engineers to formulate the methodology of construction and often seek approval/advice of the Resident Geologist for going ahead with the work.

Civil engineering is the oldest engineering in the world. Civil engineering is the mother of all Engineering. The civil engineer, in pursuing his field of endeavor, has two responsibilities: toward the society and toward hydropower … a golden thread that connects people, power and prosperity by designing and constructing stable hydro structures along with safe guarding the life of people using and passing by the structure.

The aim of this book is to present how an opportunity was explored and clean energy was extracted from a difficult natural scenario by accepting the challenge where in renewable energy was derived from the critical base of water through an integrated approach to economy, society and the environment. Opportunity meets preparedness which delivers services to meet the basic needs of human beings that would sustain for long.

This book is a complete package on Concept of Hydropower Engineering and offers unique insight, domain specific knowledge and problem solving domain to the readers (Students, Engineers and Researchers) and enable them to become hydropower engineers in the true sense.

1.2 Outcomes

Readers of this guide will acquire a comprehensive concept of hydropower and its formation through various stages starting from investigation, design, planning, estimation and budgeting, construction and up to generation of power. They will come to know how this book offers guidance on key issues pertaining to development of renewable energy, be aware of its unique characteristics compared to other energy sources, and learn about various hydropower schemes like run off river, storage, pumped storage and from small hydro to mini to micro to pico hydro plant, including their benefits and future in the world market of power generation. The readers will envisage how the hydropower reservoir renders services to mankind in different aspects like local livelihood, economic growth, water quantity management inclusive of drought control, ground water stabilization and ecosystem services inclusive of guaranteed downstream flows both for environment and human—water quality management and reduction of carbon emission.

1.3 Theme and Philosophy

Green power energy is the need of the hour to reduce impact on the environment. Hydropower is favored as renewable and alternative energy which is generated by extracting energy from moving water and depends on rain fall; with more rain fall more power will be generated. Hydropower is a golden thread of high ductility where long development and sustainable growth connects with people, power and progress.

1.4 Concept of Hydropower Project

The eternal Hydrologic Cycle has become vulnerable and susceptible to global warming and the emission of carbon phenomena. Whatever was given by Mother Earth has been stolen by Father Time due to our ignorance, carelessness and indifferent attitude towards life. Under the cloud of this global threat, renewable energy is one of the possible cool solutions to a hotter earth which is the universal need of the hour.

Hydro is a word of Greek origin which means water, and hydropower is the energy contained in water. It is generated by the water in motion and a combination of head and flow. The river water flows from the source at higher elevations to lower elevations by gravitational force, and it loses its potential energy to gain kinetic energy. The difference in water level is called hydrostatic head, which is capable of helping water to flow downstream, and in many projects; the requisite head created by building an impervious barrier across the river is called a dam, which enables the creation of a water reservoir in the upstream side of the river. Hydropower engineering taps this energy of water (potential), which is converted into kinetic energy due to gravity, for subsequently onward transmission to mechanical energy and finally into electrical energy.

This transformation of energy is done through a system known as a Water Conducting System. Stored water is released from the reservoir and flows through an intake and subsequently through a low pressure tunnel to the surge shaft and onward flow through the high pressure tunnel/penstock to rush down below to the power house (underground or surface) to rush through the turbine, which spins the shaft, and a generator is coupled with the turbine through the shaft which generates power. The water continues to flow through the tailrace channel to the river. The generation of electricity is nothing but the conversion of one kind of power to another. The turbine converts water power to rotational power at its shaft which is then converted to electrical power by the generator. It is a fact that some power will be lost through friction during conversion and efficiency is measured by the amount of power that is actually converted.

$$\text{Approximate power developed} = W \cdot Q \cdot H \cdot \eta$$

where
   W = Specific weight of water, N/m^3
   Q = Rate of flow of water, m^3/sec
   H = Hydro-static head, m
   η = Efficiency of conversion of potential energy to kinetic energy

1.4.1 Role of Hydropower in Energy Mix

Our planet contains many sources of energy like oil, natural gas, coal, water, solar, geo thermal, biogas, nuclear, and wind. The combination of these sources of energy is called energy mix. It varies in accordance with the available energy in a country.

Energy mix is of two types:

   a. *Primary energy*: When energy is converted directly from the fuel in a single stage of process, it is known as primary energy. Water is utilized as fuel in hydropower which directly produces electricity, so it falls under primary energy.
   b. *Secondary energy*: When energy is produced in a multi-stage conversion process from fuel, it is known as secondary energy. Fuel like gas, oil or coal is burnt in power plants to heat water to produce steam for generation of thermal power.

Global energy mix was controlled by fossil fuel and it has been a major contributor for global warming and carbon emission. Renewable energy is found to be one of the alternatives to have a cooler earth where hydro—a clean energy, is a possible solution which can dominate the world's energy mix as a cooler solution to hotter earth so as to change the power generation mix in the world. Hydropower is the leading renewable source for generation of electricity in the world. Global Hydropower generation has increased by more than 30% in last few years.

Hydropower is a reliable generation system in energy mix due to its constant and steady flow of water through turbines.

Hydropower plants have the ability to release water at the shortest possible notice so as to meet the immediate high demand of electricity.

Hydropower plants with reservoirs with storage provide, with a high degree of flexibility, the ability to store potential energy for usage to meet demand at any point of time.

1.4.2 Energy Storage and Its Function

Energy storage is the capture of energy during off demand to utilize this energy at the time of peak demand in the future in order to avoid the waste of energy and create a positive effect on the economy.

The energy storage system has become an important tool for the management of energy and provides different technologies to manage the power supply. There are many technologies for storage such as storage in hydrogen, electrochemical storage, thermal storage, and fast energy storage systems which are not the matter of discussion. Our focus shall be on hydro pumped storage. A dam in a hydroelectric project acts as a barrier to create a reservoir for storing water in the upstream in the form of potential energy and can be utilized to generate electricity at the time of need.

Pumped storage is a unique phenomenon for the storage of energy like a battery system. The phenomenon explains that electrical energy, already produced, is converted into potential energy when water is pumped from the lower reservoir to the higher one and stored in the form of water at the time of less demand, and later this water can be released from the higher reservoir through a pipe or any suitable tunnel to flow through a turbine to convert into electrical energy.

This plant consists of two water reservoirs at different levels. Water is pumped from the lower reservoir to the higher one with a force of magnitude equivalent to $F = mg$ against gravity when demand of electricity is low, resulting to work done ($W = F \times h = m \times g \times h$) in the process. Ability to do the work is energy, so potential energy equivalent to $m \times g \times h$ is stored at a reservoir of a higher level. The stored water is released when the supply of water is less than the demand of electricity. The water travels from the higher reservoir through pipes to flow through the turbine in order to generate electricity and store the water in the lower reservoir instead of releasing this water back to the river to complete the cycle. The potential energy becomes zero when water touches the ground and kinetic energy is generated before converting into electrical energy. Refer to Section 2.7.3 for details of scheme and Figure 2.1.

1.5 Global Status of Hydropower

1.5.1 Hydropower Installed Capacity by Region

The world around us has changed and people now understand the effect of environmental change in daily life. All around efforts are being made from all sectors to protect the environment, which has led to a focus on the development of hydropower in the world.

There has been significant growth in the development of hydropower globally as per the report of the World Energy Council, 2016. Hydropower has become the leading renewable source for electricity generation in the world. It supplied 71% of all renewable electricity at the end of 2015. Undeveloped worldwide potential is approximately 10,000 TWh/y. The world hydropower installed capacity has increased by 39% from 2005 to 2015, with an average growth of 4% per year, accounting to a total of 1209 GW in 2015, of which 145 GW in pumped storage (Table 1.1).

1.5.2 Top 10 Hydropower Producing Countries, 2016

Hydropower is being generated in the most of the countries in the world. China, as per report, is the largest hydropower generating country in the world. The 10 top hydropower generating countries are indicated in Table 1.2.

1.5.3 Ten Largest Hydropower Project in the World

Ten largest hydropower projects of the world are tabulated in Table 1.3.

TABLE 1.1

Hydropower Installed Capacity by Region, 2017

Sl No.	Region	Installed Capacity Including Pumped Storage (GW)	Pumped Storage (GW)	Generation (TWh)
1	Africa	35.33	3.376	131
2	East Asia and Pacific	468.3	66.45	1501
3	Europe	248.56	51.76	599
4	North and Central America	203.05	22.966	783
5	South America	166.959	1.004	716
6	South and Central Asia	144.71	7.451	456
	Total	1266.95	153.04	4185

Source: Adopted from hydropower status report, IHA 2018.

TABLE 1.2

Top 10 Hydropower Producing Countries, 2016

Sl No.	Country	Installed Capacity including Pumped Storage (GW)	Installed Capacity Pumped Storage (GW)
1	China	319.37	23.00
2	United States of America	101.755	22.44
3	Brazil	91.65	0.003
4	Canada	79.202	0.177
5	India	47.05	4.78
6	Japan	49.05	27.63
7	Russia	50.62	1.38
8	Norway	30.56	1.35
9	Turkey	25.88	Nil
10	France	25.397	6.98

Source: Data collected from World Energy Council, 2016.

TABLE 1.3

Largest Hydropower Project in the World

Sl No.	Name of Project	Country	River	Installed Capacity (MW)
1	Three Gorges Dam	China	Yangtze	22,500
2	Itaipu Dam	Brazil and Paraguay	Parana	14,000
3	Xiluodu	China	Jinsha	13,800
4	Guri	Venezuela	Caroni	10,235
5	Tucurui	Brazil	Tocantins	8370
6	Grand Coulees	USA	Columbia	6809
7	Xiangjiala	China	Jinsha	6448
8	Longlan	China	Hongshui	6426
9	Sayano	Russia	Yenisei	6400
10	Krasnoyarsk	Russia	Yenisei	6000

1.6 Definition of Hydropower

The hydropower project is a project where energy of water (potential) is used, controlled and converted into kinetic energy due to gravity for subsequent onward transmission to mechanical energy, and finally, into electrical energy by the help of a hydro turbine connected to a generator. This transformation of energy is done through a system known as a water conducting system, which is comprised of many structures such as a dam, dyke, reservoir, intake, low pressure tunnel, surge shaft/fore bay, high pressure tunnel/penstock, power house, and tailrace tunnel.

Hydropower is a versatile, mature and flexible energy source. A hydropower plant is unique and site-specific, which solely depends on the geotechnical, topo-graphical and hydrological features of each site and accordingly, the type of facility to be built is decided.

Hydropower plants are generally very diverse. Some plants are made with dams to create the reservoir and head, some plants are made with run off rivers, and sometimes on small rivers or on tail race channels, and even on small streams or springs.

Before going onto details of the hydropower project, let us focus on merits and demerits of the hydropower project.

1.6.1 Merits and Demerits

Hydropower is a renewable energy where potential and kinetic energy of water is utilized to generate electricity through turbines. Hydropower offers both merits and demerits as mentioned below.

1.6.1.1 Merits

- *No fuel cost*: Hydropower does not require any fuel, so there is no effect of enhancement of cost of fuel as it happens now and then in case of other energies
- Efficient, flexible and reliable because river water is not subjected to market fluctuation
- Sustainable development tool
- Economically viable as cost of production of electricity is low; payback time within 8–10 years, having long span of life of up to 100 years
- Environmentally sensible as hydropower is a clean energy and reduces emission of greenhouse gases
- Energy storage is a unique phenomena of pumped hydro storage, where energy is stored for the utilization at the time of need
- Any size hydropower project is possible such as pico, micro, mini, small, medium and large hydropower plants as per the need
- High load factor varies from 40% to 60% which is higher than other energies
- *Long life*: Varies from 50 to 100 years
- Socially responsible
- Controls flooding
- Facilitates irrigation systems
- Provides drinking water for human consumption as well as for industrial use

- Creates recreation facilities, fisheries and navigation
- Recharges ground water tables and makes the land suitable for cultivation
- Provides employment in rural areas

1.6.1.2 Demerits

- Long gestation time
- Displacement and loss of livelihood of locals and rehabilitation is required
- Wildlife is affected
- Heavy deforestation, especially in reservoir area
- Siltation
- Dam failures
- Earthquake vulnerability as construction of large dam increases the possibility of earthquake
- Geotechnical constraints in dam foundation and in reservoir area for water tightness

1.6.2 Type of Hydropower Projects

Hydropower plants are classified based on their installed capacity in MW and it is also related to the head. Hydropower facilities can be categorized in the following types of facility.

1.6.2.1 Large Hydro Project

Power is generated through either run-off-river or storage type in this category of hydro project, and is connected to the national grid of the country. The national grid is the high voltage electric power transmission network in the mainland which transmits power safely to the users through the main power station and sub stations. The installed capacity of this type of project is greater than 100 MW and feeds into a large electricity grid.

1.6.2.2 Medium Hydro Project

Power is also generated through either run-off-river or storage type in this category of hydro project and is connected to the national grid of the country. The national grid is the high voltage electric power transmission network in the mainland which transmits power safely to the users through the main power station and sub stations. The rated capacity of this type of project is between 20 and 100 MW and always feeds into a grid.

1.6.2.3 Small Hydro Project

A small hydro project is mainly based on run-off-river where the size of the dam is very small, and even in some cases, where a weir serves the purpose. It does not need any storage of water, resulting in no adverse environmental effect on locals. It is very effective to have access to rural area of developing countries. The rated capacity of small hydro projects is between 1 and 20 MW and usually feeds into a grid.

1.6.2.4 Mini Hydro Project

The rated capacity of this type of hydro project varies between 100 KW and 1 MW. The construction of this type of project is very convenient as it requires a small area and less time than other projects. It is constructed on a small flowing river, on a canal drop or on a tailrace channel. It is a standalone system but sometimes it is connected to an available mini grid or national grid, if any.

1.6.2.5 Micro Hydro Project

The rated capacity of micro hydro varies between 5 and 100 KW. It is generally constructed on a small stream or a small stream with some difference in elevation to have a sufficient flow of water. It is not connected to the national grid and it is only used in rural electrification or a small community.

1.6.2.6 Pico Hydro Project

The output of electricity of this type of hydro project is up to 5 KW. It is most suitable for a hilly area and also at rural areas. This system can use water from a source a few meters above ground level to facilitate the flow of water to pass through a small scale turbine generator for generation of power for domestic use.

Bibliography

Hydro Electric Power. A Guide for Developers and Investors. International Finance Corporation (World Bank Group).
Hydro Power Status Report 2018. International Hydro Power Association (IHA).
World Energy Resources, 2016. World Energy Council.
World Small Hydro Development Report 2016. United Nations Industrial Development Organization, Vienna, and International Center on Small Hydro Power, Hangzhou.

2

Planning, Project Cost Estimation, and the Future of Small Hydropower (SHP): Large Hydro and Its Various Schemes and Components

2.1 Global Definition and Various Rated Capacity of SHP

There is no global consensus on a definition of Small Hydropower. The definition differs from country to country and continent to continent. As per the European Small Hydro Association, the UK favors 5 KW, France likes 8 MW, and Italy settles for 3 MW. ESHA, the European Commission and UNIPEDE agree upon 10 MW or less.

Hydropower with installed capacity not more than 50 MW is considered as SHP category in China, whereas India favors up to 25 KW.

Here is the under rated power of SHP as tabulated below for different countries in the world (Table 2.1).

2.2 Type of Facilities and Its Applicability

The type of hydropower facility is classified as tabulated in Table 2.2.

2.2.1 Classification of Small Hydro by Head

Hydraulic head and specific speed are the key factors for the selection of turbine for hydro projects of any magnitude so some reference model shall be fixed for an easy solution but other parameters shall be considered as per site prevailing condition/site specific (Table 2.3).

TABLE 2.1

Rated Power of SHP of Different Countries

Sl No.	Country	Installed Capacity in MW
1	Italy	≤3 MW
2	Domenian Republic, Guatemela, Macidonia	≤5 KW
3	Mauritius	≤0.05 MW
4	Morocco	≤MW
5	Afghanistan, Burundi, Iran, Malaysia, Mali, Nepal, Norway, Sri Lanka, Tunisia, Kenya, Uganda, Zambia, Madagascar, Armenia, Austria, Croatia, Montenegro, Nigeria, Turkey, Serbia, Slovenia, Switzerland, Senegal, Azerbaijan, Cambodia, Philippines, Indonesia	≤10 MW
6	Georgia	≤13 KW
7	Bangladesh, Laos, Lesotho, Thailand	≤15 MW
8	El Salvador, Peru	≤20 MW
9	Bhutan, India, Mozambique	≤25 MW
10	Argentina, Brazil, Mexico, Benin, United States	≤30 MW
	Canada, China, Pakistan, New Zealand	≤50 MW

Source: Adopted from AHEC-IITR, "1.1-General: Small Hydropower Definitions…Ministry of New and Renewable Energy," Roorkee, September 2013.

TABLE 2.2

Type of Hydropower Facility

Sl No.	Type of Facility	Rated Power	System of Distribution	Applicability
1	Large hydro	>100 MW	National grid	Large population centre
2	Medium hydro	25–100 MW	National grid	Medium population centre
3	Small hydro	2–25 MW	National grid/ Regional grid or Stand alone	Small community and expected to connect regional or local grid
4	Mini hydro	100 KW to 2 MW	Do	Isolated communities or small factories
5	Micro hydro	5–100 KW	Rural electrification	Small isolated locals
6	Pico hydro	<5 KW	Do	3–4 houses

Source: Adopted partially from AHEC-IITR, "1.1-General: Small Hydropower Definitions… Ministry of New and Renewable Energy," Roorkee, September 2013.

2.3 How Small Scale Hydro Is Utilized World Wide

Today's world is on the verge of drastic change … a change in the low carbon era. There is an all-around effort in advocating ideas of green development of small hydropower. Environmental changes, energy

TABLE 2.3

Type of Turbine to be Selected by Head

Sl No.	Head % of Designed Head		Type 1 of Turbine to be Assessed
	Max	**Min**	
1	125	65	Axial flow, Kaplan, Propeller, Cross flow, Francis
2	110	90	Kaplan, Propeller, Francis
3	125	60	Kaplan, Turgo, Pelton

Source: Adopted from AHEC-IITR, "3.1 Electro-Mechanical–Selection of Turbine and Governing System," standard/manual/guideline with support from Ministry of New and Renewable Energy, Roorkee, June 2012.

security, and volatile fuel prices are major challenges that pose a great threat to industrial and economic development. World leaders and strategic thinkers have already decided to shift from traditional energy sources to renewable ones. It is clear that access to economically viable, environmentally sensible, socially responsible and reliable energy brings changes in industrial competitiveness while creating employment and equilibrium in the society.

It is proven that small hydropower is a befitting cooler solution to a hotter Earth. It can be used as an effective means of electrification of rural areas in the world. Hydropower is considered to be a mature technology. It can be designed, constructed, and maintained locally as it is a flexible, affordable, sensible and reliable energy source which is not subjected to the market fluctuation like other sources of energy.

Today, about 17% of the world's population is still lacking access to electricity. The United Nations has recognized that clean energy and electricity is the key to development.

The total installed capacity of small hydropower is 78 GW in 2016. It has increased about 45 GW in data from a 2013 report furnished by the United Nations Industrial Development Organization (UNIDO). Now the total estimated potential has increased to 217 GW.

SHP represents about 1.9% of world's power capacity. Seven percent of total renewable energy capacity and 6.5% (<10 MW) of total hydropower capacity. Small hydropower is fifth in development.

Despite many benefits, the potential of small hydropower remains untapped in many developing countries.

UNIDO has taken up the campaign to make awareness amongst the leaders of countries about the use of renewable energy like small hydropower in industries and small enterprises. It is needless to say that when it is supported by the environmental protection policies and concrete supervision from regulatory authority, small hydropower can be considered as the most important renewable energy technology.

2.4 Planning of Small Hydropower (SHP)

Planning of small hydropower depends on the site condition of a respective project because each site is unique. The character of the site depends on topography, geological and hydrological feature. A hydel project, large or small, needs careful study and extensive investigation before planning for any project. The conceptual footprint shall be made to take up extensive investigation and exploration of the site in order to examine all collected data for implementation. Adequate exploration of geological and geotechnical characteristics of a proposed site is one of the vital aspects of hydraulic structure safety evaluation.

The concept is put on the drawing board and a conceptual drawing is prepared so that the project authority can go for a *reconnaissance survey* of the site which includes access to the site, along with the basic concept of the site topography, hydrology such as flow and discharge of river, available head, geologic features of the site, social and environmental aspects.

A hydropower project is generally constructed in a remote area where access to the site is always a problem. If proper infrastructure along with in-plant road network is not in place from the ZERO date of the project, the project starts slipping the schedule from the zero date, so the reconnaissance survey group should emphasize this aspect to make the project more viable from the point of view of constructability.

The second phase of investigation is the *prefeasibility study* of the proposed site, where further exploration activities shall be taken up based on data collected in the reconnaissance survey. This feasibility study will educate about the first hand idea of engineering and geological as well as hydrological characteristics of the proposed site. These studies shall be interpreted in conjunction with all data being collected in the reconnaissance survey so as to establish the technical feasibility of the project.

Once the proposed project acquires the status of a technically feasible project after the prefeasibility study, the project authority gives the green signal to the project group to go ahead with the detailed *investigation and survey* of the site. It includes the following investigative activities.

- *Topographical survey:* The survey group should move to the site with topographical map made by Survey of India in the prescribed scale for the entire country which will provide the group a fairly good idea about the site topography. The group undertakes further requirement-specific surveys like a contour survey of the area, detailed route survey, layout survey of water conducting system, or power house.

- *Geological survey:* Surface and sub surface investigation including sub soil foundation investigation in the power house, water conducting systems and other important structures and ground water tables.

- *Geotechnical survey*: This survey is conducted to obtain data on physical characteristics of the soil and rock of water conducting system and power house and the stability of slope for protection for safe design of the structures.
- *Hydrological survey*: This is conducted to assess *in situ* flow determination, runoff generation, discharge and duration of flow and sediment analysis also to assess requisite head.
- *Load survey*: Load survey shall be conducted within a 4–10 km radius from the location of SHP. Data is to be collected by interacting with common villagers, teachers, doctors of health centers, and Gram Pradhan. Data should include number of villages and its houses, population and its projected connections and average consumption, number of schools, health centers, community centers, small scale industry and their demand.
- *Socio-economic survey*: The socio-economic survey shall be conducted in the influence area of the project with the latest census report which will provide details of population and sex ratios. The surveyor should cover various requirements of surveys such as community, demography, settlement pattern and housing structures, source of livelihood and income, source of water, health and sanitation, and physical infrastructure.
- *Environmental and ecological survey*: EES shall be conducted broadly on the land and land cover along with the number of displaced persons, hydro-meteorology, floristic and forest type, flora and fauna. This includes identification and prediction of impact, impact evolution and mitigation systems.
- *Survey for availability of construction material*: A survey shall be conducted for construction material like stones and sand by conducting various field and laboratory tests for the quality of materials confirming to project specification.
- Survey for nearest railways heads or highways to facilitate transportation of construction materials, equipment, and work force. The survey shall be conducted to locate nearby railheads/ports to carrying construction materials and equipment smoothly to the project area. A detailed survey shall also be made inside the road network to facilitate transportation of materials and equipment to the respective location.

2.4.1 Feasibility/Detailed Project Report

A feasibility/detailed project report is made based on data collected from the detailed investigation to assess technical feasibility and financial viability. This report is ultimately used as the main document to get approval from various authorities for implementation.

2.4.2 Scheme of SHP

Here are some critical features are to be considered during planning of the project. The type of scheme is to be selected to suit the site condition after analyzing collected data from the investigation and detailed survey.

2.4.2.1 Selection of Type of Project

Small hydropower projects are found to be suitable for the following schemes:

- Run-of-River (ROR)
- Canal Falls
- Dam Toe
- In-stream

2.4.2.2 Desilting Prevention

It is very essential for a ROR type of project to be taken up in the hilly region where rivers carry huge amounts of silt during the monsoon period. It causes erosion to the blade of the turbine and other parts of underwater equipment if sediments are allowed to pass through to the power house. It is experimented that erosion is directly proportional to the shape, size, concentration and hardness of sediment. During planning, sediment's characteristics should be checked by relevant laboratory tests.

2.4.2.3 Selection of Type of Structure

Type of structures shall be selected in such a way that construction materials available in close proximity to the site can be used in the construction and also that labor is available locally to make the project economically viable. Stone masonry work shall be used in the construction of diversion work in place of reinforced concrete as much as possible.

Usage of pipes or channels or tunnels shall be chosen depending on the site topography, and channels can be made of stone masonry and the inside of it can be lined either with shortcrete or concrete to prevent leakage. If the length of the channel is found to be longer in a terrain where the channel may encounter slips, landslides or drainage crossings result in the disruption of services during operation, so it should be wise to adopt the tunnel with free flow or under little pressure. On many occasions, a pressure tunnel is chosen over an open channel to ease the operation of the plant. A long pressure tunnel needs to have a surge shaft at the end, but a free open channel needs a fore bay at the end.

2.4.2.4 Economic Layout

The following river characteristics are to be considered for the economic layout of a ROR SHP project. Flow of water is the key to success of this sort of scheme.

- Steep slope of river to have desired flow of water
- Rapids and water falls
- Canyons and narrow valleys
- Major river bend to accelerate the speed of flow

2.4.3 Type of Scheme

The most common small hydropower projects are as follows:

- Run-of-River
- Canal Falls
- Dam based outlet and spillway
- Storage
- Pumped in-stream

2.4.4 ROR Scheme of SHP and Its Components

A small hydro project is mainly based on ROR where the size of a dam is very small, and even in some cases a weir serves the purpose of diverting water. It does not need any water storage and depends on the flow of water, resulting in no adverse environmental effects on locals. A small hydro plant consists of the following components:

- *Diversion dam/weir*: It is constructed in the river to divert water from the main flow through an intake structure. A diversion dam is constructed either with concrete or stone masonry or rock fill as per design requirements.
- *Intake structure*: Flowing water, required for generation of power, is tapped by a structure from river or canal or reservoir is known as an intake structure. It may be low head or high head and it is constructed either parallel or perpendicular to the river flow where spiral flow is strong. Raised crest or trench intake is suitable for low head diversion. It is suitable for a narrow river with flow 5 m^3/sec and above and trench intakes shall be suitable for a steep river in 1 in 20 with flow 10^{-15} m^3/sec. Its function is to ensure requisite quantity of water intake and to draw quality water from the river by segregating any floating materials, debris, vegetation, or logs and to provide smooth entry so as to ensure minimum disturbance. The intake structure consists of appurtenances like a trash rack, curtain wall, settling trap, and gate.

- *Tunnel or open channel*: It is used to carry water from the intake structure to the sedimentation tank and an onward journey to the fore bay or surge shaft. It follows the contour of terrain as per requirements to have a sufficient slope so that requisite flow of water is maintained.

- *Sedimentation tank*: The sedimentation tank is used to trap sand, silt or any suspended material from the water so as to avoid any damage being caused to the turbine. It should conform to the specifications of the turbine.

- *Fore bay*: A pond which regulates the water head before the start of the penstock. It also stores a daily or weekly load to ensure the demand of electricity. It also facilitates the gentle flow condition of water. The layout of the fore bay is governed by the topographical and geological condition of the site. The location of the fore bay and power house should be selected so that the penstock has minimum length. It provides small storage for a few minutes to supply water to the turbine. It also serves as the final sedimentation tank where floating debris is either passed through an intake or swept into the channel to be removed before entering the turbine area. In many occasions, as per necessity, a balancing reservoir is provided in addition to the forebay or in place of the fore bay with larger storage for a few hours for diurnal basis.

It is provided for meeting the demand where water in the lean period is not sufficient for the turbine.

An escape channel is provided in the fore bay to discharge any excess water being generated in case of the stoppage of the turbine due to load or grid failure. This excess water is discharged to the river or in a nearby open channel.

In a ROR scheme the surge shaft is provided, when the head is high to absorb the water hammer coming from the penstock.

- *Penstock*: A penstock is a closed conduit or pressurized pipeline made of steel is used to carry water from the fore bay to the turbine installed in the power house. There are different types of penstock used in the hydropower project such as buried penstock or exposed penstock, depending on the site condition. In some cases the powerhouse penstock is taken through the body of the dam where the power house is situated in close proximity to the toe of dam.

- *Power house*: The power house may be of two types: surface or underground, where power is generated by the help of the turbine and generator. Forces being generated by the flow of water causes the turbine to spin and the turbine converts the kinetic energy of the moving water into mechanical energy. A generator is connected to the turbine through a shaft and gear and the spinning turbine also causes the generator to spin and convert mechanical energy into electrical energy.

Layout of the power house depends on the type of turbine selected as per technical requirements.

- *Tail race tunnel*: Tail race is a passage through which water is pumped out of the power house and put back into the river after generation of electricity.
- Access road.
- Operator's residence (optional).

2.4.5 Canal Falls Scheme

Canal Falls Schemes are planned along a canal having head from 0.5 to 15 m with dependable flow. In an agrarian country, cultivating the land is the prime source of wealth where large no of irrigation canal will be available with several falls. These type of falls are also available in a cooling water return channel in thermal plants, drinking water supply system and sewage channel after treatment. These schemes can be planned easily due to their assured flow of water and these locations are easily accessible and plain.

These type of projects shall be planned considering following constraints:

- Power house shall be planned without interfering irrigation and drinking system
- Hydropower scheme shall not affect the safety of existing structures

A Canal Falls Scheme consists of the following:

- A bypass canal with or without a regulating system
- Compact intake cumulative power house
- Tail race channel joining canal down stream

2.4.5.1 Layout of Canal Falls

Layout shall be planned considering the prevailing foundation condition and access to the site, and it should be planned on which side of the canal the power house and diversion channel shall be located. Hydraulic design shall be done in such a way that head loss is minimized.

There are varieties of Canal Falls Schemes. These variations depend on the location of falls and type of falls whether new or old.

- *Old canal*: If the fall exists at some distance from canal head the power house shall be planned and located on a new bypass channel.
- *New channel under construction*: The power house can be planned on a canal bed and if the situation demands, canal width may be widened to accommodate the power house and bypass arrangement.
- In some cases, falls or drops may be available at a distance from the head regulator of the branch canal where in such condition a link

bypass channel shall be planned and be connected to branch the channel with the main channel. The power house shall be planned to be located on the link channel.

- All this planning shall conform to the requirements for topography of the canal side, availability of land, proper access to the proposed site and evacuation of power arrangement.

2.4.5.2 Component of Canal Falls Scheme

Canal Falls Scheme consists of many components that are described below.

- *Intake structures*: Intake structures shall be fitted with a trash rack.
- *Water conducting system*: No conducting system is required as water *is directly taken to turbine.*
- *Power house*: Power house in this scheme is generally planned as a surface power house. Location of the power house shall be planned considering the constructability of the power house. It should be kept in mind that there will be huge difficulty in excavation work due to close proximity of canal where ingress/seepage of canal water will make the foundation soil unstable, the development of uplift pressure, and a reduction in frictional force.
- Safety assessment of existing structures shall be adhered to without compromise as will interaction with irrigation for smooth operation.

2.4.6 Dam/Weir Based Scheme

A Dam/Weir is constructed across a river to create a reservoir from which water is utilized for irrigation, water supply for human consumption and controlling the flood with no power generation.

Power can be generated at the time of releasing water for the purpose of irrigation or supply of drinking water. Under this circumstances, a Dam Toe power house can be planned for generation of power with the existing system where no power generation was planned earlier.

For the planning of SHP in an existing facility the following data and information are to be gathered for effective planning of the scheme:

- Assessment on discharge and head obtained from available record from archive.
- Use of consumptive water.
- Long term river flow data.

Alternative study to be carried out for effective planning of the scheme:

- Provision of additional storage in monsoon by raising the height of the dam/weir.

- Possibility of usage of pilled water for generation of power may be examined by raising the height of the gate or replacement by a new one keeping in view of safety of the structures.
- Head for generation of power depends on location of power house which is to be located in the toe of the dam so head may be examined.

After examining all this data, planning of this scheme shall be taken up and installed capacity is to be calculated.

2.4.6.1 Component of Dam Toe Scheme

A Dam Toe scheme consists of the following components:

- Intake structure
- Short penstock including bypass
- Power house
- Tailrace

2.4.6.2 Layout of Dam Toe Scheme

Layout shall be planned taking the practical aspects of the site into consideration for optimal management.

The power house can be located at the toe of the dam if both topography and geology of the site are found to be suitable for construction. This location shall conform to the existing outlet from where water shall be supplied to the power house. The distance between the power house and dam toe shall be sufficient enough so that excavation of the foundation of the power house should not be affected during construction.

In the case of a concrete dam having no outlet, an intake structure shall be planned away from the dam and a channel shall be planned and laid through one of the abutments to supply water to the power house.

In the case of a composite dam made of concrete and earth in the flanks a new intake shall be through one of the flanks at suitable height.

In the case of a barrage the power house can be located away from the barrage in the downstream. The possibility of usage of an existing intake for the channel shall be examined. Since the location of the power house is away from the barrage, a new channel shall be made from existing channel by providing a head for transferring water to power house as an additional feed.

2.4.7 Pumped Storage Scheme

Pump storage is a scheme where water works between two reservoirs at different levels. Water is stored in the upper reservoir and is released from there when power is needed. Water flows through the discharge inlet and power is generated from the turbine coupled with generator; water then flows through the discharge outlet to lower reservoir. Water is pumped back to higher level reservoir when power is again needed. It is like recharging of battery system and supports operational flexibility. Please refer to Section 2.7.3 for more details pertaining to the scheme.

2.5 Benefits of Small Hydropower (SHP)

Hydropower is a globally accepted source of clean and green energy. Small hydropower consists of the following unique benefits in its favor to capture the global market as an energy source:

- Provides mature technology
- Inducts minimal impact on environment
- Economically efficient
- Solution to rural electrification
- Generates sustainable industrial development
- Can be installed and run concurrently with a larger system of dams and reservoirs constructed for irrigation, drinking and industrial water supply
- Turbine spins oxygen into water
- Segregates recreational and other wastes from water
- No fuel is required to run
- Provides predictable and reliable energy
- Life expectancy is much more than other energies
- Does not indulge greenhouse gas emission
- Does not obstruct fish migration
- Provides stability to power grid
- Promotes employment to locals
- Generates a high energy payback ratio

2.6 Project Cost Estimation and the Future of SHP

There is a global consensus of trends to exploit maximum possible resources for the development of small hydropower. China is the leader in this field and has taken an initiative to promote and sell the idea of SHP technology by giving incentives for base construction, and have established the idea of a public private partnership, thereby becoming the leader in this field.

To become a leader in any front, technical and economic viability of the project should be clearly understood by computing a project cost estimate based on the investigation, planning and feasibility report. It is essential to consider all necessary items that are taken into consideration and that none of these items are scantly estimated at the time of preparation of estimation. An exhaustive list of activities is to be made in relation to the specific site for making the estimate and it may not match fully with the other site due to the uniqueness of each site, and the degree of detailed activities should be estimated depending on the magnitude of the project which can be taken as guidelines for SHP.

Small hydropower is generally made under the ROR scheme where water is diverted from a river or stream through a water conducting system to a turbine installed in a power house and finally discharges the utilized water back into the river. No reservoir is created in this scheme and generation of power depends on the flow of water.

2.6.1 Break Up of Activities of SHP

Activities involved in SHP can be divided in to three parts:

- Enabling works
- Civil works
- Electro-mechanical works

2.6.1.1 Enabling Works

It covers activity of site preparation and development such as creation of infrastructure like road and site establishment.

2.6.1.2 Civil Works

Major activities that are required to complete civil works of SHP are narrated below:

- Diversion/weir/intake
- Open channel from intake to sedimentation tank and onward journey to fore bay
- Sedimentation tank
- Any bypass arrangement
- Penstock and its anchorage system
- Power house and equipment foundation
- Tailrace channel
- Switch yard
- Mise work like roads, etc.

There will be some change in the combination of activities with respect to specific site and type of scheme adopted.

2.6.1.3 Electro-Mechanical Work

Electro-mechanical work of SHP is made up of the following components:

- Turbine
- Generator
- Transformer
- Switch yard equipment and station accessories including mechanical and electrical auxiliaries
- Hydro-mechanical equipment such as gates, valves, trash racks, etc.

2.6.2 Preparation of Estimate

Estimation of a project has a two-fold requirement—one is to calculate the estimated cost, and the second one is to find out the time line of the project, which means how long it takes to complete the work. Both components are very crucial for any project. Faulty or inaccurate estimates will delay the work beyond scheduled completion and overrun the cost, making the project unsuccessful.

A work breakdown structure is to be prepared with tasks and sub tasks. Each item of work is to be estimated from which time schedule of each activity can be envisaged. The estimation process consists of two stages as indicated below.

2.6.2.1 Preliminary Estimate

A preliminary estimate shall be prepared on the basis of preliminary drawings, specifications, and standards. It is essential to complete the preliminary survey, investigation and project report including appointment of a Project Management Consultant (PMC) before taking up estimation work. Each project is unique and site specific. All professional companies maintain a data bank acquiring data from successful projects with the intention of using this data in any project of the same kind in a similar geography. Under this stage, estimation is done on per KW installed capacity using the trend of collected data from previous successful projects. The following guidelines may be adopted for the purpose of estimation:

- Cost of civil and ancillary work can be computed by considering materials available locally along with local workforce and on the basis of costs of similar items in the same area.
- Quotes shall be invited from leading competent vendors for electro-mechanical works.

- Cost of channel pipe, tunnel, penstock can be found out on the basis of cost unit length of such components available in the area.

2.6.2.2 Detailed Estimate

A detailed estimate shall be prepared after engineering drawings and specifications of the work are finalized. Engineering drawings and specification can be finalized after having administrative approval and expenditure sanctions from a competent authority. After having this approval a detailed estimate can be prepared with the support of all drawings such as architectural, general civil engineering, structural, piping, mechanical equipment drawings and electrical. The detailed estimate is a time consuming activity but it gives a fairly realistic idea of the cost expenditure of the project and estimated cost of the project shall be within $\pm 10\%$ of cost as executed.

2.6.2.3 Formula Based Estimate

There are many formulae available which can be used for the assessment cost based on influencing parameters. These formulae can provide a preliminary idea of the cost of the project.

2.6.3 Detailed Estimate for Evaluating Budget of the Work

The total estimated cost can be divided in to three heads such as direct costs, indirect costs and others.

2.6.3.1 Direct Costs

- Works
- Establishment
- Tools and Plants (T&P)
- Receipts and recoveries on capital accounts

2.6.3.2 Indirect Costs

- Land revenue
- Audit and account charges

2.6.3.3 Other Costs

- Financing cost
- Local area development
- Interest during construction

Various heads as mentioned below shall be considered for computing project cost:

- Preliminary cost
- Cost of land
- Infrastructure costs
- Cost of civil work
- Electro-mechanical equipment costs
- Maintenance costs
- Switch yard costs
- Transmission line costs
- Plantation costs
- Special tools and plants costs
- Communication costs
- Miscellaneous

This estimate excludes impact of government subsidies and incentives, system balancing costs associated with variable renewable or any CO_2 pricing, etc.

2.6.3.4 Preliminary Cost

The preliminary cost of a project includes the cost of many activities like detailed survey and investigation, which include topographical survey, geological and geophysical investigation, and a survey for the availability of construction material model test, if any. Preliminary cost should not be more than 2% of the total cost.

2.6.3.5 Land Cost

Land required for project establishment is acquired on purchase or lease. The rate for land of each category shall be assessed to arrive at a realistic cost. A compensation package shall be made maintaining different provisions like existing structures, standing crops, prospective mineral deposit, fertility of land, types of plantation or trees and processing charges for transfer as an actual.

2.6.3.6 Infrastructure Cost

The provision being kept under this head includes temporary and some permanent works like offices, stores, buildings for residential purposes for operation and maintenance staff, and roads.

2.6.3.7 Cost of Civil Works

The capital cost of a hydropower project is controlled and dominated by civil and electro-mechanical works which can represent between 75% and 90% of total cost. The cost of civil work come to about 58%–62%, although the cost of civil work is very much site specific.

Execution of all civil activities such as diversion structures, intake, desilting basin, water conducting system, fore bay, power house and tail race shall be put under this head.

The quantity of each civil component shall be computed as per drawings, specification, standards and site data available. Quantities thus obtained shall be multiplied by the rate of each item being fixed by rate analysis of each item as per standard norm. The rate analysis of each item shall be done with the prevailing cost of ingredients like sand, aggregates, cement and other construction materials along with transportation, carriage or haulage of materials, testing of materials, quality control and safety, labor cost, tools and tackles (T&P), and water charges (1%). The coefficient of labor and material required per m^3 or square meter or RM or quantity shall be considered as the prevailing norm of the states and countries for arriving at a unit rate of item, and overhead and profit at 10%–12% shall be added to compute unit rate of each item. The rate thus obtained for each item shall be multiplied with the quantity of each item to find out the cost of each item. Contingency provisions of 3% shall be considered as agreed upon by the project authority.

Probable major items involved in execution of a hydropower project:

- Preparations of site inclusive of cleaning and stripping of vegetation up to specified thickness, felling of trees up to specified girth and transportation of surplus material up to specified distance and dumping the material in demarcated dumping area with all leads and lift as per direction of engineer in charge.
- Earthwork foundation excavation to specified depths including levelling and dressing the founding level with all lift and lead with regards to specification and drawing as per direction of engineer in charge.
  - Excavation in ordinary soil, morrum, etc.
  - Excavation in hard soil
  - Excavation in soft rock
  - Excavation of hard rock requires blasting
  - Excavation of hard rock requires chiselling
- Back filling with excavated material including spreading, compacting in layers as per specification up to 95% of MDD.
- Providing and placing of plain cement concrete of approved grade as per specification.

- Providing and placing reinforced cement concrete of requisite grade as per specification duly compacting, curing, etc., as per the direction of engineer in charge (excluding shuttering and reinforcement).
- Providing reinforcement of required grade of steel including cutting, bending as per bar bending schedule and placing in position duly tied with binding wire.
- Providing and placing shuttering in position with plywood/steel shuttering including required staging and scaffolding.
- Fabrication and erection of structural steel in position in trash rack, gates, and trusses, including surface preparation by grit blasting, and painting.
- Providing of brick work or stone work with mortar of approved proportion.
- Protection work such as stone pitching and energy dissipation work.
- Grouting work.
- Various item of building work.
- Carting of surplus material to the demarcated dumping area along with levelling the area with specified lead.
- Diversion work if any.

2.6.3.8 Electro Mechanical Work

The cost of electro-mechanical equipment depends on the discharge capacity and head. Proper selection of a turbine depends on many factors such as plant capacity and head, speed and setting of the turbine with regards to tail water, weight, and size of runner and shaft. All these factors shall be considered carefully to produce power at the least cost. It incurs about 35% of total cost.

The components are mentioned as follows:

- Turbine with governing system
- Generator with excitation system
- Electrical and mechanical auxiliary
- Transformer and switch yard

2.6.3.9 Transmission Line

The cost of transmission lines depends on the amount of power to be evacuated from the powerhouse, voltage level, types of conductor, types of poles and the length of transmission line. The cost of executing the transmission line has to be shown as a separate item, as providing a transmission line for power evacuation may be the responsibility of the state utility.

2.6.3.10 Plantation Cost

During execution of the project, huge damage will occur in the forest, so in order to compensate the greenery, new plantation has to be done in the identified green belt as approved by the forest department and expenditure shall be limited to 0.50% of the project cost.

2.6.3.11 Maintenance Cost

Completion of construction of a small hydro project may be accomplished in 1–5 years depending on many factors such as site condition, accessibility to project area, remoteness of the project, size of components and availability of resources. Some work already completed needs some maintenance if work prolongs for more than a year or two. Provision of maintenance of completed structure shall be kept at 1% of the cost of civil work.

2.6.3.12 Plants and Erection Equipment

Most of the construction equipment is generally mobilized by business associates as per contract. But some equipment like batching plants, transit mixer, concrete pump or boom placer, hydra, roller, and cranes are procured by the department as an added facility to accomplish the work. One to two percent of the project cost may be allotted under this head.

2.6.3.13 Communication

Effective communication is a key to success, so an effective communication system shall be in place which will help to accomplish the project in the scheduled time. Provision under this head shall be kept.

2.6.3.14 Miscellaneous

During construction of a project, many activities of a small nature are required to be carried out for the benefit of the project:

- Transport facility
- Power supply, which is within the purview of department
- Photography and videography of the progress of work
- Medical help and sanitation
- Worker compensation
- Security and safety arrangements
- Safety competition at site and awards to workforce

2.6.3.15 Receipts and Recoveries

Equipment, tools and tackles, and buildings are sold after completion of work, so an amount equivalent to 10% of cost of T&P shall be deducted from total cost of the project.

2.6.3.16 Indirect Cost

The following expenditures are to be considered as indirect charges:

- *Capitalized abatement on land revenue*: Charges for capitalization of abatement on land revenue shall be 5% of cultivable land cost, or 20 times of the annual revenue cost.
- *Audit and account charges*: 0.5% of project work cost can be considered under this head.

2.6.3.17 Other Cost

2.6.3.17.1 Land Areas Development Assessment Cost

During construction of the project, local inhabitants face a lot of inconveniences and hazards related to construction activities like sound pollution, air pollution, and obstruction in roads and open areas. One percent of the project cost shall be allotted for the development of the area.

2.6.3.17.2 Financing Cost

The project authority secures a loan from a different financing institution for the execution of the project on the terms and conditions of the institution. The financial institution charges some amount towards processing fees for approval. One percent of the project cost can be kept under this head.

- *Interest during construction (IDC)*: Interest accrued on the loan during construction period shall be added to the cost of project
- *Escalation of cost*: Escalation is a provision being kept in the estimate for an increase of price over time. It is used to estimate the future cost of the project. Construction of a hydel project consumes more time. During construction, the price of labor, materials, POL (Petrol, Oil, Lubricant) and essential commodities become elevated. This elevated price shall be calculated as per the price index issued by the authority and is to be considered in the cost of the project by establishing a relationship between the previous price and current one through approved formulae for various commodities.

The project authority deals with this issue in two different ways. They consider price quoted by the contracting agency as fixed and not subject to any price escalation during the timeline of the contract. For the second method, price elevation is payable through agreed formulae within the contract period as mutually agreed upon, subject to a max 5% of the contract value. In this case, the basic price of labor, POL, and other materials shall be specified in the contract so that escalation can be calculated with respect to basic and current price.

2.6.4 Format for Detailed Estimate

Detailed estimates consist of the following:

- A report in the prescribed format of the project authority.
- Abstract of Cost duly indicating total estimated cost of each item with the requisite Bill of Quantity (BOQ) and rate of each item as per approved Schedule of Rate (SOR).
- Drawings and specification of work.
- Detailed measurements as per drawings and design calculation if any.

2.6.5 Type of Estimates

Type of estimates, as indicated below, shall be prepared in case of any changes required to be incorporated after approval of the detailed estimate.

2.6.5.1 Recasting of Estimate

If it is felt that some changes are required in condition or method contemplated earlier, an estimate is sanctioned. In this case, the original abstract shall be recast as per the established guidelines of the project authority. The cost and quantity already approved shall be rearranged and approved by the competent authority.

2.6.5.2 Supplementary Estimate

During execution of a project, some supplementary works need to be executed for the sake of the project which were not envisaged at time of making the detailed estimate. These works are totally independent of the works already considered in the detailed estimate; it may be due to change in design/specification/addition and alternation due to change in drawing as per site condition. The cost and quantity of the already approved estimate shall be rearranged while indicating the original estimate and total amount for which sanction is required, including supplementary amount, and shall be placed with clarification for approval to the competent authority.

2.6.5.3 Revised Estimate

A revised estimate is prepared when an excess beyond the permissible limit of 10% of the project cost is envisaged, and it shall be submitted for approval with proper justification for variation.

2.6.6 Economic and Financial Cost Analysis

The purpose of this discussion is to get an idea about financial terms along with financial and economic analysis so as to determine the generation cost and necessary tariff of electricity of a small power plant.

Development of small hydropower plants has gained momentum on our planet. Construction of SHP have been taken up by various state bodies where they are made of all available investments and no repayment of this funding is required. Since generation of power was opened to private and corporate sectors as well as government developers, in various countries repayment of capital is a prime concern. Financial analysis is a must for evaluation of a project for financial viability.

2.6.6.1 Requisite Tools for Financial Analysis

Any prospective developers or potential investors would like to know the detailed cost components/related cost factors in order to find out the ability of the project to repay the investment so as to understand whether the project is financially feasible and worth the investment.

The tools include capital cost, capacity factor, interest during construction, fiscal incentives, Debt-Equity Ratio, working capital, discount factors, present value, net present value, Internal Rate of Return, installation cost, unit generation cost, Benefit-Cost Ratio, plant availability, and payback.

We will discuss the aforementioned financial terms, costs and parameters before making any analysis.

As narrated above in the Sections 2.6.3.7 and 2.6.3.8, it can be summarized that there are two major cost components of small hydropower plants indicated below:

- Civil works of SHP consist of all components (diversion work, intake, silting basin, water conducting system (low pressure and high pressure tunnel), forebay tailrace, power house, etc.), inclusive of all infrastructure development and necessary project development cost (indirect and other cost). The project development cost includes planning and feasibility assessments, environmental impact analyses, licensing, fish and wildlife/biodiversity, mitigation measures, development of recreational amenities, historical and archaeological mitigation, and water quality monitoring and mitigation.

- Electro-mechanical cost

    Economic analysis: Economic analysis is a quantitative evaluation of the economic feasibility of the project and gives a comparison between the benefits and costs of the projects over the lifetime of the project.

    Financial analysis: Financial analysis is a quantitative assessment of the ability of the project to repay the investment on a self-liquidating basis. For a project to be financially feasible, the anticipated revenue receipts over the lifetime of the project should be more than the project disbursements.

Financial analysis of the project compares benefits and costs to enterprise, and economic analysis of the project compares benefits and costs to the economy. Economic and financial analysis are complementary to each other. An economically viable project must be financially sustainable and financial sustainability should not be a matter of concern as long as the project is economically sound.

2.6.6.2 Capital Cost

Capital cost is the total cost of installation of a project which includes all direct, indirect and other costs required for the accomplishment of a project. For a hydel project, capital cost is dominated by civil works. The cost of civil works is influenced by numerous site related factors. Hydropower is very site specific. Around three-quarters of the total investment costs of hydropower projects are driven by site-specific elements that impact the civil engineering design and costs. The proper selection of site and hydro scheme will play a key role for the reduction of capital costs of the project.

The site-specific factors that influence the cost of civil works include geological and hydrological characteristics, remoteness and topography, distance from existing structure and transmission. It also depends on labor cost and commodity prices of the country.

Electro-mechanical equipment cost is correlated with the capacity of the plant. The designed capacity of the plant can be achieved by adopting a combination of large and small turbines and generators. If one generator of large capacity is used and it goes offline due to some reason, then generation drops to zero. So, designers should design the system to minimize the cost and maximize the efficiency.

The cost of electro-mechanical equipment is a function of plant size and head. A relationship has been established between the cost, power and head of SHP.

$$\text{Cost (per Kilowatt)} = \alpha' P^{1-\beta} H^{\beta 1}$$

where
P	= Power in kW of turbine
H	= Head in meters
β & $\beta 1$	= Coefficient of power and head, respectively

2.6.6.3 Funding and Its Means

Finance is the oxygen of any project. Financiers are to be attracted for the development of a small hydropower project. Many developers around the world have put their best efforts to get financiers through community finance, public funding, equity investments, grants and loans from financial institutions.

In Africa, SHP development has been achieved through soft loans from foreign development institutions and other countries. Efforts are being put in place to create an environment that is financially attractive for private institutions. Suitable grid coverage and infrastructure are required to enhance facilities to make a robust electricity sector that attracts investors. Although long time benefit lies with SHP, in high level investment SHP is still perceived as high risk by the investor.

Availability of funds in time is the key to success where a project can be completed within the schedule timeline. Financial institutions aim to finance projects, but before the sanction of any loan they like to examine the proposal of the project in the light of legal eligibility, loan repayment capability as well as managing ability and health of the organization as per procedure of extending loan financing. Govt/Noda/Corporates have a system of participation in equity up to 20%–30% of the project cost. (Government sometimes consider direct or indirect equity participation in a project. It assures government support for the execution of the project. This participation sometimes also helps to secure public support for any strategic and politically sensitive project.) Generally, funding can be had from:

- Promoters
- Financial institutions
- Equity developers
- Government aids/subsidy

2.6.7 Phase-Wise Distribution of Budget and Expenditures

Time phased budget allocation and distribution of expenditures shall be done proportionately with respect to the proposed time schedule of the project based on the terms and conditions of the financial institutions.

2.6.7.1 Capacity Factor

Capacity factor is the ratio of an actual electrical energy output over a given period of time to the maximum possible electrical energy output over the same amount of time. Capacity factor generally lies at 50% for both SHP and large hydro projects with a varying capacity factor from 25% to 80% for most of the projects. This wide range depends on the design flexibility due to inflows and characteristics of the site. Resources and capacity factors in some countries in South America for new large and small hydropower are 63% and 66%, and 66% and 52%, respectively. In most of the region, capacity factors of large hydropower are higher than small hydropower projects.

2.6.7.2 Interest During Construction (IDC)

Interest during construction shall be deduced depending on the withdrawal schedule of the loan.

2.6.7.3 Fiscal Incentives for Small Hydropower

Many countries have introduced fiscal incentives in order to promote small hydropower projects, such as:

- Income tax exemption
- Rebate on income tax
- Soft loan from financial institutions
- Import duty and sales tax exemption on purchase of equipment
- Capital subsidy
- Clean development mechanism benefits
- Renewable energy certificate benefits

2.6.7.4 Debt to Equity Ratio

Debt to Equity ratio is a financial liquidity ratio that compares a company's total debt to total equity. It can be computed by dividing total liability by stockholder's equity. A ratio of 1 means that the creditor and stock holder equally contribute to the asset and a less than 1 ratio indicates that the portion of assets provided by stockholders is greater than the portion of assets provided by creditors. Generally, a creditor likes to have less Debt to Equity ratio. In SHP, it has been considered that a 70:30 ratio is normal. If equity is more than 30% of capital cost, then equity in excess of 30% shall be considered as a loan. If equity is considered less than 30%, then actual equity shall be considered for determination of tariffs.

2.6.7.5 Working Capital

The working capital of a company includes both current assets and current liabilities which are convertible into cash or equivalent within a year and those which are required to meet day-to-day operations of the company.

The working capital is bifurcated in four categories such as gross, net, variable and permanent working capital.

Gross working capital: The gross working capital is capital that is invested by a company into its current assets, and these assets can be converted in to cash within an accounting year.

Net working capital: The net working capital is the difference between current assets and current liabilities. These liabilities are expected to mature within a year. Working capital will become positive when assets exceeds current liabilities. It teaches the company how much amount is left for day-to-day operation of the company.

Permanent working capital: Permanent working capital is an amount that is required to run day-to-day business of the company in the leanest season of the year. This amount changes from year to year depending on the growth of the company.

Variable working capital: Variable working capital is an additional amount to current assets particularly in cash, accounts receivable or inventories which are required in very active business season of the year.

Working capital that is required in a hydel project for day-to-day operations can be computed in accordance with the following methodologies:

- Operation and maintenance expenses as decided
- Account receivables equivalent to specific duration's energy charges as decided
- Maintenance spares at 15% of the M&O expenses

2.6.7.6 Discount Factors

It is a factor that is needed to find out the present value of predicted future cash flows. It is the factor by which future money received is multiplied to obtain net present value. A discount factor signifies the time value of money, which means the revenues that may be received for money over the time. Money may depreciate or appreciate in value over time, but it won't earn any revenue unless it is invested.

The following inputs are required to find out discount factor of a cash flow:

- Amount of cash flow
- Interest rate
- Time

Suppose discount factor to be calculated for one year from now with the rate of interest at 10%

The discount factor (DF) can be found by dividing 1 by 1 plus the rate of interest $= 1/1 + 0.1 = 1/1.1 = 0.90$ or 90%. The formula indicated below can be used to find DF.

$$DF = 1/(1 + i)^n$$

where
   i = Rate of interest
   n = No. of years

2.6.7.6.1 Present Value

Present day value of an amount that is received in future date with specific discounted rate. It can be calculated by the formula,

$$PV = Amount/(1 + rate\ of\ discount)^2$$
$$= \$500/(1 + 0.08)^2 = \$428.66$$

So present value of $500 is $428.66.

2.6.7.6.2 Net Present Value

Net present value (NPV) of an investment is sum of all positive cash flows minus present value of all negative cash flows.

$$NPV = \Sigma\{(Periodic\ revenue - Periodic\ expense)$$
$$\times\ Compound\ interest\ factor\} - I$$

If net present value is more than zero it means that the benefit is more than the expenses of the project. A project cannot be taken up if NPV is negative. NPV should be positive for economically viable project.

2.6.7.7 Internal Rate of Return

Internal Rate of Return (IRR) is the interest at which the NPV of all cash flows (both positive and negative) from a project or investment is zero. IRR is used to evaluate the attractiveness of a period or investment. If IRR of a project exceeds the company's required rate of interest, then the project is desirable. If IRR is below the value of rate of interest, then the project is rejected.

2.6.7.8 Installation Cost

Installation cost is the cost in dollars per Kw determined by dividing the project cost from the installed capacity of the project.

2.6.7.9 Unit Generation Cost

Unit generation cost is obtained based on the annual cost by annual salable energy generation. The generation cost includes operation and maintenance cost, depreciation cost, insurance cost, equipment cost, interest on capital borrowed, and interest on capital cost.

2.6.7.10 Energy Available for Sale

Auxiliary consumption, plant availability and water royalties shall be deducted from the energy generated in order to find energy available for sale.

1% of gross energy generated may be taken as auxiliary consumption which also includes transformer losses.

5% of gross energy generated shall be considered for plant availability towards energy shutdowns and unplanned maintenance.

12% of gross energy generated shall be deducted for water royalty for the plant capacity of more than 5 MW, but no deductions will be made for plants having a capacity of less than 5 MW.

2.6.7.11 Sale Price of Electricity

Sale price is a price fixed as per a power purchase agreement. In absence of it, leveled tariffs can be taken as the sale price.

2.6.7.12 Payback Method

The payback period is the length of time required to recover one's investment. The payback method is used to evaluate the project investment proposal and to calculate the payback period of the investment.

$$\text{Payback period} = \frac{\text{Net investment}}{\text{Net annual income from the investment}}$$

Net investment includes purchase cost plus installation cost minus anticipated future salvage value.

Net annual income is the annual financial benefits expected from the assets.

2.6.7.13 Benefit Cost Ratio

Benefit cost ratio is the ratio of the present value of benefit and present value of expenditure.

$$\text{Benefit cost ratio} = \frac{\text{Present value of benefit}}{\text{Present value of expenditure}}$$

This ratio provides a value of benefits and costs that are represented by actual dollars spent and gained. The expenditure includes O&M, depreciation and interests. The project becomes viable when the ratio lies above 1.

2.6.7.14 Debt Service Coverage Ratio

The debt service coverage ratio (DSCR) is obtained by dividing net operating income by total debt service.

$$\text{DSCR} = \frac{\text{Net operating income}}{\text{Total debt service}}$$

If calculated DSCR is more than 1, this indicates that there is sufficient cash flow available to pay back the debt. The financiers, generally, consider that their investment is safe and secure if DSCR is above 1 and if anything goes wrong it will protect their interest.

2.6.7.15 Financial Analysis

Financial analysis is to be conducted to select, evaluate and interpret financial data of SHP so as to assess the financial viability of the project. This analysis is done based on various costs such as installation cost, benefit cost ratio, net present value, internal rate of return with some maximum and minimum criteria in NRV and IRR, installation and generation cost, respectively.

Various aspects already narrated above such as capital investment, phase wise distribution of expenditure during construction period, expected life of the project, amount of debt, rate of interest and repayment time are to be considered for the analysis.

Viability or feasibility on the financial point of view can be ascertained by the help of many approaches and accepted methodologies like the Payback Method, Net Present Value, Internal Rate of Return Method, Debt Service Coverage Method, or Benefit Cost Ratio Method.

Each method has its own advantages and disadvantages, so viability can be assessed keeping in consideration all these factors.

The payback method ignores the time value of money and it focuses on short term profitability. This method fails to consider the inflow of cash that occurs beyond the payback period. It gives importance to liquidity and ignores profitability. Any good or alternative project may be overlooked in this method.

Net Present Value requires guess work to find the total cost of capital, and it is also not useful in comparing two companies because it results in dollars.

Many financial institutions consider IRR and DSCR as the most acceptable methods for evaluation of financial viability of a project.

2.6.8 Due Diligence

In small hydropower, as per the international standard, due diligence shall be conducted to obtain reliable information and data pertaining to the proposed project prior to making any decision for purchase/merger/implementation of technically viable projects, and also for the purpose of investment in the project if it is found to be financially feasible through the process of systematic research work.

After conceptualization of the project the project authority conducts exploration and a detailed investigation as per norm such as the reconnaissance survey, prefeasibility study and feasibility/detailed investigation in order to find the proposed project technically feasible and financially viable.

Due diligence through a third party shall be initiated to assess the technical health of the project on the following issues:

- Authenticity of parameters being obtained by geological, geotechnical and topographical and hydrological investigation
- Review of hydrological analysis
- Review of project layout and technical design of civil structure and electro mechanical equipment and relevant specifications

- Review of BOQ/SOR so as to justify the cost analysis
- Social cost benefit analysis
- Check the approval of environmental and social impact assessment
- Financial feasibility of the project

2.6.8.1 Due Diligence of Financial Feasibility of Hydropower Project

One of the most important stage of due diligence is to find out the capital expenditure of the project.

2.6.8.1.1 Empirical Hydro Cost Equation

There are many ways to calculate the cost, and every step of a calculation brings its own perception. The cost of a hydropower project is very much site specific. The cost depends on various cost components such as equipment cost, replacement cost, financing cost, total installed cost, fixed and variable cost maintenance and operation, and levelized cost of energy (LCOE).

Capital cost is the dominating factor for any major hydel project. About three fourths of the cost of the project is driven by site-specific elements like site condition, proper selection of site, etc., which have an impact on design and engineering of the project, and the proposed capacity of the plant can be achieved by combining larger and smaller turbines to find a solution for the economy.

Capital expenditures (CAPEX) include cost items like project development cost, engineering cost, environmental cost, civil work, electro-mechanical cost, equipment cost, and grid connection. Studies found that main cost items show a wide range of cost, and is reflected by a coefficient of variation like 0.47 for civil work, 0.49 for electro-mechanical work and 0.47 for project development.

Many studies were conducted that analyzed the cost of electro-mechanical equipment as a function of total plant size and heads. A formula was derived to establish a relationship between cost, power and head of a hydropower project:

$$\text{COST (per kW)} = \alpha P^{1-\beta} H^{\beta 1} \tag{2.1}$$

where
P is power in KW of turbine
H is head in meters
α is a constant
β and $\beta 1$ are coefficient of power and head

Mr. Gordon studied and developed an empirical formula[*] to find out the cost of electro-mechanical equipment considering the relationship between the head, hydrological characteristic of site and water flow level.

---

[*] Gordon's formula is adopted with permission from research work (The cost of small scale hydro power production etc.) by Mr. G.A. Aggidis, Engineering Department, Lancaster University, Lancaster, UK.

$$C_{EM} = 9000 \times kW^{0.7}/H^{0.35} \qquad (2.2)$$

where
  kW is the installed capacity in kW
  H is the hydraulic head in m

Gordon developed a site factor which has an influence over the cost of project. Site factor (S) is derived by dividing the total cost of project by the cost of electro-mechanical work and the overall cost of the project can be obtained.

$$Cp = 9000 \times S \times kW^{0.7}/H^{0.35} \qquad (2.3)$$

where
  Cp is overall cost
  S is site factor

He further developed and updated his empirical formulae by considering civil site cost, engineering cost, equipment cost and administrative cost for a different range of head and installed capacity

$$Ch = k \times L(kW/H^{0.3})^{0.82} \qquad (2.4)$$

where
  Ch = Cost of overall project
  k = System factor load
  L = Location or country

Value of k depends on Head, for Head <350 m, $k = 3 \times 10^6$, and for Head >350 m, $k = 5 \times 10^6$

After some time, Mr. G.A. Aggidis pointed out that some caution is needed to operate Gordon's formula by modifying the empirical formula for the assessment of project cost

$$C_P = 25000 \, (kW/H^{0.35})^{0.65} \quad \text{for Head between 2 and 30 meters} \qquad (2.5)$$

$$C_P = 45500 \, (kW/H^{0.3})^{0.6} \quad \text{for Head between 30 and 200 meters} \qquad (2.6)$$

For canal based scheme, civil works comprise of diversion canal, spillway and power house

Civil components such as intake, penstock, power house and tailrace were involved in the scheme of the dam toe. For the ROR scheme, civil works comprised of diversion weir, intake, desilting basin, LP tunnel, fore bay, HP tunnel or penstock, power house, and tailrace were included in the cost estimate. But electro-mechanical components were the same for all schemes.

TABLE 2.4

Initial Project Cost of SHP

Sl No.	Type	Installed Cost per Kw in US $	O&M Cost	Capacity Factor (%)	LCOE
1	Large hydro	1050 to 7650	2–2.5	25–90	0.02–0.19
2	Small hydro	1300 to 8000	1–4	20–95	0.02–0.27

Source: Adopted from IRENA Hydropower report.

A turbine with a governing system, governor with excitation system, and switch gear were considered for the cost estimate.

As per Cost Analysis of Hydropower by IRENA, the initial cost of a small hydropower project with varying head is tabulated for idea (Table 2.4).

Project cost per kW (ECU per kW) with varying head is depicted in Figure 2.1.

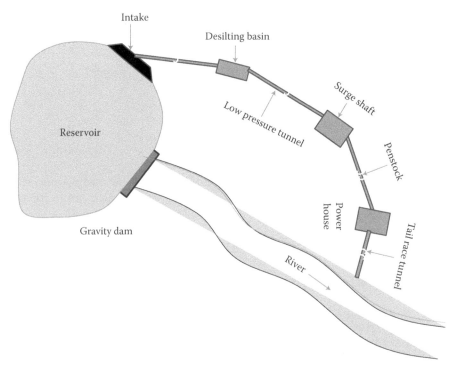

FIGURE 2.1
Storage hydro project.

2.6.8.1.2 *Capital Cost Breakdown*

Capital cost is a one-time expenditure incurred for the construction of a project that represents the all-in-plant cost. It includes engineering, procurement and construction, and also interest during construction. Capital cost is site specific, but a capital cost breakdown is created by collecting data from three different project reports. This cost does not include grid connection cost, regulatory cost, engineering cost and other indirect costs.

It is observed that civil works and electro-mechanical equipment are two major components. The cost of a turbine is a major cost component in electro-mechanical equipment in a new hydro project.

Civil works in a high head project incur about 47% of the total 60% cost of civil works on penstock, and a low cost project will incur about 38% of 47.5% total civil cost on a dam and power house. This means penstock is the main cost driver for high head pump house and dam in a low head project.

Approximate cost (%) of SHP as indicated in Figures 2.2 and 2.3. Worldwide hydropower installed capacity is indicated in Figure 2.4.

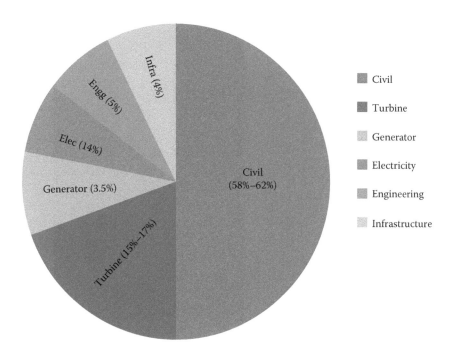

FIGURE 2.2
Cost (%) of small hydropower.

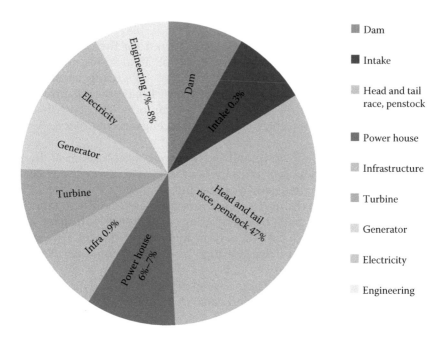

FIGURE 2.3
Cost (%) of all components of SHP.

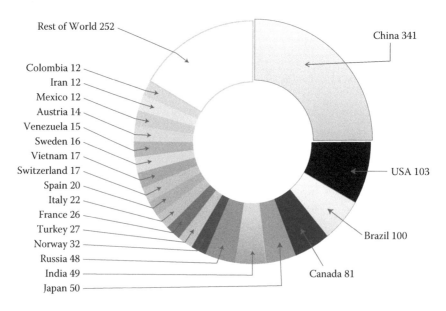

FIGURE 2.4
World wide power installed capacity in GW. (Adapted from Hydro Power Status Report, 2018.)

2.7 Major Hydropower Project

The types of major hydro projects is divided into three types in accordance with their operation.

2.7.1 Run-of-River Project

This type of project is the simplest and the cheapest hydropower project to develop. This type of hydro project does not require a reservoir by constructing a dam across the river, so major construction cost is avoided. Geotechnical problems generally encountered in dam foundation areas during construction are avoided too. But some sort of diversion structure is required to divert water from the river to the power house. Water is directly diverted from the river through an open or closed channel and pipework to the turbine being installed in the power house, situated in the downstream of the river, and finally water is discharged through the tail race tunnel into the river. This type of project takes advantage of a drop in elevation in the entire route to have a requisite flow rate of water to generate electricity. The location of the project is selected based on some geological features such as requisite flow rate of water and tilt in the river so as to speed up the flow of water. The suitable location of the power house may be obtained far away from the source in order to have the requisite flow of water, and then the geology of the entire route needs to be studied carefully.

Small hydro, micro hydro and mini hydro are examples of ROR schemes. Refer to Figure 2.5 and also refer to Figure 2.4 for the world's installed capacity by region.

2.7.1.1 Advantage and Disadvantage of This Project

ROR project has advantages and disadvantages that are narrated below.

2.7.1.1.1 Advantages

- Dam is not required to create a reservoir
- Geological problems during construction are avoided
- Construction cost is minimized
- Shorter timeline
- No rehabilitation of local people dwelling close proximity is necessary
- Less emission of carbon and methane due to no flooding or pondage
- Source of clean and green renewable energy
- Provide employment opportunity in remote area

2.7.1.1.2 Disadvantages

- Generation of power depends on flow of river
- Power generation is not steady as it fluctuates with flow of river

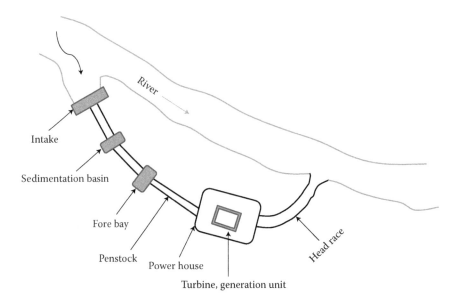

FIGURE 2.5
Run-of-River Scheme.

- Power generation is low during dry season
- Water cannot be stored and flows without utilization of excess water during monsoons

2.7.1.2 Ten ROR of the World

Detailed Run-of-River scheme is featured in Figure 2.5 and world's 10 ROR project is reflected in Table 2.5 for reference.

TABLE 2.5

Ten Run-of-River Projects of the World

Sl No.	Name of Project	Country	Installed Capacity (MW)
1	Santo Antanio	Brazil	2498
2	John Day Dam	USA	2160
3	Teles Pires Dam	Brazil	1820
4	Nathapa Jhakri	India	1500
5	Lime Stone Generation Stn	Canada	1340
6	La Grande-1	Canada	1436
7	Outrades-3	Canada	1226
8	Kettle Generating Stn	Canada	1020
9	Tala Bhulan		1020
10	Longspruce	Canada	1010

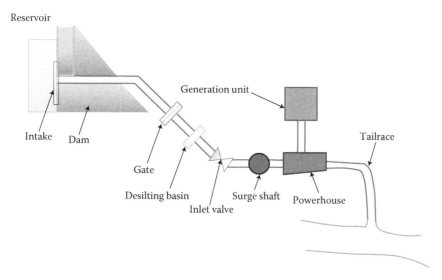

FIGURE 2.6
Section of storage scheme.

2.7.2 Storage Project

In this type of project, a dam is constructed across the river and a reservoir is created on the upstream, which will generate a requisite head of water for the generation of firm power. The power house is located much farther away from the dam. The fluctuation of water can be controlled and any surplus water during the rainy season can be discharged by opening sluice gate or radial gate.

It is, of course, a costly one because it needs a detailed geological study of the dam and reservoir site and if any fault zone/shear zone is detected, it should suitably treated for stability of the dam; simultaneously the reservoir site shall also be tested for water tightness which will incur massive cost and delay the construction. In addition to this, there will be additional expenditures for construction of a coffer dam, diversion tunnel, low pressure tunnel, surge shaft, and penstock and also includes rehabilitation of people affected due to the formation of a reservoir; it also affects ecology because any vegetation/ forest falling within the periphery of the reservoir shall be destroyed.

Refer to Figures 2.1 and 2.6 for Storage Project.

2.7.2.1 Advantages and Disadvantages

The following are the advantages and disadvantages of Storage Project.

2.7.2.1.1 Advantages

- Water can be stored as it consists of a reservoir
- Ensures steady generation of electricity when season is dry

- Can control flood by regulating water
- Renders help to irrigation system
- Water can be used in human consumption
- Facilitates recreation and fishery
- Provides navigation
- Provides employment opportunities in rural areas
- Recharges close water table

2.7.2.1.2 Disadvantages

- Construction cost is very high
- Geo-technical problems are encountered in dam foundation, low pressure and high pressure tunnel and the huge reservoir area
- Care to be taken to ensure water tightness of huge reservoir area
- Causes ecological imbalance
- Destroys forestry, vegetation and wildlife
- Needs a lot of rehabilitation of people residing in reservoir area as well as in the area in downstream area in close proximity to dam

2.7.3 Pumped Storage

Every country has its national power grid that connects different parts of the country, and in addition to it many countries have regional and local grids as per their policy of power transmission and distribution systems. So, constant and regular power supply to these grids at any time of the day is a prime requirement inclusive of times of high demand. The demand of supply of power fluctuates during different time zones of the day as per the consumption of electricity. The demand is at its peak from morning to afternoon and it is significantly lower in the night. A storage power plant is developed to promote energy efficiency and to avoid waste of energy during low demand hours. Pumped storage energy is a proven and reliable source of electricity that can ensure a steady supply to the national grid so that clean and affordable energy can be made to reach the people.

Pumped hydro storage is an extended version of the conventional hydro project where already produced electrical energy is converted to potential energy in the form of water, and later on this stored potential energy is converted into electricity. In this type of project water is pumped from a lower level reservoir to store as potential energy in the higher level reservoir during low demand hours, and later it can be utilized at the peak demand of power. During demand hours of electricity, water is released from the higher level reservoir thorough pipe ways or tunnels to flow into the system through a turbine to generate electricity and get the water stored in the

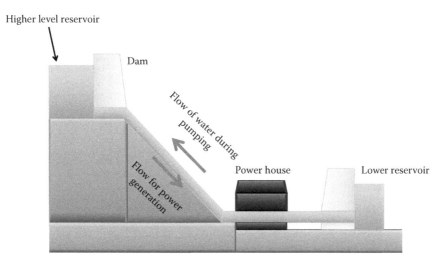

FIGURE 2.7
Schematic of pumped storage.

lower reservoir instead of discharging the used water back into the river. Afterwards, water is pumped in the storage at a higher level from the lower level reservoir when demand for power is low. This pumping back process is done by the help of a single reversible turbine. This is a cycle and can be considered as a battery.

Refer Figure 2.7 for pumped hydro storage.

2.7.3.1 Advantages

- No wastage of water
- Energy conservation
- Ensure stable supply to grid

2.7.3.2 World's 10 Pumped Storage Project

10 Pumped Storage projects of the world are tabulated in Table 2.6.

2.7.4 Multipurpose Project

Hydropower projects being developed with reservoirs render various services like water supply, irrigation, flood control and drought management, navigation, fishery, and recreational activity in addition to the generation of electricity. This sort of hydropower project is called a multi-purpose project as it falls under the storage scheme and can be termed as a multi-purpose use of the hydropower reservoir.

TABLE 2.6

World's 10 Pumped Storage Project

Sl No.	Name of Project	Country	Installed Capacity (MW)
1	Bath County PSH	USA	3003
2	Huizhou PSH	China	2448
3	Guandong PSH	China	2000
4	Dinorwig PS	UK	1728
5	Castaic PS	USA	1566
6	Chaira HP Cascade	Bulgaria	1455
7	Baoquan PSH	China	1200
8	Bailianhe PSPH	China	1200
9	Blenheiam Gibea PS	USA	1160
10	Bed Creek HE Stn	USA	1065

The World Water Council agreed to work on "the Multi-Purpose Water uses of Hydropower Reservoirs." The hydropower is used for water management and the reservoir can control and regulate water flow for drinking water supply, water supply for industrial uses, flood control, mitigation, navigation and recreational activities; the hydropower reservoir is a multi-use reservoir.

Refer Figure 2.8 for Multi-Purpose Dam and Figure 2.9 for Cost Breakdown Structure for Major Green Field Hydropower Project.

As per ICOLD's register, there are at least 38,542 large dams in the world. Only 9,568 (25%) of large dams in the world have hydropower as one of their purposes.

In additional to power generation, the other activities of this type of project are:

- *Irrigation*: Water is an essential ingredient for cultivation. Farmers in developed countries get water for cultivation through different systems of irrigation. They bank on rivers, available reservoirs or underground water. Irrigation is an artificial supply of water to land or soil. It helps to grow crops, maintain landscape, and provide a lifeline to arid zones. In many places irrigation canals are constructed to supply water from hydropower reservoirs in the dry land and make the area irrigated to grow crops to meet human demand.

- *Drinking water supply*: The entire world is suffering from the scarcity of water—both quantity and quality. Fresh water for drinking as well as commercial purposes are being provided from hydropower reservoirs through cross country pipelines. These pipelines run parallel along highways, having tapping points in designated locations where from requisite water can be tapped for different uses.

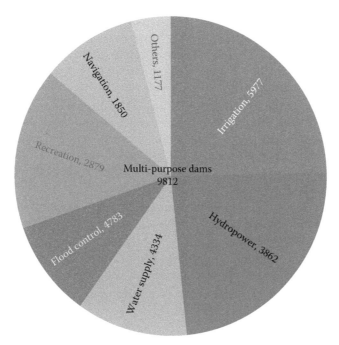

FIGURE 2.8
Usage of hydro reservoir services.

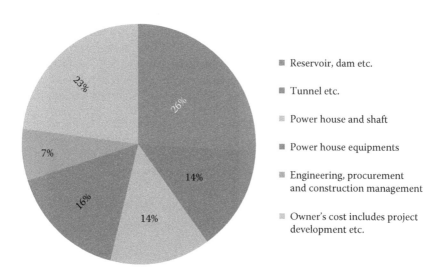

FIGURE 2.9
Cost break down structure for major green field hydropower project.

- *Navigation*: Construction of dams across the river creates a reservoir in the upstream side and it also retards the flow of the river in the downstream side. Navigation systems can be improved in the upstream and water can be released through spillways in the downstream during dry periods to have sufficient water depth for navigation.

- *Flood control*: Flow of the river is controlled by the construction of a dam that creates a reservoir to store water during monsoon times up to the high flood level of the area, and any excess water above the designed high flood level shall be released over the spillway to the downstream with safety to avoid flooding of the area where people are residing in adjacent areas downstream.

- *Drought mitigation*: Reservoirs created by dam supply water in dry seasons which will have an impact and influence on agricultural systems in the area.

- *Recreation*: Reservoirs are a large water body which provides facilities for boating, fishing, swimming etc., and it attracts tourists to the spot.

- *Commercial fishery*: Some reservoirs support commercial fishery in addition to recreational fishing.

2.7.5 Component of Hydel Project

It is essential to draw water from a reservoir to the power house for the generation of power through the components indicated below, and this entire process is known as a water conductor system.

Hydel projects are comprised of several components such as the reservoir, dam, dyke, intake structure, sedimentation tank, low pressure tunnel, surge shaft, fore bay, high pressure tunnel or penstock, power house with requisite turbine and generator, tailrace tunnel/channel, and draft tube, all of which constitute a water conductor system. Water is drawn from the reservoir through various structural arrangements to discharge back to the river with two major objectives.

a. Water passing through the conducting system shall be free from any undesirable material such as silt, sand, gravel, boulder, vegetation, and logs so as to avoid any possible damages to turbines or getting the system clogged. Coarse sediments cause huge abrasions and cavitations in the blades of the turbine.

b. The flow of water, also known as energy, shall be kept intact for optimal utilization of kinetic energy for the generation of power.

Depending on the type of hydro project based on topography, two types of water conducting systems are generally adopted:

a. Open channel system
b. Pressure tunnel system

The pressure tunnel system consists of a low pressure (LP) tunnel and high pressure (HP) tunnel or penstock. The LP tunnel made of reinforced cement concrete (RCC) starts from intake to surge shaft, and the HP tunnel or penstock made of steel starts from surge shaft to power house.

2.7.6 Difference between Dam and Weir

Dam, weir, barrage and embankment are four different structures by names, and are quite often used in water management engineering but create confusion amongst engineers to differentiate between dam, weir, barrage and embankment.

- *Dam*: A dam is an impervious barrier built across the river so as to create the head of water by the formation of a reservoir on the upstream of the dam. Surplus water during monsoons can be discharged over the spillway which is controlled by radial gates. An opening of adequate size through the body of the dam is provided at a lower level than that of the crest of the spillway in the overflow section, which is called the sluice. Water can be released through the sluice at the required time by controlling the sluice gate. The top of the dam is used as a road way. Dams are classified in accordance with function, material and on technical aspects such as concrete dam, rock fill dam, storage dam, coffer dam, check dam, gravity dam, and arch dam.

- *Weir*: A weir is constructed across the river to change the characteristic of the flow of water. It may be overflow or non-overflow. Water starts overflowing when more water accumulated than is the requirement. Weirs are classified in accordance with the shape of the opening and crest.

- *Barrage*: A barrage is also constructed across the river, having a series of gates through which the flow of water can be controlled by opening or closing as per requirements so as to use the water for irrigation or as a supply of drinking water. The top of the barrage is also used as a roadway.

- *Embankment*: An embankment is an artificial barrier made by either earth, rubble or a mix of both in an area to protect the area from flooding. A portion of marshy land raised by a filling to take road over it is also called an embankment in highways construction. A non-overflow earthen barrier made across the river is called an embankment dam. If any portion of the reservoir in its periphery is lower than HFL, an earthen barrier that is made in that portion of the reservoir to retain water within the reservoir is called an earthen dyke. These are also called non-rigid dams because they are made of material like earth or rock fill.

2.7.7 Classification of Dam, Other Components and Its Function

Dams are classified in accordance with their functions, materials and technicalities. Dams are basically water-retaining structures.

- *Storage dam*: The purpose of this type of dam is to store water in the reservoir during monsoons and utilize this stored water when it is required in the dry season. Its function is multipurpose such as power generation, irrigation, recreation, supply of drinking water, and fisheries, and it consists of a system of discharging surplus water.
- *Detention dam*: This type of dam is constructed to control floods and reduce the impact of flowing water in the downstream area.
- *Coffer dam*: A coffer dam is a temporary structure made across the river to divert the water during construction of the main dam. The purpose of construction of this dam is to facilitate easy working by keeping the working area fairly dry.
- *Check dam*: A check dam is constructed across the channel to reduce the flow to eliminate erosion and allow more water to percolate into an adjacent area, and also store water in patches so that this water can be used during dry climates.
- *Diversion dam*: This type of dam is a low height dam constructed to divert the water through a channel for irrigation or any industrial purpose.
- *Gravity dam*: This type of dam resists water thrust and other overturning moments by its self-weight and ensures its stability by the Middle Third Rule. No tension is allowed to develop at the bottom of the dam. It is made of concrete and stone masonry. A gravity dam is generally made straight and triangular in shape. It is a solid profile without any openings except a gallery and sluice being provided in the body of the dam.
- *Arch dam*: An arch dam is a curved structure in plan, having convexity against the flow of water which transfers pressure through arch action to the abutments. It is economical, but a complex structure. It is most suitable for a narrow gorge with strong and firm flanks so that loads being transferred through the arch action to the abutments are sustained.
- *Buttress dam*: A buttress dam is a hydraulic structure having an RCC flat slab in the upstream side that is supported by a series of buttresses in the downstream to transfer the load. Buttresses are compressive members. They may be either curved or straight in plan and made of reinforced concrete. It is also called hollow dam.
- *Embankment dam*: Embankment dam is a non-rigid dam because it is constructed with soil that has specific properties. It is built

by laying earth in layers of specified thickness, and each layer is being compacted by roller up to the designed MDD and OMC. The upstream slope is protected by stone pitching with the arrangement of wire net so as to prevent erosion and a suitable filter arrangement is made in the downstream slope to control the seepage line against possible erosion.

- Gravity, arch and buttress dams are known as rigid dams because these types of dams are constructed with materials like concrete or stone masonry.
- Embankment dams or dykes are known as non-rigid dams because they are constructed with material like earth, rock fill or debris.
- *Reservoir*: Water is impounded in the area on the upstream of river after a dam is constructed. It stores water during the period when there is excess flow for the utilization of this water in the lean period of the year.
- *Dyke*: Impervious earthen barrier that is constructed in the area lower than HFL in the periphery of the reservoir to restrain water in a designated reservoir area.
- *Intake structure*: Intake is a structure built to draw water from the source such as reservoir, canal, or river. The location of an intake structure shall be fixed in such a way that it provides smooth curves like a bell mouth to avoid water loss in entry and entry of air should not affect the operation and safety of tunnel.
- *Sedimentation tank*: Hydel projects, situated in a hilly region, are made on rivers that carry large quantities of silt that make the hydel project Vulnerable and may harm the equipment being used for generation of power.
- *Low pressure tunnel*: Water passes through this tunnel made of concrete to the surge shaft or fore bay.
- *Surge shaft*: It is a tank whose main purpose is to reduce the water hammering effect on the high pressure tunnel/penstock and simultaneously act as storage.
- *Power house*: The power house is a structure where water energy is converted into mechanical energy and electrical energy by the help of a turbine and DG. It is comprised of a turbine room, generator room, erection bay and service area.
- *Tail race channel*: It is a channel where turbines discharge the water that ultimately discharges in the main river.
- Overflow and non overflow section, core wall, sluice, gallery, bucket, crest, ogee, inspection chamber, training wall, radial gate, sluice gate, energy dissipation.

2.7.8 Profile and Various Sections of Dam

A dam is an impervious structure constructed across the river to create a necessary water head for generation of power, irrigation, recreation, fishery, and supply of drinking water. It is basically triangular in shape with a widened top crest used as a passage for a roadway.

It is divided into two sections:

a. *Non-overflow section*: It is a section of the dam over which water is not allowed to pass. A dam has non-overflow sections on both ends of the length. Sometimes, both end blocks are designed as composite sections with concrete and earth where concrete in the core of the section is known as the core wall.

b. *Overflow section*: It is a section that comprises an appropriate spillway to release excess water during monsoons, and it covers most of the width of the river and exists in the middle of the stream. It looks different than the non-overflow section of a dam. On the top of the spillway, there will be a number of piers which will house radial gates over the entire width of spillway to hold water up to HFL, and any excess water during monsoons will be released by operating these radial gates over spillway. There will be another passage of water within the body of the overflow section below crest level called the sluice, which enables the release of a requisite quantity of water through the sluice. When water levels are below the crest level, the sluice gate is operated from the inspection chamber.

2.7.8.1 Non-Overflow Section of Dam

Refer Figures 2.10 and 2.11 for non-overflow section.

2.7.8.2 Overflow Section of Dam

Refer Figure 2.12 for overflow section.

2.7.9 Energy Dissipation, Hydraulic Jump, Stilling Basin

Hydraulic jump and energy dissipation is a matter of prime concern in hydropower engineering, especially in the case of designing a spillway and bucket of the dam where care is to be taken to dissipate kinetic energy of water at the toe of the overflow section of the dam and to protect the bed or bank, and even to protect any structure like bridges that exist downstream of the dam that may be affected by the jump. So before finalizing the design, a detailed model study is also taken up.

Before the discussion of hydraulic jump, let us discuss open channel flow and some of it parameters to understand the phenomena of hydraulic jump

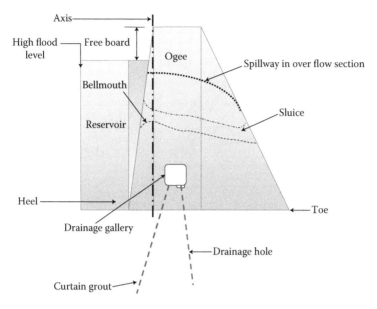

FIGURE 2.10
Non-overflow section of gravity dam.

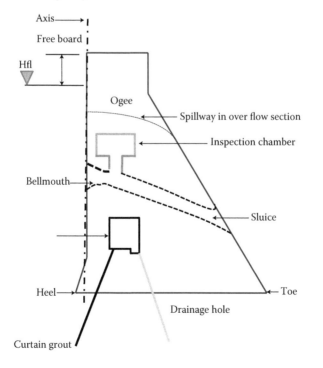

FIGURE 2.11
Non-overflow section of gravity dam with an inspection chamber.

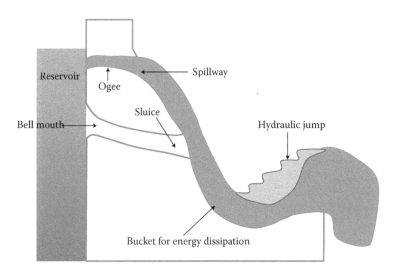

FIGURE 2.12
Overflow section of gravity dam.

in a better way. Hydropower deals with an open channel, stream or river and falloff water from a height of spillway to a bucket or stilling basin below to create a jump.

- *Open channel flow*: Open channel flow is the flow of fluid with a free surface exposed to atmosphere. A conduit which involves flow of fluid but not completely filled falls under the same category. The main driving force is gravitational force, which is balanced by equal and opposite shear force between fluid and channel base.
- *Depth of flow*: It is the depth at which water flows above the bed of the channel.
- *Critical depth*: It is a minimum depth of flow at which the river or stream attains high velocity with the creation of some turbulence, and the velocity of the channel at this depth is called critical velocity.
- *Super critical depth*: The depth of flow of water at this point is less than the critical depth, and velocity at this depth is called super critical velocity. Flow is fast and rapid.
- *Sub critical depth*: The depth of flow of water is more than critical depth and velocity at this depth is called sub critical velocity. The flow is slow.
- *Froude number*: It is a ratio of velocity of the fluid and speed of wave.

$$Fr = V/gD$$

where
  $D =$ Hydraulic depth
  $g =$ Gravity

2.7.9.1 Hydraulic Jump

Hydraulic jump occurs in open channel flow due to the sudden transition of supercritical flow to sub critical flow. When water of high depth falls onto water of lower depth it creates hydraulic jump. The hydraulic jump is generally observed in an open channel or river where a flow of higher velocity meets with lower velocity resulting in increased height of water as the water with high velocity gets reduced.

Hydraulic jump is classified with respect to Froude number:

- *Undular jump*: In this jump the surface of water maintains undulation and Froude number for this jump is 1–1.7. Refer to Figure 2.13 for undular jump.
- *Weak jump*: Weak jump is formed when Froude number varies from 1.7 to 2.5 and velocity is uniform where a small amount of energy is dissipated. Some rollers are formed on the surface of the jump.
- *Oscillating jump*: When Froude number lies between 2.5 and 4.5, the jumps occurred is termed as an oscillating jump. Flow of water in this jump oscillates back and forth from the bed of the stream.
- *Steady jump*: When the Froude number lies between 4.5 and 9, the jump that occurs is known as the steady jump. It is stabilized and has no back and forth oscillation.
- *Strong jump*: A jump is called strong when Froude number lies at or above 9 (Figure 2.13).

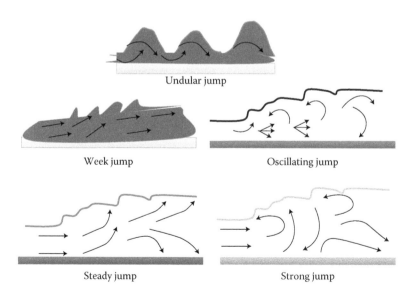

FIGURE 2.13
Hydraulic jump.

2.7.9.2 *Energy Dissipation*

Surplus water from the reservoir of a hydropower system is released and allowed to flow over the spillway of the overflow section of the dam, and water rushes down the spillway and falls near the toe of spillway where potential energy of the water is converted into kinetic energy. This huge energy acquired, if not controlled, will scour the toe of the dam and adjacent bank. The proper and adequate arrangement is made to dissipate this huge energy, which is known as an energy dissipater, and the process is called dissipation of energy. Stilling basin buckets, both flips and rollers, are generally used as a dissipater.

When water falls from the crest of the spillway to a low level where hydraulic jump occurs due to the abrupt transition of super critical flow to sub critical flow, the velocity of water gets reduced in this zone resulting in the dissipation of kinetic energy, and water is directed to pass through a structure like a bucket and forced to drop this water away from the bucket so as to protect the toe of the dam from scouring. This type of arrangement is called an energy dissipater.

- *Stilling basin*: Water falls from the crest of the spillway down near the toe of spillway of the dam where energy is dissipated in a structure designed in such a way that velocity of flow is controlled. Length of jump is regulated so that the flow leaving the area can be guided safely to the river below without any scouring effect on the toe of dam, and the sill of the designed structure and adjacent bank of the channel is known as stilling basin.

   The stilling basin may be used for energy dissipation when the head of water is less than 50 meters. A stilling basin is not suitable for water head that is more than 50 m because cavitations will form on the river bed that will affect the basin and adjacent bank of the channel. A concrete apron is provided considering the length of the jump. In order to avoid a huge excavation for the construction of the apron to suit the length of jump, numbers of chute block, buffer blocks and sills are provided to stabilize the jump and also to control the length of the hydraulic jump, resulting in a reduction of length of the jump and the cost of stilling basin.
- *Chute block*: Chute blocks are triangular in shape with a horizontal top surface. These blocks are installed in the starting point of the stilling basin. These blocks stabilize the jump, improve the performance and decrease the length of hydraulic jump.
- *Baffle blocks*: Baffle blocks are placed between chute blocks and the end of the stilling basin to stabilize the jump and dissipate energy.
- *Bucket*: A bucket is provided to dissipate energy where the tail water level is low and the downstream rock is strong enough to resist erosion. Many drainage holes are also provided in the area of the bucket to reduce uplift. This type of bucket is called a flip bucket.

The roller type of bucket is provided when downstream water depth is greater than the requirement of formation of jump (Figure 2.2).

Bibliography

Cost Analysis of Hydropower by IRENA, 2012. Cost Analysis Series, Vol 1, Power Sector, Issue 3/5. Renewable Energy Technologies, Abu Dhabi, UAE.

Hydro Electric Power, A Guide for developers and Investors, International Finance Corporation (IFC) World Bank Group.

Hydro Power Status Report 2018. International Hydro Power Association (IHA).

Standards/manuals/guidelines for small hydro development. 1.2 and 2.1 General/Civil Works – Planning and Layouts, Ministry of New and Renewable Energy, Govt. of India. Alternate Hydro Energy Center.

Standards/manuals/guidelines for small hydro development. 1.4 General: Reports Preparation: Reconnaissance, Pre-feasibility, Feasibility/Detailed Project Report and As Built Report, Ministry of New and Renewable Energy, Govt. of India. Alternate Hydro Energy Center, Indian Institute of Technology, Roorkee.

The costs of small-scale hydro power production: Impact on the development of existing potential by G.A. Aggidis, D.C. Howard. Journal Renewable Energy.

World Energy Resources, 2016 World Energy Council.

World Small Hydro Development Report 2016, UNIDO. United Nation Industrial Development Organisation. ICSHP, Vienna, and International Center on Small Hydro Power, Hangzhou.

3

Concept of Forces Acting on
Gravity Dam and Its Stability

3.1 Terminology and Some Definitions Related to Gravity Dam

Terminology, nomenclatures and definitions of various components of dams are described below for the understanding of readers.

- *Plan*: A plan is a view from the top of a dam structure being cut at mid height in a horizontal plane, showing the main features of the dam and its accessories.
- *Profile*: A profile is the side elevation of a dam with designed height up to the original ground surface, rock surface, or excavation surface along the axis of the dam duly indicating up and down stream safe slope.
- *Free board*: Vertical distance between HFL in the reservoir and the top of dam.
- *Height of dam*: Vertical distance between the top of the dam and the lowest point in the excavated foundation area.
- *Axis of dam*: A surface, plain or curved, is selected by the designer for the measurement of horizontal dimension of the dam.
- *Hydraulic height*: Hydraulic height is the height to which the water level rises in the upstream of the dam. It is the difference in level between the lowest point in the original river bed and the maximum level of controllable water behind the dam.
- *Abutment*: The part of the valley against which the dam is constructed. Sometimes when a natural solid valley does not exist, artificial abutments are made.
- *Length of dam*: The length of the dam measured along the axis of the dam from abutment to abutment.
- *Core wall*: A wall of impervious material such as concrete is made within the body of the embankment dam in order to prevent seepage of water.

- *Crest of dam*: The crown of the overflow section of a dam is called the crest.
- *Crest length*: The length of the spillway is called the crest length.
- *Spillway*: It is constructed on the overflow section of a dam in order to safely release the excess water downstream of the dam. The top or crest of the spillway is made in a smooth S-curve profile called an ogee for smooth downward flow of flood water into the river. The purpose of providing a spillway is so that the reservoir of each hydel project has a designated capacity to store water. If the reservoir is full and flood water enters the reservoir increase of the water level may overflow the dam. In order to avoid this situation, a spillway is provided to discharge the surplus water in the river downstream. When flood water is controlled by the gate it is called a controlled spillway and if water is allowed to pass over the crest naturally is called an uncontrolled spillway.
- *Sluice*: The sluice is a water channel being provided within the body of the dam to discharge water as required by controlling the sluice gate when the level of water is below the crest level. It is provided to meet the requirement of the lower reaches of the river. It helps to divert water when the power house shuts down, desalting the reservoir by draining out silt laden water.
  - The shape of sluices are rectangular or circular. The horse shoe shape of the sluice is also provided as per the discretion of the designer. The selection of rectangular or circular sluices shall be taken based on economic and hydraulic consideration as it does not perform better on partial openings of a gate. The horse shoe section gives satisfactory performance on partial gate opening.
- *Bell mouth entrance*: A bell-like opening is provided at the entrance of the sluice to minimize head loss and to avoid development of cavitation pressure that will ensure the smooth flow of water through the sluice.
- *Aeration of sluices*: Water is allowed to pass through the conduit either operating the gate fully open or partially open. There are possibilities of the creation of sub atmospheric pressure immediately downstream of the gate when the gate is partially open and hydraulic jump occurs in the sluice.
  - High velocity water tends to absorb air due to sub atmospheric pressure. Generally, an air vent is provided so that air is discharged into sluice where low pressure exists.
- *Pier*: Piers are constructed upstream of the dam to facilitate of making a roadway and also to provide support to the radial gate being constructed over spillway.

- Training wall
- *Drainage gallery*: The drainage gallery is either rectangular or circular in shape and is provided within the body of dam. It is parallel to the axis of the dam and provided throughout the length of the dam.
  - The main objective of the drainage gallery is to facilitate inspection and arrest seepage occurring within the body of the dam. Seepage water is collected in a sump provided in the lowest level of the gallery. A collection drain runs along with the gallery and it is connected to a sump at the lowest elevation.
  - Drainage holes of the required depth are drilled in regular spacing along the drain up to a designed depth to reduce the uplift force so as to release internal pressure of the dam. The entirety of the water collected in the sump is being pumped out suitably from the drainage gallery.
- *Silting basin*: Huge amounts of water come down from the spillway with high velocity downstream of the dam that may scour the foundation, resulting in damage to the foundation and causing severe erosion downstream of the dam. A silting basin is provided to dissipate this huge energy from the water falling down from the spillway in order to stop severe scouring downstream of the dam. This scouring will obviously damage the foundation. It is provided at the end of the spillway, and the primary objective of energy dissipation is to create a hydraulic jump.
- *Control gates*: There are many types of control gates that are provided so that water passing through the gates are not obstructed when the gate is fully open. These control gates are installed on top of the spillway and sluice to control the water flow and include gates such as the slide gate, high pressure lift gate, radial gate and jet flow gate.
- *Consolidation or area grouting*: Dams are constructed in hills that have a rocky foundation. The rocks consist of fissures, cleavages, cracks and clay seams within the structures of rock. These features weaken the rocks and make the dam structures vulnerable. Geologists make geological mappings of the foundation area, duly indicating fissures, cracks, clay seams, faults, fold dips and strikes, and make drawing prescribings drill holes of desired size, spacing, depth and direction. These holes are to be grouted in order to fill the fissures and cleavages to consolidate and strengthen the rock and make it water tight under the dam foundation, which will enhance the stability of the rock by preventing relative movement between them under the designed loading. This phenomenon is known as consolidation grouting of foundation.

- *Blanket grouting*: Blanket grouting is done in the reservoir area to reduce permeability and subsequently to decrease uplift forces under the structure due to the effect of hydro static head.

- *Curtain grouting*: It is a high pressure grouting done to reduce seepage through the dam foundation by means of sealing all geologic constraints such as fissures, cracks, and clay seams that will act as a barrier to the seepage while increasing creep length of the seepage. The longer creep length will reduce the uplift force at the bottom of the dam. A series of holes with greater depth are drilled along the gallery as per the direction of the geologist. Grouting is taken up in various stages, for example, primary, secondary, and tertiary depending on the grout intake in each hole and every stage up to the acceptable grout intake as prescribed.

- *Drainage hole*: Drainage holes are also drilled along the length of the gallery of the dam. The holes are drilled in the alignment of the drain in the gallery so that the seepage water comes up through these holes in the drain and is collected in the sump for effective removal of water. The purpose of providing drainage holes is to reduce the uplift force to the dam.

- *Dental treatment*: Rocks generally consist of cavities, fault/shear zones and other imperfections/weak zones. These geologic imperfections/ shortfalls or weak zones of foundation are vulnerable for any sort of water retaining structure.

  - The methodology adopted is to strengthen this weak zone of the foundation by doing some specific treatments such as excavating fault breccias, clay, and weathered rocks from the foundation until solid rock is reached at the bottom level. Grouting shall be taken up in the area excavated and the entire area shall be filled up with the requisite grade of concrete with reinforcement as per the designer. This entire activity is called dental treatment or fault zone treatment of the dam foundation.

3.2 Concept of Forces Acting on Gravity Dam

A gravity dam is a solid impervious structure made of concrete and built across the river to create a reservoir upstream for the purpose of the generation of power, irrigation, recreation, water supply, fishery, and flood control. It is designed and shaped in such a way that its weight is sufficient enough to ensure its stability against the effects of all imposed loads.

The design and construction of a gravity dam is a critical phenomenon. The forces that act on the dam are of such a nature that exact magnitude, direction and location cannot be assessed correctly. It should be considered based on data being obtained by site exploration and to some extent it depends on experience, engineering judgement and common sense. Post design exploration shall also be taken up during the construction stage in order to assess the unique problem or unusual geological constraints, especially in the foundation, for stability. There are adequate instances and experiences which indicated that the stability of a dam becomes vulnerable when critical attributes of dam foundation are not explored and unknown.

The stability is the prime criteria for any structure. The stable gravity dam must have a sufficient cross section so that horizontal and vertical forces will pass through the middle third of the structure at any horizontal plane. The dam will be subjected to cracking due to the development of tension downstream of the structure if resultant passes through beyond the middle third. The greater the distance of resultant from the middle third, the greater likelihood of the occurrence of cracking due to greater tensile forces. The structure will overturn if resultant a pass beyond the downstream face.

Though design of a dam is not within the scope of this book, for the benefit of reader, the basic concept of design as per IS 6512 is described.

3.3 Requirement of Stability

The design shall satisfy the following requirement of stability:

- The dam shall be designed safe against sliding on any plane within the foundation
- The dam shall be designed safe against overturning at any plane within the dam
- The safe unit stresses in the concrete of the dam or in the foundation shall not be exceeded
- Crushing

3.4 Forces Acting on Dam

- *Dead load*: Dead load of a gravity dam is equivalent to the self load plus the weight of appurtences of the dam such as weight of radial gate, sluice gate, bridge, etc.

- *Reservoir and tail water load*: These loads are hydrostatic pressure applied on the upstream and downstream faces of the dam. These loads are obtained from hydrostatic data like reservoir capacity.

- *Uplift pressure*: When the reservoir is full with water and water starts percolating through dam foundation due to hydrostatic head and exerts pressure upward on the base of dam. It is developed at the contact between the dam and the foundation. Drainage is provided in the drainage gallery in order to reduce the intensity of uplift pressure on the body of the dam.

- *Earthquake force*: Earthquakes occur due to tectonic causes which create many fault zones within the rock structure, and movement of crust along the fault generates an earthquake. In any dam site the resident geologist will inspect and examine the rock foundation to identify the presence of any fault and establish whether the fault is active or not so as to find out the length of fault, fault material/breccias, width of fault, etc. The geologist should find out the seismic record of the area along with the intensity, magnitude and location of the earthquake. Earthquakes generate both horizontal and vertical acceleration. Analysis of response of a dam to earthquake force shall be done as per standard method.

- *Earth and silt pressure*: It is not necessary that all dams will be subjected to silt pressure. Hydrological data shall be consulted by the designer before considering the effect of silt pressure on the dam. The magnitude of silt pressure varies directly with the depth.

- *Ice pressure*: Intensity of ice pressure depends on the climatic condition of the site where the dam is proposed to be constructed. Ice pressure is generated by thermal expansion of ice, and is also exerted when an ice sheet floating on the surface of water is dragged by wind. Intensity of pressure depends on thickness of sheet of ice, coefficient of thermal expansion and the strength of ice.

- *Wave pressure*: Wave pressure is exerted on the dam due to wind. The area of the reservoir of any hydropower project is very vast where a wave is formed due to the blowing of wind and waves hit on the top surface of the dam on the upstream face.

- *Thermal load*: A gravity dam is comprised of a huge mass of concrete which is subjected to volumetric changes due to the change of temperature of concrete. The dam body generally has contraction and construction joints. The load generated is transferred to a grouted transverse contraction joint. The designer should study weather conditions and temperatures of reservoir water which will affect the temperature of the concrete of the dam. A temperature gradient occurs in the concrete surface due to differences of temperature between ambient temperature and temperature of water surface, which generates secondary stress on the surface of dam or in the opening which form cracks.

- *Internal hydrostatic pressure*: A dam is a water retaining structure and made of concrete which will have pores. Water will seep through these pores and percolate in the body of the dam which will generate internal pore pressure through pores, cracks, and joints within the body of dam. Distribution of pressure varies linearly from full hydrostatic head upstream face to zero in downstream.

3.5 Category of Forces to Be Resisted by Dam

The categories of forces to be resisted by dams are as follows:

- Forces, such as weight of the dam and water pressure, are directly calculable from the unit weight of material and property of fluid pressure.
- Forces such as uplift, earthquake load, wave pressure, silt pressure and ice pressure which can only be assumed on the basis of varying degrees of reliability.

3.6 Assumptions for Consideration of Stability

The following assumptions are made for the consideration of stability:

- That the dam is composed of individual transverse vertical elements each of which carries its load to foundation without transfer of load from or to adjacent elements.
- That vertical stress varies linearly from upstream face to downstream face on any horizontal section.

3.7 Load Combination

The design of the dam shall be done considering the most adverse loading conditions. The combination of loads are as follows:

- *Usual load combination*: Under this condition, design is done with normal reservoir water level, dead load, uplift, silt pressure, minimum temperature load, tailwater.

- *Unusual load combination*: Under this condition, design is done with maximum reservoir level, dead load, uplift, silt, temperature and tail water pressure.
- *Extreme load combination*: Under this condition, the design is done with normal reservoir level, dead load, uplift, ice, wave, temperature tail water and effect of earthquake as per zone applicable to the area.

3.7.1 Dead Load

The dead load of a dam is comprised of its self weight plus the weight of appurtenances such as the pier, gates, and bridges. The self weight shall be calculated by taking the unit weight of concrete, that is, 2400 kgs/m^3. The weight of appurtenances shall be obtained by considering the unit weight of each material used for the job. Small openings like the drainage gallery is not generally deducted unless the designer feels that this opening accounts for significant quantity of dam volume.

The basic shape of a dam is triangular and the top is widened to accommodate the roadway.

Dead Load Acting on Dam (Figure 3.1):

1. W1 will act in CG of block 1 at the center
2. W2 will act on CG of block 2 at 1/3rd height from base

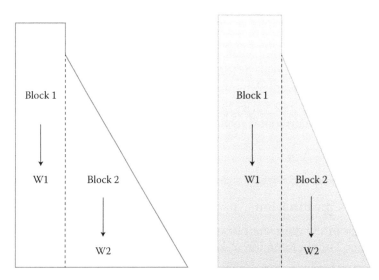

FIGURE 3.1
Dead load acting on dam.

3.7.2 Reservoir and Tail Water Load

It is understood that the weight of water varies a little with the temperature, but this variation is ignored. The mass of water is considered as 1000 kgs/m³. The water pressure increases linearly with the depth. The linear distribution of static water pressure acts normal to the surface of the dam. The hydrostatic pressure at any point on the dam is equal to the hydraulic head at that point times the unit weight of water.

Tail water pressure will be taken at full value for the non-overflow section and the reduced value for the overflow section depending on the type of energy dissipation arrangement that is made.

A gravity dam is comprised of a spillway in an overflow section over which excess flood water is released during monsoons. The flood water glides over the crest and downstream of the spillway which creates a complicated hydrodynamic effect. Hydrodynamic force occurs due to change of speed and direction over spillway. This force may be neglected in stability analysis when it is insignificant but it should be considered when it becomes significant (Figure 3.2).

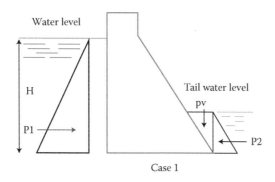

Case 1

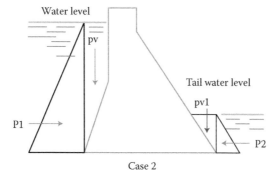

Case 2

FIGURE 3.2
Reservoir and tail water pressure.

3.7.3 Uplift Pressure

After completion of the construction of a dam the reservoir is filled up with water and a hydraulic head will be generated and water will enter in any cracks, pores, or fissures available in the structure of the foundation under hydrostatic pressure. Simultaneously water will enter in the space between the foundation and the bottom of the dam. Further, water will ingress in the pores of the body of the dam under the same influence. This water seeps and moves downstream from upstream due to the hydraulic gradient and generates upward pressure which is known as uplift.

Uplift pressure is nothing but internal pressure that occurs in the pores, seams, fissures and cracks of concrete at the contact of the dam and bottom of the foundation as well as within the structure of the foundation.

Effective drainage will limit the uplift at the toe of the dam to the tail water and drainage holes drilled after the curtain grout and all along the length of the drainage gallery within the body of the dam will relieve uplift pressure under the body of the dam.

The following criteria shall be considered:

a. Uplift pressure distribution in the body of the dam shall be assumed, in the case of both preliminary and final designs, to have an intensity which at the line of the formed drains exceeds the tail water head by one-third of the differential between the reservoir level and tail water level. The pressure gradient shall then be extended linearly to heads corresponding to the reservoir level and tailwater level. The uplift shall be assumed to act over 100% of the area.

b. Uplift pressure distribution at the contact plane between the dam and its foundation, and within the foundation, shall be assumed for preliminary designs to have an intensity which at the line of drains/drainage holes exceeds the tail water head by one-third of the differential between the reservoir and tail water heads. The pressure gradient shall then be extended linearly to heads corresponding to the reservoir level and tail water level. The uplift shall be assumed to act over 100% of the area. For final designs, the uplift criteria in the case of dams founded on compact and unfissured rock shall be as specified above. In the case of highly jointed and broken foundation, however, the pressure distribution may be required to be based on special methods of analysis taking into consideration the foundation condition after the treatment proposed. The uplift shall be assumed to act over 100% of the area.

c. For the loading combination f and g, the uplift shall be taken as varying linearly from the appropriate reservoir water pressure at the upstream face to the appropriate tail water pressure at the downstream face. The uplift is assumed to act over 100% of the area (Figures 3.3 and 3.4).

d. No reduction in uplift is assumed at the downstream of spillways on account of the reduced water surface elevation (relative to normal tail water elevation) immediately downstream of the structure that may be expected.

e. It is assumed that uplift pressures are not affected by earthquakes.

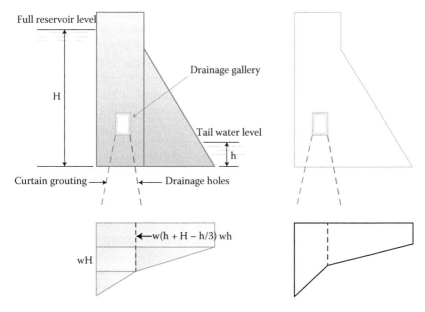

FIGURE 3.3
Uplift pressure distribution.

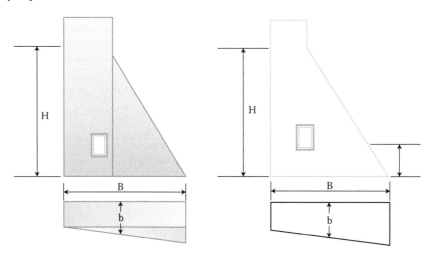

FIGURE 3.4
Vertical stress on foundation (without considering uplift).

3.7.4 Earthquake Force

Earthquake is any vibration of earth crust due to natural and artificial causes. There are three types of earthquakes: tectonic cause, valconic cause and superficial cause. The most severe is tectonic, which is considered for structural design.

The core of the Earth is comprised of molten metal of very high temperatures, which generates elastic strain and get stored in the rock structure. In the passage of time, when this stored energy becomes more than the strength of rock structure, the rock structure releases this energy and extends in all directions around its focus. This elastic wave travels to the Earth's surface, which creates a trembling of the Earth called an earthquake.

Earthquakes cause random motions of ground, which can be resolved in any three mutually perpendicular directions. This motion causes the structure to vibrate. The vibration intensity of the ground expected at any point depends upon the magnitude of the earthquake, the depth of focus, distance from epicenter, and the strata on which the structure stands. The predominant direction of vibration is horizontal.

Seismic force calculation is necessary when the dam structure is subjected to extreme load combination. The designer should review the seismic record of the site along with the geologist to determine the seismicity of the site where the dam is proposed to be constructed. It can be done either with a probabilistic or deterministic method.

The designer and geologist are to conduct a geologic study of the area to find out the existence of any fault, its length and fault material, and also to find out the time when the detected fault was active.

The designer and geologist are to further study and find out the magnitude and location of the last earthquake recorded in the proposed site. Based on available seismic data, the designed magnitude of earthquake force is considered greater than the historical magnitude of the earthquake force. This earthquake is considered to be more severe than the earthquake that occurred at site or close vicinity of the proposed site. This is called the Maximum Credible Earthquake (MCE).

The load exerted on a dam due to horizontal acceleration where the dam is having a sloping or vertical face on the upstream.

$$P_E = C_\alpha wz$$

$$C = \tfrac{\varsigma}{2}m\left[\frac{h}{z}\left(2-\frac{h}{z}\right)+\sqrt{\frac{h}{z}\left(2-\frac{h}{z}\right)}\right]$$

where
P_E = Pressure normal to the face of dam
w = Weight of water
Z = Depth of water

C = Pressure coefficient

$$\propto = \frac{\text{Horizontal Earthquake Acceleration}}{\text{Acceleration of Gravity}}$$

h = Vertical distance from reservoir

C_m = Maximum value of C for a given slope

Silt pressure shall be considered as saturated cohesion less having full uplift. It is proven that silt and water pressure exist together in submerged fills.

Horizontal silt and water pressure shall be 1300 kgs/m² and vertical silt and water pressure shall be 1925 kg/m².

3.7.5 Ice Pressure

Ice pressure may not be applicable to all dams, as ice pressure is created by thermal expansion of ice and wind drag. Ice pressure is dependent on the temperature rise of ice, thickness of the ice sheet, coefficient of expansion, elastic modulus and strength of ice.

3.7.6 Wave Pressure

A dam is subjected to the impact of waves in addition to static water pressure. The pressure of a wave depends on the velocity of wind, configuration of water surface and depth of reservoir.

3.7.7 Thermal Load

Stresses in the dam are affected due to temperature variation in the body of the dam, temperature of the reservoir, air temperature, and solar radiation. The magnitude of deflection is maximized in the morning and minimized in the evening and it varies with operation of the spillway. While calculating thermal load, the temperature gradient is assumed based on location, orientation and surrounding topography.

3.8 Reaction of Foundation

A dam is placed across the river and raises the water level on the upstream which develops hydrostatic force. This force tends to slide the dam horizontally and causes over turning about the toe downstream. The upstream water causes seepage through the dam foundation and generates an uplift force acting at the base of dam which is vulnerable for stability against overturning and sliding. The dam foundation shall be treated properly by consolidation grouting and shall be strengthened by dental treatment in cases of any fault/ shear zone in the foundation existing.

3.9 Resistance against Over Turning

If the resultant of all forces acting on the dam passes through the toe of the dam or away from the toe, the dam will rotate and overturn and the dam will fail in cracking before this situation arises. No tension is developed if the resultant of all forces passes within the middle third of the base of the dam. The factor of safety against overturning is defined as the ratio of the resisting moment to the overturning moment about the toe of the dam and it should not be less than 1.5.

3.10 Sliding Resistance

Horizontal forces acting on a dam tend to slide the dam along the horizontal plane about the base or any other joint along the horizontal plane towards downstream. This horizontal force will be resisted by frictional and shear forces developed in the base or any other joint plane in the body of the dam. The dam will fail if the horizontal forces causing sliding are more than the frictional forces developed in the plane. The ratio of the summation of horizontal forces to vertical forces is called sliding resistance. This sliding resistance is a function of cohesion and the angle of internal friction of the material of the surface where sliding occurs.

For a gravity dam the shear friction factor of safety should be greater than 3, 2 and 1 in the case of usual load combination, unusual load combination and extreme load combination, respectively.

In practice, upward slope of the foundation is generally kept towards downstream and foundation is kept in bench to develop resistance against sliding.

3.11 Factor of Safety

Factor of safety is a mathematical expression and a useful tool for designers for practical purposes. This calculated value is compared with the available reference value. If this calculated value is greater than the reference value, then the structure is considered as safe; if not, then it is termed as unsafe.

3.12 Discussion on Earthquake Engineering of Dam

A concrete gravity dam is a huge mass of concrete with no reinforcement except some skin reinforcement which is provided to take care of temperature stress being developed during the process. A concrete gravity dam is generally constructed in a wide river valley having a long length and height. It resists all loads by its self-weight. Water is filled in the upstream of the dam which exerts horizontal pressure on the body of the dam. These forces tend to slide and overturn the structure towards downstream. These actions are countered and resisted by its own weight to establish the stability of the dam.

If the designer during computation finds the presence of any tensile force, then the section shall be redesigned with a condition that no tension is developed in the body of the dam by keeping the resultant of all forces passing through the core of the section having zero tension.

Seismic force is one of the components of consideration for stability of the structure, and the magnitude of vertical and horizontal acceleration of an earthquake to be considered for design should be assessed in advance for the particular site of construction. It is very important to assess the seismic behavior of the particular site where a gravity dam is proposed to be constructed, so a detailed investigation needs to be conducted to estimate the seismic parameters, that is, fault/shear zone, thrust, etc., which are of vital importance for the safety of the dam.

The dam shall be designed to withstand extreme seismic effects and the designer should have data and information collected from basic seismic studies. These are preliminary values of ground motion which can be useful for evaluation of the effect of earthquake force in all seismic zones as specified in relevant codes of various countries and it may be suitable for the final design of the dam in a low intensity seismic zone and having low consequence of failure. When a large multi-purpose project of great national interest is planned where seismic loading is critical, then detailed seismic studies are required to be evaluated. Because an earthquake is a multi-hazard natural phenomena, it can cause damage to the dam to any extend due to its severe shaking or tremor and it also can cause landslides and floods. It may pose a threat to the potential of liquefaction of the foundation.

A huge mass of water is stored in the reservoir which may cause impact on the face of the dam due to the motion of the Earth and may strike severely on the face of the dam, and even huge waves created by an earthquake make a huge flow of water over the free board and create havoc in the downstream, resulting in disaster. This huge wave may hit and damage the radial gate installed on the spillway.

Fault or shear zones existing adjacent to the dam foundation shall be identified and investigated because they may be a source of an earthquake as energy release is associated with rapid movement on causative fault and may be the reason for many earthquakes.

Reservoir triggered seismicity is a matter of concern for many engineers. This phenomenon is a physical response to reservoir impounding. This impounding may cause an earthquake due to the existence of an earthquake producing fault (causative fault) near by the dam site. Weight and pore pressure is increased due to impounding which can trigger the release of seismic energy, but it is clear that it cannot change the underlying tectonic process and seismic hazards at the dam site. After locating and identifying a fault near the site, the added water weight cannot increase the seismic energy release and this secondary reason may be ignored at the time of the structure's design because this energy release is much less than the actual seismic energy release.

The question is how does an earthquake act upon a dam? An earthquake is created due to a sudden release of energy. It produces ripples and trembles on the Earth's surface. It generates various waves like P-waves, S-waves, etc., in all directions. A P-wave is faster and has higher intensity than an S-wave. The action of seismic force on a dam is like a man who stands on the floor of a bus and the bus starts abruptly; this sudden movement of the bus will try to throw the man in front towards the direction of movement of bus, his legs will try to stick to the floor of the bus which will try to bring the upper part of his body back to original position. The force which tends to bring him back into position is known as inertial force. The same action is generated in the body of a dam during seismic acceleration. In other words, a man stands on a mat spread on the floor and a second man pulls the mat suddenly from behind. Imagine the effect of this sudden pull of the mat on the body of the man is the same action being generated on the body of the dam due to seismic force (Figure 3.5).

It means that when a wave moves beneath the dam, the dam moves forward with the wave and the inertia force generated will move in the opposite direction of the earthquake direction. The magnitude of an earthquake

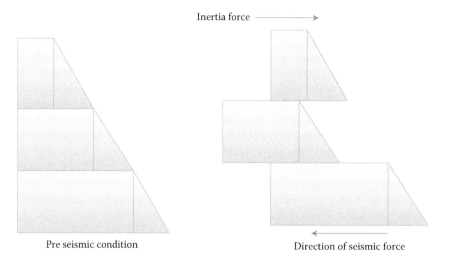

FIGURE 3.5
Action of seismic force of gravity dam.

depends on severity, weight of the dam and elasticity of the material. Earthquakes may come from any direction and the structure shall be designed for worst conditions. The design of the dam shall be done in reservoir empty and reservoir full conditions to evaluate the worst one for the consideration.

In reservoir full conditions, the worst case occurs when seismic acceleration moves in an upstream direction and the corresponding inertia force moves in a downstream direction.

In reservoir empty conditions, the worst case occurs when the seismic force moves downstream and the corresponding inertia force moves in an upstream direction.

Horizontal force due to seismicity is the product of the mass of the dam, and horizontal acceleration will act through the center of gravity of the dam.

Vertical acceleration acts downward and as a normal condition, the inertia force being generated acts upwards, which will reduce the weight of the dam and water resulting in a reduction in stability. The storage water of a reservoir comes under the influence of horizontal acceleration of seismic force. This horizontal acceleration creates a hydrodynamic pressure which exerts against the upstream side of the dam. This hydrodynamic force acts in addition to hydrostatic pressure. The direction of this force is opposite to the direction of seismic acceleration.

3.12.1 Integrity Check of Concrete Dam

Security and safety of a dam are a critical need. Many people have lost their lives due to the failure and sudden collapse of dams in many countries. Failure occurs due to many reasons such as structural failure, natural calamity or change of human behavior and development in downstream area. Dams are built to serve people through the supply of drinking water, agriculture and irrigation, power supply, and flood control, but they do not serve any purpose if adequate safety and security preparations for the people are not made to save their lives from disaster. So, an integrated approach should be adopted for dam safety and security in the following procedures.

- Identification of hazard as a source of risk
- Possible flaws in construction
- Possible dam structure weaknesses due to aging
- Probable flood
- Earthquake
- Landslide
- Human activity in potential inundation zones
- Structural failure due to overtopping of high floods
- Foundation failure

- Piping and seepage, etc.
- Safety program
- Review of existing data
- Regular visit of dam site by regulatory authorities
- Inspection and surveillance
- Assessment of observance and its consequence
- Decision for remedial and corrective measure
- Procedural guidelines shall be strictly implemented
- Inspection procedure
- Operational procedure
- Maintenance procedure
- Procedure for instrumentation
- Procedure for emergency action
- Monitoring and feedback system

Bibliography

Design of Concrete Gravity Dam: US Department of the Interior, Bureau of Reclamation. Design Manual for Concrete Dam. A Water Resources Technical Publication Denver, Colorado 1976.

4

Contract and Administration Management

4.1 Introduction

Contract management is a process where a contract, mutually agreed upon by two parties for a common goal, is created lawfully and executed efficiently while adhering to terms and conditions in delivering the product within a stipulated timeline and as per the desired financial performance of the contract.

Contract management invites some adoptability and flexibility from both parties for some situations not envisaged during contract negotiation, which should be sorted out amicably without any dispute within the purview of the contract so as to maintain the sanctity of the contract.

4.2 Component of Contract Management

When two companies want to do some business with each other having a common goal and intention/perception, and the common goal is communicated through expression of interest to form some obligations between the companies, this is known as an agreement. The agreement that is enforceable by law is called a contract.

Existing trends in the professional world suggest that companies' preferences to enter into very highly competitive situations are on the rise. This eagerness of adopting high competitiveness has put organizations under tremendous pressure to accomplish any assignment with optimal cost expenditure and a compressed timeline which can be achieved through effective and structured financial performance.

This sort of demand has increased the importance and benefit of effective and structured contract management. Proper contract administration, effective service delivery, persuasive relationships between the companies and solid financial performance are the keys to success of contract management.

Some important ingredients of the contract:

- Promisor is the organization that makes the offer of the work
- Promisee is the organization that receives the offer from promisor
- A particular act is performed by the Promisor for the Promisee, which is known as consideration
- Offer and acceptance together make an agreement
- Both parties should be competent to enter an agreement

4.2.1 Significance of Contract Management

A contract is made to form a relationship between owners, vendors, customers, project management consultants, and business associates. Contracts stipulate every aspect of key strategies, specifications, various terms and conditions to be fulfilled under the purview of the contract. It is a methodology that enables all related parties to meet their obligations in order to deliver the product within the stipulations of the contract. It is an effective tool between clients, customers, vendors, suppliers and contractors for the accomplishment of work.

4.2.2 Challenges in Contract Management

Contract management systems consist of the following challenges which should be identified and sorted out for an effective management system.

- Non-compliance of terms and conditions of the contract
- Revenue leakage
- Extension of time limit
- Disputes
- Termination of contract due to unsatisfactory performance
- Arbitration
- Affected relationship with business associates

4.2.3 Contract Administration

Contract administration consists of numerous activities starting from preparation of the Bill of Quantities (BOQ), Schedule of Rate (SOR), estimate of the work, preparation of tender documents along with specification, standard and various clauses, invitation for quotation with credentials, negotiation with bidders and finalization to accomplishment of the contract without any disputes.

Contract administration is the process where the concept of work is understood, the cost of the work is measured in accordance with drawings, specifications and standards, and the entire work is worded compactly with the least possible gaps and flaws in a tender document and is placed before a qualified party for an agreement. The contract manager is to ensure that the party has the capability, both technically and financially, to satisfy his contractual obligations after the contract is awarded.

4.2.4 Service Delivery

Service delivery is management that ensures the completion of awarded work in accordance with the terms and conditions as stipulated within the contractual time limit and adhering strictly to the essence of quality and safety.

4.2.5 Relationship Domain

The contract manager is to interact with vendors effectively and keep an avenue always open for any discussion or conflicts which arise from time to time during execution of the contract, and sort out issues amicably within the purview of the contract. If not possible, he should ensure possible solutions subject to approval of a competent authority.

4.2.6 Financial Analysis

The contract manager is to monitor periodically the status of the contract with departmental engineers as well as with vendors for the assessment of quantity completed along with billing position of the work, and this exercise shall be done with due interaction with the planning department to avoid any errors in terms of the amount of major activities that have been completed.

4.3 Function of Contract Management

The function of contract management is divided into two categories.

- *Function of pre award of contract*: Function of this category has already been narrated under Contract Administration.
- *Function of post award of contract*: The contract manager's responsibility is to dispatch the Letter of Intent (LOI) with Zero Date of the contract to the vendor and a copy to be endorsed to the project manager. Contract documents duly signed by the competent authority shall be sent to all concerned. The contract manager will arrange a kick-off meeting between the vendor, project manager, planning manager, store &

procurement, quality manager, safety manager, HR and inspectors to appraise the vendor about the working system of the company. They will be briefed on details about quality, safety and other HR related activities like labor licenses, etc. All relevent formats/documents shall be handed over to the vendor by the respective department. The project manager will take a commitment from the vendor regarding starting work along with the mobilization plan of manpower and machinery as stipulated in the contract, and this plan will be linked to the payment of the mobilization advance on the submission of the requisite bank guarantee; actual mobilization is done at site in accordance to furnished plan. A minutes of meetings shall be made and signed off by all participants for records.

4.4 Type of Contract

As mentioned below, there are many type of contracts that are executed during construction projects. The type of contract is selected based on nature and condition of work. In mega projects a combination of contracts are used due to various nature of construction activities.

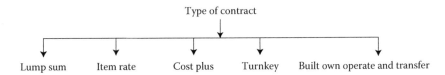

4.4.1 Lump Sum Contract

A lump sum contract is a contract where the bidder is asked to quote a lump sum price of the contract in accordance to scope, drawings, specifications, and operational requirements being furnished along with the tender document. The vendor is expected to complete the work within the lump sum contract value. This type of contract is not suitable for a large project. It is suitable for small projects or works well with defined projects like a cooling tower or waste water treatment plant. Break up of payment is submitted along with a bid based on milestone completion of work. The percentage of payment of each milestone is fixed and agreed upon by both parties.

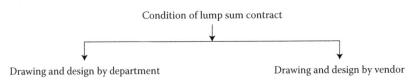

4.4.2 Drawing and Design by Department

All the detailed drawing and specifications, etc., along with a detailed scope of work pertaining to the contract shall be furnished by the department and will be part of the tender document. The entire work shall be carried out per available drawings, and it will be the basis of payment of completed work, any work not envisaged earlier that is executed shall be paid separately as extra, and any deviation from the tender drawing without prior approval shall be recovered and will be done by another agency at their own risk and cost.

4.4.3 Drawing and Design by Vendor

Under this category, vendors are supposed to design the structure as per guidelines provided by the client and prepare all relevant drawings pertaining to the contract. Vendors in this case first submit a technical bid comprised of a detailed design, plan, and structural drawing along with calculation, BOQ and other technical parameters for scrutiny and review. Technical bids shall be evaluated by the approved committee so as to avoid any future complications. A pre bid conference will be called where clarification will be sought from vendors and the technical requirements of the job will be finalized. All vendors will be asked again to submit a revised technical bid in accordance with the discussion held in the pre bid conference.

Vendors shall submit revised technical and price bids containing the following:

- General and special conditions of contract
- Detailed design calculation
- Plan, architectural and structural drawings, etc.
- Detailed specification of work
- Bill of quantities, etc.
- Various milestones of work and percentage wise payment break up for each milestone
- Deviation from original document, if any

All revised technical bids shall be reviewed, and finally price bids of all vendors shall be opened and action will be taken as per the norm.

4.5 Item Rate Contract

In this type of contract, the entirety of work is executed as per drawings issued for construction under relevant item codes for which rates were quoted during the tendering process. These rates were approved and reflected in the Schedule of Rate (SOR) which is a part of the contract document. Any item executed but not available in the SOR shall be considered as an extra item and the rate of which shall be finalized jointly as per the terms and conditions of the contract.

All large projects are being executed under this contract. Hydroelectric projects are generally executed under an item rate contract due to its huge activities and its undefined and unpredictable geological and geotechnical characteristic.

4.6 Cost Plus Contract

Cost plus contracts are generally awarded to renowned, reputable, reliable and highly cost plus quality conscious contractors. In this contract, rates and quantities are not indicated but a limit of reimbursement is mentioned in order to have control over the contract. Reimbursement of expenditures incurred for deployment of workforce to be done on the job are accomplished in accordance to IFC drawings issued to them with agreed overhead and profit.

Cost plus contracts ensure better quality of work because materials and machinery are being provided by the client. It requires strict day-to-day supervision and monitoring of each activity up to completion. It consists of certain clauses like maximum limits of reimbursement of cost expenditures and incentives for accomplishment of work within stipulated time lines.

Incentive schemes are being made from time to time as per requirements of the project on a particular set of activities with a fixed date of completion so as to expedite the starting of the next set of activities. Incentives are paid to the contractor as agreed upon on completion of the work. Any leftover incomplete work can be taken up under this contract without contractual hazard. This contract is not effective for fixed budgeted projects.

4.7 Turnkey Contract

A turnkey contract is taken up when specifications and all functional or operational parameters are finalized with a defined scope of work. No alteration and addition is expected to crop up during execution. In this type of contract, the contractor has the overall responsibility of design, execution, erection, procurement, quality control, safety, etc., up to successful commission of the system for a fixed lump sum price and fixed date of commissioning. The contractor has to ensure that the facility conforms to the performance of the system as per contractual obligations.

In this contract, a performance guarantee is to be furnished by the contractor based on certain stipulations of the contract such as the quality of material used and defined operating conditions with desired working efficiencies.

4.8 Estimation and Tendering Process of Concrete Gravity Dam

Construction of a dam is comprised of multiple activities. Some activities are identified, and some are not identified and defined but unpredictable and change during the construction process. These activities primarily depend on the geologic and geotechnical characteristics that cannot be envisaged accurately in advance. This information can be had from a similar neighborhood project or data gathered through experience or by the help of exploration or investigation. A tender can be called successful if the tender is executed and completed without any extra items and deviation of the contract. This status can only be achieved by proper design, identified and well estimated activities, correctly budgeted and logically well scheduled through an experienced professional.

- *Estimation*: Estimation is a process to forecast the cost of work in which the entire work is fragmented into manageable identified activities. Each activity is quantified technically and rated analytically to sum up the total cost of the work in accordance to specification, drawings, standards, etc.
- *Tender*: A tender is a fair priced competitive offer by a company to accomplish a series of activities of a particular work in accordance to specification, drawings and conditions, both general and special, within a defined timeline and agreed upon costs.

The tendering process for the construction of a dam is a very complex activity for both clients and business associates. Both have to feel the pulse of the geological and geotechnical characteristics of the dam site correctly to get competitive prices while the bidder can furnish a realistic price bid to accomplish the work within the scheduled time frame.

A reconnaissance survey of the site shall be conducted by the bidder before quoting the price bid in the following guidelines:

- Location of site and its access
- Topographical, geological, geotechnical features of the site
- Availability of construction materials
- Environmental and social conditions

- Communication and logistics arrangement
- Infrastructure, etc.
- Nearest railway head
- Nearest airport
- Nearest city/town, etc.

4.8.1 Preparation of Cost Estimate

An effective cost estimate can be prepared with sufficient details of the work. The adequate details of the work can be found from an effective planning technique and work breakdown structure which provides ample details of manageable and identified smaller activities with the requisite level of information so that an effective and realistic estimate of work can be made.

A cost estimate is comprised of two components as shown below:

4.8.2 Bill of Quantities

Detailed estimate of quantities of each activity/item is calculated from drawings prepared by designers and also based on data collected during the reconnaissance survey in regard to topography, geological and geotechnical features of the site.

4.8.3 Schedule of Rates (SOR)

BOQ is prepared taking all components of work into consideration. A description of all work items shall be framed for the completion of each activity as per specification and drawings. Any item of similar category available either in the company's SOR or in any previous contract can be adopted, or the rate shall be derived by analysis from the client. Analysis of items shall be done with the following components:

- Cost of materials
- Cost of carriage of materials
- Cost of labor including skilled, unskilled and ordinary
- Hire charges of equipment and machinery
- Water charges

- Sundries
- Contractor's profit and overhead

The coefficient of materials and labor required per unit of a particular item shall be taken from the approved coefficient of consumption per unit of each activity. The rate of material shall be per current market rates.

4.9 Type of Estimates

After completion of the above two activities, a detailed estimate of the work can be completed.

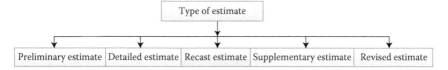

4.9.1 Preparation of Report of Detailed Estimate

A comprehensive and transparent report shall be enclosed with a detailed estimate about the scope of work, drawings, specification, methodology of analysis of rates, condition of site, type of equipment and machinery considered and time limit for technical sanction of the project.

4.9.2 Technical Sanction of Estimate and Recasting of Estimate

The detailed estimate shall be technically sanctioned by the competent authority; if it is technically sanctioned with some comment, the recasting of the estimate shall be done while duly incorporating the comments.

4.9.3 Preparation of Notice Inviting Tender (NIT)

NIT shall be prepared carefully with the following information:

- Name and scope of work shall be clearly mentioned
- Details of Earnest Money Deposit (EMD) and process of deposit
- Eligibility criteria
- System of tender
  - 2 bid system (document related to criteria and commercial bid). 2 bid system is adopted when specification of work has already been finalized and indicated in NIT

- 3 bid system (document related to criteria, technical bid and commercial bid). It is adopted when specification has not been finalized and it will be finalized on receipt from tenderer.
- Detailed procedure of submission of each bid
- Rejection criteria of bid
- Last date with time of receipt of bid
- Opening date with time of bid

4.9.4 Preparation of Tender Document

The tender document prepared shall include the following:

- Notice inviting tender
- Schedule of quantities of work
- Set of drawings related to quantities of work
- Technical specification of work
- General condition of contract
- Instruction to bidders

4.9.5 Publicity of Tender

The prepared tender shall be put in the press or be published on website.

4.9.6 Tender for Sale

The tender document shall be prepared and shall be kept ready for sale. The tender document shall only be sold to eligible contractors.

4.9.7 Receipt of Tender

The tender shall be furnished within the stipulated time.

4.9.8 Opening of Tender and Procedure of 2 Bid and 3 Bid System Tender

As per instructions of NIT, contractors are required to furnish their offers in the laid down process of 2 bid or 3 bid system on the prescribed time and date after submission of EMD to the client. Sealed offers shall be opened in the presence of authorized representatives of bidders in the stipulated time frame. The detailed process is explained below.

- 2 bid system when specification has already been finalized. A pre bid conference is conducted between intended tenderers if any clarifications in connection with specification is required. Minutes of meeting shall be made and circulated to all tenderers.

- Documents related to criteria shall be opened in presence of tenderers/representatives and criteria shall be evaluated. Party shall be qualified or rejected as per evaluation by the competent authority. Financial bid of qualified bidders shall be opened at notified time and date in presence of tenderers.
- 3 bid system when specification has not been finalized.
- Documents related to criteria shall be opened and shall be evaluated first in the presence of tenderers.
- Technical bids of qualified tenderers shall be opened at the given time and date in the presence of tenderers.
- A technical conference shall be conducted at notified dates for finalization of technical specifications. A MOM shall be made and circulated to all to incorporate alterations and deletions of technical specifications.
- All qualified bidders shall be asked to modify their commercial bid and to submit at notified date.
- Modified commercial bids shall be opened at the notified date and time in the presence of tenderers or their representatives.

4.9.9 Comparative Statement

A comparative statement shall be made in the prescribed format of the company and it should be elaborate and comprehensive. All items quoted by the various bidders shall be checked for feasibility of executing each item under the quoted rate. It should be examined whether any rate is quoted excessively low. The lowest bid in the statement shall be examined thoroughly and should be tallied with a departmental estimate of the work and the percentage difference shall be calculated to find out the executability of the quoted price.

4.9.10 Acceptance of the Contract

The lowest bidder from the comparative statement shall be selected and all credentials, resources, equipment and machinery, etc., all are in order and in the line of requirement of the tender document. Finally, the tender of lowest bidder shall be accepted.

4.9.10.1 Issuance of Letter of Intent

A letter of intent (LOI) shall be issued to the successful bidder including the name of the work, amount of work and completion time with zero date of the contract reckoning from the date of issue of the LOI. The LOI shall be signed in duplicate and second copy shall be sent back as a token of acceptance of the offer.

4.9.10.2 Issuance of Work Order

The LOI will be followed by a work order containing the following document:

- Letter of Intent
- Notice Inviting Tender
- Schedule of Rate
- General Condition of Contract
- Special Condition of Contract
- Addendum

4.9.10.3 Glimpses on General Condition of Contract (GCC)

Some salient feature of GCC is narrated below for the benefit of readers:

- *Earnest money deposit (EMD)*: Earnest money at 2.5% of contract value is deposited by the tenderer to ensure their sure participation in the bidding and that they won't back out and refuse to do the work when work is awarded to them. Earnest money deposited is returned to unsuccessful bidders and it is held by the client for the successful bidder.
- *Security deposit*: Security at 10% of the contract value shall be deposited to the project authority in the form of a demand draft or bank guarantee with a validity up to the end of the contractual completion time of the awarded work. EMD deposited at 2.5% shall be considered as the initial security deposit and balance of 7/5% shall be deducted at 10% of each running bill during execution of the project as retention money and shall be held up to the end of the defect liability period.
- *Mobilization advance*: 10% of the contract value is paid to the contractor as per terms and conditions of the contract as an advance to facilitate the mobilization of manpower and machinery on submission of the bank guarantee with validity up to the end of the contract. 10% of each RA bill shall be deducted towards this advance until the entire advance amount is recovered.
- *Composite bank guarantee*: If the authority and contractor agreed to have one bank guarantee known as a composite bank guarantee for both mobilization advance and security deposit/retention money at 10% of the contract value up to end the of the defect liability and an agreed deduction of 10% from each RA bill, the same amount stands credited to security/retention money account until recovery of the full advance. From the date of full recovery of the mobilization advance,

the composite bank guarantee will automatically be converted and towards security/retention during the stipulated defect liability period.

- *Secured advance*: Contractor may be allowed to have secured advance on materials brought to the site for the execution of the items as per the rate schedule. Advances in such cases shall be given in the extent of 75% of the value of materials and an agreement shall be made between the parties. Recovery shall be made from each RA bill where the material used in the item is measured and billed. Complete recovery shall be made well before the work is completed.

- *Zero date*: Zero date of the contract is considered from the date of issue of the Letter of Intent to the contractor.

- *Contractual completion time*: Contractual completion time is the duration in which the entire scope of work is expected to be completed by the contractor.

- *Extension of contractual time*: If a contract is overrun due to reasons attributable to the authority like delay of release of drawings, work fronts or any other holds which are within the scope of the owner, then the contractual time limit is extended proportionately to complete the work. This additional time limit is known as an extension of time without any financial implication on both sides.

- *Defect liability period*: It is the time period after the completion of work where any defects occurred/ liable to occur shall be rectified by the contractor at his own risk and cost. This defect liability period is kept 1 year or 1.5 years as per discretion of contract and BG held shall be released after this period is over.

- *Completion certificate*: The engineer in charge will issue a completion certificate to the contractor after successful completion of the work in accordance to specification and drawings in the prescribed format, duly mentioning the name of work, work order number, contractual contract value, actual value, contractual date of completion, actual date of completion, etc.

- *Final completion certificate*: After defect liability time of the contract is over, an extensive checklist shall be prepared jointly by the department and representative of the contractor and this checklist shall be issued to the contractor for liquidation of the checklist. A final completion certificate in prescribed format shall be issued to the contractor after liquidation of the checklist and observing successful performance of the plant. The security deposit shall be released after issuance of final completion certificate.

- *Liquidated damage (LD)*: Liquidated damage shall be imposed on the contractor due to delay in completion of work beyond contractual completion times. Liquidated damage is imposed at some fixed percentage for each week delay maximum up to 10% of the contract value.

- *Cancellation/termination of contract*: Owner is entitled to cancel the work at any time at his discretion and for stoppage of work if owner feels it is necessary due to any cause indicated below. A notice in writing from the engineer in charge to the contractor of such cancellation shall be issued as a proof of cancellation. Contractor shall immediately stop the work and assess the status of completed work jointly.
  - If the contractor breaches any of the conditions of the contract and fails to remedy the breach within the stipulated time
  - In case of death of tenderer
  - Completion period is not adhered to beyond three months of scheduled time as agreed upon
  - If contractor or any partner of the firm is convicted of any criminal offences
  - Factors as specified by owner

- *Suspension of contract*: The engineer in charge may at any time suspend the work or any part in writing stating that the contractor shall not proceed with the work until he receives a written order to proceed with the work. In this suspender order, the contractor shall not be entitled to claim any compensation for any loss or damage sustained by him due to this suspension and an extension of time shall be granted to him, as suspension was not a consequence of any default or failure on his part. If any suspension order is served for more than three months, then contractor shall option to terminate the contract.

- *Force mejure*: Force mejure means war, revolution, civil war, tidal wave, fires, major flood, earthquakes, epidemics, etc.

- *Misconduct*: If any personnel of the contractor or any sub agency or sub agent are found to be guilty of misconduct or be incompetent or under qualified or negligent in performance or duties in the opinion of the engineer in charge is final and binding. The concerned personnel shall immediately be removed from the site and substitutes shall be put in place without hampering the progress of the work.

- *Labor laws and regulations*: The contractor shall be responsible for strict compliance of all laws, rules or regulations for all agents and agencies working with him. The contractor should comply with the provision of payment of wages, minimum wages, workmen's compensation, etc., in accordance with the prevailing acts or laws in the country. The owner shall be entitled to carry out checks or inspections of contractor's facility, records, etc.

- *Indemnity and insurance*: The contractor shall indemnify and keep indemnified the owner from and against all third party claims including but not limited to property loss, damage, personal injury, or death of any personnel.

4.9.11 E-Tendering

At present, the professional environment is highly competitive and dynamic. The contract department of any organization is under immense pressure to reduce cost expenditures and they have to locate, identify and select efficient and qualified contractors who will be able to deliver the quality of work within the stipulated time. The traditional tendering process is mostly manual, time consuming and cumbersome. The responsibility of the contract department is to find a qualified contractor with a very competitive offer who can deliver the work with quality so as to start a project. A tender is released in many occasions with a few bidders by compromising many aspects in order to avoid delay of re-invitation in the elaborate process of traditional tendering.

Industries are dominated by a wide range of technologies and e-activities. Exchange of electronic information is an order of the modern professional world. Information and communication technology plays an important role of exchanging information between owner and contractor during the tendering process. Transformation of this information is accurate and efficient whereas the traditional tendering process is manual and mostly paper based which consumes more time and is more costly.

E-tendering is an internet based tendering system where the entire tendering process can be performed online from invitation, uploading bid documents and subsequent activities like submission of bid documents, opening, evaluation and award of contract. It provides a centralized process that enhances the efficiency and subsequently reduces the cost expenditure.

4.9.11.1 Different Phases of E-Tendering

Hassle free and effective, E-tendering systems where tenderer can download the tender and submit a bid online through the company's portal. An indicative E-tendering system is described below.

- Authorized representative of owner shall obtain digital signature certificate from approved certifying authority.
- E-mail shall be sent updating authority and indicating name, user ID, and digital signature certificate number.
- After logging in for first time, the system will ask to change the password and the password shall be changed for safety.
- Scanned copies of bid documents that include eligibility criteria, technical bid and price bid shall be uploaded in the system.

- All relevant drawings and specifications in prescribed format shall be uploaded as a virtual library for reference.
- Invitation of bid shall be uploaded and notice shall be sent to all intended bidders to download and respond online.
- Enrollment as new contractor
  - Enter user ID, name of applicant, password and confirm password, company's name, email ID. Unique company's ID will be generated.
  - Use digital signature certificate/E-token when it is asked.
- Contractor will download entire document for electronic bid response.
- Document related eligibility criteria uploaded by contractors shall be opened first and a list of qualified bidders shall be published.
- Documents of technical bids shall be opened and qualified contractors will be called for a conference to discuss queries, if any, and finalize the specifications. New technical bid documents shall be prepared in accordance with MOM made in the conference. If there are no changes in specification, a revised commercial bid need not to be made.
- New bid documents shall be uploaded with revised financial BOQ, date of submission and date of opening of revised financial bid.
- Intimation shall be sent to qualified bidders about last date of submission and date of opening of revised commercial bid. Submission of revised bid is essential; earlier submitted bids shall be null and void.
- Revised financial bids shall be opened.
- Schedule of quantities duly filled up by each bidder and comparative statement shall be downloaded by the authorized person and signed off on.
- Bid documents of lowest bidder shall be downloaded and signed off.
- The comparative statement is an auto generated document which shall be checked by regulatory authorities carefully. The schedule of rate duly filled out by each bidder shall be checked and verified.
- Processing of bid: The following documents shall be submitted to accepting authority
  - Hard copies of all documents of lowest bidder
  - Schedule of quantities with quoted rates of all bidders
  - Comparative statement
  - Approved NIT
- LOI shall be issued to successful bidder as approved by acceptance authority.

4.9.11.2 Work Breakdown Structure in Hierarchical Structure

The work breakdown structure follows the principles of hierarchical decomposition where project works are broken into smaller and manageable tasks for achieving the completion of the project in the scheduled timeline. This decomposition process starts from a top down perspective in light of the whole system. Size and scope is decided as per requirements suited to the nature of the project which can be easily understood and managed by each member of the project team. The higher level represents project deliverables and the next and subsequent level contains smaller manageable sub deliverables/tasks with more details so as to enable each member of the team to perform in a way to achieve the target within stipulated timeline.

4.9.11.3 How to Prepare WBS

Project, planning, construction and contract groups jointly create the breakdown structure of the project. A major role is played by the planning group who identify the major functional deliverables for the project and break it down it into sub deliverables and subsequently to further smaller tasks and calculate the weightage of each task in terms of percentage of the whole to draw an S-curve of the project schedule. It is an effective tool for planning, scheduling, and monitoring the performance of construction activities so as to create work status reports. All works considered in the preparation of WBS are identified, estimated, scheduled and budgeted. Any work excluded from WBS is to be considered beyond the scope of work. WBS is a multi-level supporting tool of management for the entire project life cycle, and acts as an eye of management that displays the status of the work and emphasizes on slippages in schedules, flaws, action required and areas of concern.

Let us take the example of construction of a dam, which is a major functional deliverable and can be broken into sub deliverables such as foundation treatment, concreting, fabrication and erection of radial and sluice gate. Sub deliverables like foundation treatment can further be broken down into smaller tasks such as drilling of grout holes, flushing and washing of grout holes by application of air and water alternatively to cleanse the clay seam from grout holes, and consolidation grouting of holes in the foundation. Next, sub deliverables like concreting can be broken into different smaller tasks like placement of shuttering, reinforcement, PVC strip in construction joints and layer wise concrete in each lift, curing, etc. These sub deliverables are further broken into smaller segments with more specific details, size, scope, time line, and cost, which can be called a work package and is to be assigned to each engineer of the group.

Every concerned department will have this sort of work package to complete the work and each deliverable can be budgeted and the entire project can track financial performance along with project performance by integrating cost expenditure from the organization breakdown structure and project breakdown structure.

4.9.11.4 Type of WBS

WBS is generally classified in two categories:

- The project authority develops WBS considering the total project in multilevel development. Each vendor participating with a specific contract can develop contract specific WBS based on this overall project WBS.
- Vendor's WBS is developed by individual vendor considering all deliverables that are part of their awarded contract.

4.9.11.5 Work Breakdown Structure Diagram and Outline with Work Packages

Broken down activities of civil and structural work of construction of dam are identified. Each package is placed in sequence to have a fair idea of all activities involved in the construction of a dam at a glance in the diagram below.

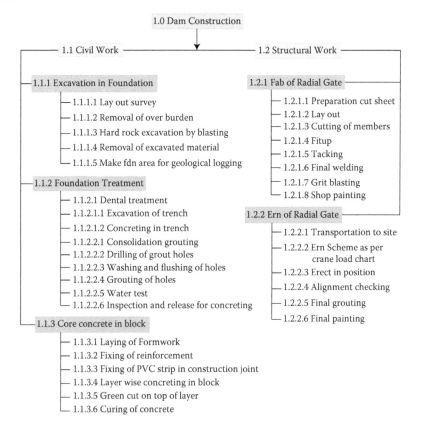

4.9.11.6 WBS Dictionary

The WBS dictionary is a document which contains all information about deliverables, activities, schedules, resources, cost, and scope of each work element being considered in WBS. It provides complete details of work to team members to understand the work package without any misconception and miscommunication, and team members can use it to control the work and ensure that no changes are incorporated in the package so as to complete the work in the stipulated timeline.

Role of WBS in project planning: Planning is a systematic project and time management technique where major project deliverables are broken down into smaller and manageable sub deliverables so that a logical network with sequential activities can be drawn based on available resources to achieve the desired target within the stipulated timeline.

WBS provides all the required information for planning and scheduling of any project. The work package of WBS provides the work, duration, and costs for the tasks required to produce the sub deliverable.

4.9.12 Network Planning

A large project contains numerous deliverables that are to be identified, estimated, budgeted and scheduled with a desired date of completion. A goal without proper planning of a project is a mere wish. If the project authority doesn't know where they are moving, they land somewhere else far away from the goal and face a disastrous situation. In order to avoid this situation, structured and elaborated planning is required at this very important phase of project management.

In this stage all activities are to be identified, estimated, scheduled with the timeline and all sequences are to be defined, interdependency, interlinks and concurrency are to be established for reaching the desired goal.

Based on these exercises, a graphical representation of all activities are drawn with logical interference with the timeline of each activity in sequence and due consideration of interdependency and concurrency for achieving the goal. This entire graphical representation of accomplishment of activities in a logical mode is called Network Planning.

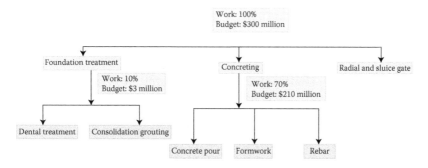

A project consists of three arms: scope, time and cost. These are called the three constraints of project management. These three arms make a triangle with quality as the core. In modern project management, safety is added as another component as a constraint of the project management system. The project manager should have 100% clarity about these constraints before making any schedule.

Main Components of Project Management:

- *Scope of the project*: Personnel engaged in project management should have clarity, conception and clear understanding about the work to be undertaken.
- *Estimation*: Project deliverables shall be broken down into various manageable activities both independent, interrelated or concurrent to estimate the quantum of each activity.
- *Timeline*: Total time within which entire work is expected to be completed. Time is the essence of project management.
- *Resources*: Resources are a source of support and strength in the form of money, manpower and machinery which shall be deployed in accordance to scope and timeline to complete a work.
- *Schedule*: A list of planned activities that are placed in network logically and sequentially to accomplish a work.
- *Risk*: Assessment of predetermination of situation where schedule may go awry and its corrective measures.
- *Quality control*: Systematic implementation of quality measures to complete a work without any flaws.
- *Safety*: Safety is not negotiable at the work site. Safety in the work place is measured by the records of accidents, injuries, illnesses, fatalities, near misses, etc. Safe work practices must be ensured at the site to save human life as well as project time losses.

4.9.12.1 Concept of Planning and Scheduling of a Project

A large-scale project having multi-task activities is a challenging job for any project manager. A systematic and structured approach is necessary to handle this magnitude of work. The entire project shall be broken into smaller, manageable activities and all activities are to be identified, quantified, estimated and scheduled within the timeline.

A detailed study shall be taken up to find out the relationship between predecessor, successor and concurrent activities. A timeline of each activity shall be defined based on relationship and quantity which shall be in a suitable planning tool to predict a probabilistic completion time of the project.

Three methods shall be discussed where numerous details shall be considered to coordinate all these activities and to make a realistic project schedule to complete the project and also to monitor the progress of work effectively.

Methods are as flows:

- Bar Chart
- Program Evaluation and Review Technique (PERT)
- Critical Path Method (CPM)

Basic information required for all three methods as indicated below:

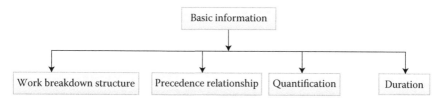

The project manager who is responsible for the completion of the assigned project within the alloted timeline. He is aware that time is directly proportional to the cost of project. Time is money, so his mind is flooded with many queries and questions which should be solved so as to complete the work within the time limit.

- How to assess all activities small or big, manageable or difficult, critical or non-critical, independent or interdependent, sequential or concurrent from the assigned project that play a crucial role in the progress curve of the project?
- How to present all these activities in one medium so that performance of the entire project by monitoring each activity can be measured?
- How to quantify all activities?
- What is the time required for each activity and entire project for completion?
- How and when to start and finish each activity?
- Type of resources required to complete each activity within the time limit?
- What are the activities that are to be undertaken concurrently?
- How much time can be permitted for the delay of one activity so that it won't affect the start of next activity?
- If delay of a critical activity postpones the start of next critical activity, what are the remedial measures to be taken to recover the loss of time so that the schedule remains unaffected?
- How to identify preventive controls so as to reduce the effects of disruption of work due to accidents or bad quality?

- How to create a contingency plan that ensures losses incurred can be recovered?
- What are the possible bottlenecks that may arise during execution?
- How to find out shortfalls and areas of concern that may affect the progress of the project?
- How to monitor cost expenditures of project with respect to the progress curve?
- Whether any cost vs time or man hour vs time progress curves can be made?
- How can a project be expedited by investing more for speedy completion?

Most of the queries invading the mind of a thoughtful project manager will have their answers in the network planning of PERT and CPM which will be discussed in this chapter.

Here under the advantages of the network system:

- All project deliverables can be displayed in the network in sequence
- It facilitates effective performance monitoring of each activity
- It helps show how to utilize resources to achieve targets
- It emphasizes critical activities and how their delay affects the time schedule of the project
- It identifies the activity lagging behind the schedule and slippages occuring in the schedule
- It identifies the shortest time required to complete the project
- It provides information regarding manpower, material and capital requirements
- It points out areas of concern in advance to take appropriate action prior to commencement, including shortages of manpower due to big local festivals, delayed deliveries of long lead items, and non-availability of adequate work front at site
- It advises about interdependencies of activities and task relationships
- It generates hold up report due to non-availability of IFC drawings, materials, etc.
- It provides facts and figures upon which decisions can be made
- It helps to do risk analysis
- It can suggest alternative plans to accomplish the work
- It can assess the effect of shifting resources from a less critical activity to an activity causing the constraint or bottle neck

- It eliminates crisis management
- It provides adequate data for effective monitoring and coordination for expediting the work

4.9.12.2 Components of Network

PERT/CPM is a project management tool which is used to draw a probabilistic network to accomplish any project within stipulated timeline. Various components of the related network are defined below.

- *Event*: An event is comprised of the start and finish of an activity without specific duration and represented in a circle with a short description of the activity.
- *Activity*: Activity represents a task that needs to be completed.
- *Duration*: Total time required to complete an activity in the network system.
- *Critical path*: Shortest time required to complete a project.
- *Predecessor activity*: An activity that must be completed before a particular activity starts.
- *Successor activity*: An activity must start immediately after a specific activity is completed.
- *Earliest start time (ES)*: Earliest start time is a time when an activity can start.
- *Earliest finish time (EF)*: Earliest finish time is a time when an activity can finish.
- *Latest start time (LS)*: Latest start time is a permissible time when an activity can start.
- *Latest finish time (LF)*: Latest finish time is a time when an activity can finish.
- *Float or slack*: Float or slack can be defined as the difference between the latest start time and earliest expected start time of an activity, so float $= LS - ES$.
- *Dummy activity*: It is an imaginary activity which establishes interdependency between two or more activities in the network and denoted in a dotted line.

4.9.12.3 Construction of Network

In the PERT network circles represent events and arrows represent activities.
ES = Earliest Start of an Activity, EF = Earliest Finish of an activity and t = duration

$$EF = ES + t$$

TABLE 4.1

Activities and Duration for Drawing Network

Activity	Symbol	Title	Immediate Predecessor	Duration (Days)	Remarks
1–2	A	Survey and Layout	–	10	
1–3	B	Infrastructure Dev	–	5	
1–4	C	Installation of BP	–	20	
2–6	D	Fdn Preparation	A	15	
3–5	E	Setting of Lab	B	3	
4–7	F	Calibration of BP	C	4	
5–8	G	Mix Design	E	28	
6–8	H	Fdn Treatment	D	15	
7–8	I	Trial Run BP	C	5	
8–9	J	Blinding Concrete	D,E	2	

LS = Latest Start of an activity, LF = Latest Finish of an activity

$$LS = LF - t$$

Draw a network with the following activities from A to J indicated in Table 4.1 and refer Figure 4.1.

Draw a network with float as indicated in the Table 4.2 and refer to Figure 4.2. Calculate float in each event from the network. Main events of a project shall be identified, where the earliest starting and earliest finishing events are fixed for any project. Let's take event number 1, which is the starting point of the project. Earliest and latest start is 0/0. Here, the critical path is 1-2-6-8-9, where float is zero. There is no difference between the earliest start time and latest start time, that is, LS − ES = 0. Events 3, 5, 4 and 7 are concurrent, and all these events are supposed to be over within 40 days before but not later. Event number 4 is a concurrent one but it is not on the critical path. The earliest time of this event is 20 and that of event number 7 is 35. The activity between 4

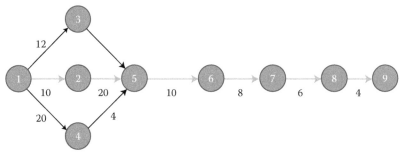

Critical path (1-5-6-7-8-9) = 58 days

FIGURE 4.1

Critical path network as per Table 4.1.

TABLE 4.2

Activities and Duration for Drawing Network with Float

Activity	Symbol	Title	Immediate Predecessor	Duration (Days)
1–2	A	Drilling of Grout Holes	–	10
1–3	B	Procurement of Cement	–	12
1–4	C	Batching Plant Installation	–	20
2–5	D	Consolidation Grouting	A	20
3–5	E	Procurement Aggregates	–	14
4–5	F	Calibration of BP	C	4
5–6	G	Blinding Concrete	D	10
6–7	H	Placement of Shuttering	G	8
7–8	I	Placement of Rebar	H	6
8–9	J	Block concreting	I	4

and 7 requires only 4 days, so event number 4 gets a permissible start time (35 − 4 = 31 days) without affecting the progress of the project. The earliest start time of the event is 20 and latest start time is 31, so float is 31 − 20 = 11 days, and an additional 11 days do not affect the event numbers 7 and 8 and critical path. Event numbers 3 and 5 are also concurrent ones. Event numbers

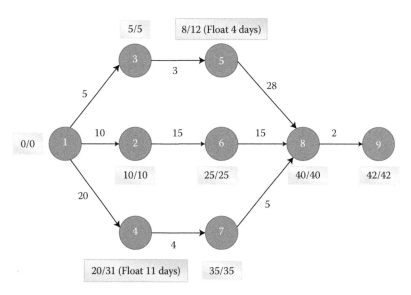

Critical path 1-2-6-8-9 = 10 days + 15 days + 15 days + 2 days = 42 days
Critical path where float is zero

FIGURE 4.2
Critical path network as per Table 4.2.

having a float of 4 days do not affect other activities on the critical path of 42 days.

4.9.12.4 Explanation of Dummy Activity

A dummy activity is an imaginary activity that does not consume any time and it is required to be present in the unique identification of the activity. It is used in the network to show dependency relationships or connectivity between two or more activities. It is shown in the network with a dotted line.

> **EXAMPLE 4.1**
>
> In a dam construction network, activity C (block concreting) is dependent on activity A (dental treatment) and activity B (consolidation grouting). A & B are two concurrent activities. Activity C can only be taken up after completion of A & B. So, activity D can be represented with a dotted line to show the connectivity between A & B (Figure 4.3).
>
> **EXAMPLE 4.2**
>
> A batching plant was being installed and calibrated in a project for taking up concreting work while design of concrete was being taken up simultaneously after the trial run.
>
> - Installation and calibration of the batching plant (Activity A)
> - Conducting mix design in the lab (Activity B)
> - After successful calibration of BP, train run was taken up (Activity C)
> - Concreting is taken up in block (Activity E) (Figure 4.4)
>
> Activity A and B start concurrently. Activity C and E are two parallel activities. Activity C is preceded by A, and activity E cannot be started

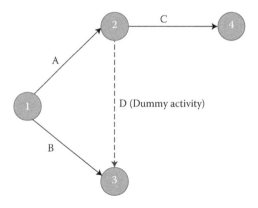

FIGURE 4.3
Dummy activity.

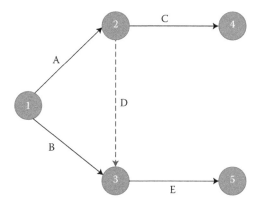

FIGURE 4.4
Dummy activity.

until activities A and B are completed. Trial run (activity C) can be taken up after A is completed and concreting can be taken up after the installation of BP and mix design is completed and connectivity is represented by a dummy activity D with an arrow pointing to event number 3.

4.9.12.5 Estimation of Time

Time is the essence of any contract or commitment. The project manager is responsible for delivering the project on time and he is supposed to closely monitor the project to find any lapse between the events. The project manager shall consider three possible times in a PERT network:

- Optimistic completion time, where everything will happen in the way everything is expected to happen without any obstacles and difficulties. It is the least amount of time to be taken for completion of a project.
- Pessimistic completion time, where everything does not happen as per the plan and the project faces lots of difficulties and obstacles. It is the longest time expected to take to complete a project.
- Most likely completion time, which is neither optimistic nor pessimistic but is assumed to go in the normal way to complete the work by putting effort again and again towards accomplishing the work.

Considering everything, a responsible manager is to find a possible but acceptable time for completion of the project by combining all three times into what is called the expected time. A standard deviation indicating a certain amount of uncertainty is considered in the estimation of time.

Let us express optimistic time as a, pessimistic time as b and most likely time as m

$$\text{Expected time, Te} = \frac{a + 4m + b}{6} \qquad (4.1)$$

Standard deviation should be calculated before expected time is calculated.

$$\text{Standard deviation, Sd} = \frac{b - a}{6} \qquad (4.2)$$

The project manager should be careful enough to select expected time of the project.

As for example, a = 6 months, b = 10 months and m = 8 months.
Expected time 1 = (6 + 4 × 8 + 10)/6 = 8 months.
2nd example, a = 5, b = 15, m = 7.
Expected time 2 = (5 + 4 × 7 + 15)/6 = 8 months.

Expected time of completion is 8 months for both cases, but Example 4.2 contains more uncertainties.

4.9.12.6 Probable Completion of the Project

The project manager got all data and the list of tasks along with critical path of completion time of 12 months. He can judge the probable completion of the project, and he can also find the probability of completion if he would like to complete the project in 11 months instead of 12 months. Project probale completion time can be assessed by use of standard normal probability Z-table.

4.9.12.7 Time-Cost Trade Off

Projects consist of three arms: scope, time and cost. These three are known as the constraints of the project activities. The project triangle is formed with these three arms, with resources at the core of the triangle. Resources are to be utilized effectively by a project manager to accomplish the project within the time and estimated cost, which is known as performance. Resources, time, cost, quality and safety are five fingers of performance that are respected by any sensible and responsible project manager.

The project manager should utilize available resources in a systematic and structured manner so as to complete the work and to achieve the target without losing any time. He should monitor all activities in such a way that all activities are completed maintaining proper quality of work and strict adherence to safety norms to save unnecessary wastes of time. Strict adherence to safety and quality norms save time and money. An efficient project manager should do a time-cost trade off analysis to reduce the expected completion time, which will lead to an increase in the cost of the project. The following activities shall be taken up for effective application of the analysis:

- Complete the project in the scheduled time frame
- Recover any slippage that occurred in the S-curve
- Avoid any compensation due to delay in completion
- Make any left over resources free for next project
- Improve fund liquidity
- Enjoy any incentive scheme launched by client for early completion

The project manager should apply the following to reduce the time of completion:

- Offering time bound incentive schemes
- Working overtime in multiple shifts
- Inducting additional resources
- Releasing work fronts at site
- Removal of hold from IFC drawings
- Removal of material hold
- Strict supervision for achieving quality and safety to avoid duplication of work and work stoppages due to accidents

Normal time: Normal time is the time that is required to accomplish work under usual condition.

Crash time: Crash time is a time when a normal time is compressed to complete work, and this crash phenomenon will increase the cost expenditure of the project (Figure 4.5).

Cost slope: The slope of line AB from the normal time to crash time is called the cost slope of activity.

$$\text{Cost slope} = \frac{\text{Crash cost} - \text{Normal cost}}{\text{Normal duration} - \text{Crash duration}} \qquad (4.3)$$

4.9.13 Preparation of S-Curve

The S-curve is a project management tool by which progress of the project can be determined and the status of the project can be visualized. This curve can be drawn as cost vs. time, quantity vs. time, and man hour vs. time. Actual and target S-curves can be drawn simultaneously to indicate actual progress and shortfalls that occurred in the curve. The slippage can be ascertained from the curve.

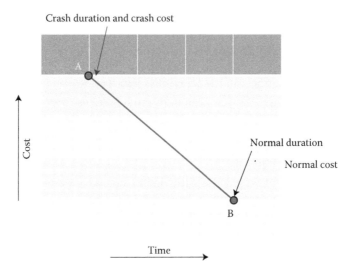

FIGURE 4.5
Crash time.

4.9.13.1 Monitoring Mechanism

An effective monitoring mechanism shall be in place to reduce time and cost overrunes. The main objective of this mechanism is not to reduce the time of completion but to arrest the slippage likely to occur in the schedule of the work by timely identifying the constraints and bottlenecks so that work can be completed in the scheduled time.

The following steps are to be taken to complete the work within the scheduled time:

- Aggressive monitoring systems in all levels
- The S-curve shall be used for monitoring progress and to determine status
- Weekly planning shall be done along with client/construction group/ contractor
- Front availability against programs during the week shall be studied
- Material, manpower and machinery availability vis-a-vis planned programs shall be checked
- Weekly material demand shall be placed to procure and store in advance
- Delivery of long lead items shall be expedited
- Engineering hold up and material hold up reports shall be made. Close interaction with engineering shall be done for early release of drawings and drawing status reports shall be sent
- Areas of great concern shall be indicated

- Planning personnel will go to sites regularly to assess the reality and physical progress at the site so as to avoid over reporting of progress
- Weekly progress review meetings shall be conducted between the client, construction group and contractor. Actual progress vs. programmed shall be checked; if there are any shortfalls, the reasons for the shortfalls shall be identified and corrective action shall be taken
- Fortnightly meetings shall be conducted between procurement, planning and construction group for the status of materials
- Geologists shall be posted at site to sort out constraints pertaining to geology to quickly release the work front
- Weekly meetings shall be conducted with quality and safety
- Smooth fund flow to contractors

Program of construction of hydropower project: Breakdown structure (WBS) is prepared for the entire work, and a detailed bar chart is made for all activities related to civil, eletro-mechanical electrical and start-up, and commissioning activities; based on these, a detailed PERT/CPM network shall be prepared to monitor all activities including critical activity to avoid any slippage in the schedule (Table 4.3).

4.10 Start-Up, Testing, Pre-Commissioning, and Commissioning

Introduction: Effective project management is a process by which a huge project can be planned, programmed, executed, tested and handed over to an operation group for smooth operation. During the construction phase, many small but critical activities are overlooked due to work pressures from top management for speedy completion of work. All those flaws ultimately become a bottleneck during the commissioning phase that delays the commissioning.

The project group is supposed to develop a commissioning group along with operations so that early monitoring of critical activities can be initiated which will address uncalled for delays in commissioning the project. This activity should start when the project is almost over more than 85% of the project complete.

4.10.1 Planning for Commissioning

Commissioning activities shall be planned meticulously and in a structured way so as to make it a great success.

The Operation and Maintenance (O&M) group along with QA/QC prepare an overall schedule of commissioning. It will cover all components such as

TABLE 4.3

Bar Chart for Construction of Dam

S. No.	Activity	Months														
		4	8	12	16	20	24	28	32	36	40	44	48	52	56	60
1	Report preparation, reconnaissance, pre-feasibility, feasibility and DPR	▮														
2	Preparation of general layout hydrological aspect, foundation engineering, seismic aspect, geological features	▮														
3	Preparation of DPR	▮														
4	Review of DPR and submission to authority		▮													
5	Engineering															
a	Preparation of technical specification, standard, preliminary drawing etc. of Civil works		▮													
b	Electro-mechanical work		▮													
c	Electrical works etc.		▮													
d	Long lead items		▮													
6	Preparation of tender document															
a	GCC, SCC, tech spec etc.		▮													
b	Development of schedule of rates		▮													

(Continued)

TABLE 4.3 (*Continued*)

Bar Chart for Construction of Dam

		4	8	12	16	20	24	28	32	36	40	44	48	52	56	60
c	Computation of cost estimate		▮													
d	Finalization of tender document			▮												
7	Preparations of detailed engineering drgs and IFC drawings															
a	Civil IFC drawing		█████													
b	Electro-mechanical IFC drgs		█████													
c	Electrical IFC drawings			███												
d	Reviewing vendor drgs and specification of E&M works				███											
	Tendering process															
	Notice inviting tender															
	Finalization of civil work order and issue of LOI			███												
	Finalization of E&M WO and issue of LOI				███											
	Finalization of elec WO and issue of LOI				███											
8	River diversion and coffer dam		████████													
9	Construction starts				█████████████████████████████████████											
a	Civil and structural work				████████████████████████████████											
a.1	Intake and low pressure tunnel fab and erection of intake structure					████████										

(*Continued*)

TABLE 4.3 *(Continued)*

Bar Chart for Construction of Dam

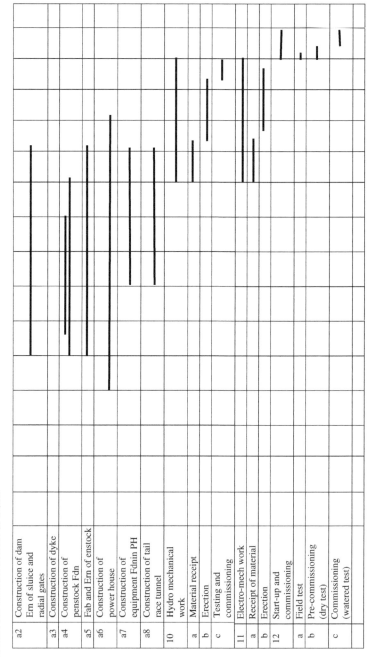

Source: Adopted from AHEC-IITR, "1.8 General– Project Management of Small Hydroelectric Projects," standard/manual/guideline with support from Ministry of New and Renewable Energy, Roorkee, December 2012.

water retaining structures, water conducting systems, hydro-mechanical equipment, electro-mechanical equipment, transmission facilities, etc., along with the sequence of testing and expected duration of each activity.

Time consuming activities like deforestation, rehabilitation of locals, relocation of wild life, etc., shall be taken up much in advance to facilitate a smooth commissioning process.

An organogram shall be prepared duly indicating assignments of individuals with specific responsibility and accountability where each test activity/test shall be documented and recorded in the prescribed formats as per the commissioning documents, and are to be signed off by all participants.

A network shall be developed duly indexing and identifying all activities. Concurrently, parallel and sequential activities shall be identified and each activity shall be allotted a timeline in logical order to connect the next activity.

4.10.2 Engineering Diligence

Each hydel project is unique in nature. It has its own difficulty and constraints. Care shall be taken from design to installation of any equipment like hydro mechanical, electro mechanical, etc., otherwise it may affect the performance due to faulty equipment and it will be a difficult exercise to rectify during start-up which may cause a huge delay in commissioning.

Proper engineering diligence shall be applicable to every stage starting from design, shop test, installation, field testing, commissioning and startup.

4.10.3 Punch List Categorization

A system shall be developed to identify and record incorrect installation, erroneous work done, incomplete tasks, damaged parts or any work completed not conforming to specifications. All these defects need to be rectified and completed to facilitate the commissioning of the project. This list of balance activities is called the punch list. The expert group will categorize the punch list items as per requirements.

- *Category A*: Items must be completed before the start of pre-commissioning
- *Category B*: Items shall be cleared during pre-commissioning
- *Category C*: Items for commissioning

The following are the general guidelines for diligence before start-up:

- Mechanical work
  - Hydrostatics test on each system

- Factory and field testing on all major equipment like turbines, governors, draft tubes, spillway gates, intake gates, shafts, etc.
- Unit alignment, bearing setting and centering, gap, etc.
- Functionality checks for brakes, cooling systems, oil pressure units, top cover drainage systems, excitation systems, etc.
- Check all dimensional sheet
- Electrical work
  - Generator
  - Main power transformer
  - Switch gear, both medium and low
  - Unit control switch board
  - Ground resistance certificate
  - Surge protection of power house and switch yard
  - All relays, station battery and charger, etc.
- Civil work
  - As-built drawings shall be prepared indicating coordinates, level, dimension and center line, etc., of all structures
  - Layout and dimension of power house and its inside space provided

4.10.4 Formation of Commissioning Group

A commissioning group shall be formed by the competent authority

- Experts from operation group
- Experts from maintenance group
- Representative from QA/QC
- Representative from construction group
- Representative from contractor
- Representative from HSE

The entire group shall be led by the lead of the operations group because ultimately he has to run the plant after handing over the system. He should guide the group in the desired direction. The group shall make a plan for commissioning, and the roles of all participants shall be identified and clarified for the smooth handing over of the system. Day-to-day interaction and evening meetings shall be conducted with all concerned to discuss the status of the commissioning.

4.10.5 Documents Required

The commissioning group requires the following documents before planning start-up and commissioning activities.

- All as-built and IFC drawings
- Results of all construction tests and inspection reports of electro-mechanical equipment, pipelines and its accessories, instrumentation, etc.
- Quality control documents as per company's quality manual, etc.

4.10.6 Planning for Start-Up and Commissioning

The commissioning scheme shall be prepared by experts who possess thorough knowledge and field experience of power generation. It requires very careful coordination and the strict generation of documents of each activity in every stage to ensure safety and reliability of commissioning activities.

4.11 Start-Up

All activities that are required to be performed after pre-commissioning and prior to trial operation inclusive of pre-commissioning inspection and checkout of all equipment, system and enable commissioning/trial operation.

4.11.1 Pre-Commissioning

All activities are to be taken up before launching the start-up, commissioning and trial run of the plant.

4.11.2 Commissioning

In this final stage, all major plant systems and equipment such as turbines, generators, governors, excitation systems, etc., shall be put into service in accordance with the approved operation manual and also as per procedures recommended by the designer/process licensor and suppliers approved by the operations group after the successful trial run of the plant.

4.11.3 Responsibility of Each Group

Experts who prepare the overall schedule covering water retaining structures, low and high pressure tunnel, electro-mechanical equipment, transmission, etc., and the contractor works out detailed procedures. Each step of the process shall be documented as required. All groups working shall report to one single entity who is overall responsible for commissioning.

- *Operation group*
  - Review the system as per design guide line, IFC drawings and operational manual, etc.
  - Witness all tests and activities as per commissioning plan
  - Make checklist of all system or equipment after inspection
  - Issue the checklist to construction group for liquidation
  - Acceptance of system and equipment after 100% liquidation of checklist
  - Operate all equipment to support start-up activity
- *Construction group and contractor*
  - Perform tests in all equipment/systems as per guidelines
  - Prepare records of test data and other tests conducted in support of commissioning and signed off by all participants
  - Mark the line/system/equipment cleared
  - Offer checklist for inspection
  - Provide skilled manpower for conducting tests, etc.
- *QA/QC group*
  - Check quality of work and all documents pertaining to the particular work
- *Manufacturers and suppliers*
  - Some tests are to be performed during the commissioning of equipment per the terms and condition of the purchase order. These tests are to be done in addition to the tests being conducted in the shop.

4.11.4 Start-Up Test

Before starting of machine, start-up tests of generating units and all other systems and sub systems should have been done during field tests. These tests are divided in two categories as indicated below.

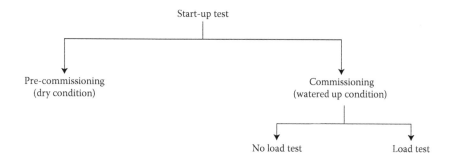

4.11.4.1 Pre-Commissioning Test

It is conducted to ensure that work has been carried out per design parameters and dry tests. All equipment is tested to verify and to confirm that all equipment/systems have been constructed, erected and aligned in accordance to design specifications their and to ensure the smooth functionality in accordance to the contract document.

4.11.4.2 Commissioning Test

It is conducted to verify the correct operation of the equipment/system after it is installed.

Dry tests on mechanical equipment shall be carried out in advance if the project has a reservoir or storage facility. A dry test must be completed before the reservoir is filled up with water and also before the high pressure tunnel and low pressure tunnel get filled.

Tests and checks are to be conducted per requirements:

- Connectivity of auxiliary equipment
- Correct operation of all instrumentations and safety devices
- Insulation resistance test and dielectric test for generator
- Fire protection system
- Tightening of penstock man hole, draft tube man hole with proper gasket and O ring
- Hydro mechanical gates and valves to be checked
- Function of all unit auxiliary and station auxiliary

4.11.4.2.1 Preparatory Work for Wet Test

Much preparatory work shall be carried out before conducting the watered up test. This test is time consuming and needs a lot of resources to make the project ready for a watered up test. The following activities are to be taken up before conducting the commissioning activity.

- Deforestation of the entire reservoir area. Logs, vegetation, etc., shall be removed
- Wildlife shall be relocated to a safe place
- Removal of any archaeological structures
- The groundwater table shall be measured in surrounding area
- Leakage and deformation of dam, LP and HP tunnel shall be observed
- Transmission lines and connected switch yard must be in operation
- Effective interaction, coordination with third party like grid operator for smooth commissioning
- Rehabilitation of affected local people

All water retaining structures are to be inspected and tested dry before taking up wet and load tests on generating equipment.

- Water leakage from closed gates, emergency closure valves, penstocks, DT manholes, coolers, shaft seals, etc.
- Pressure test on tunnel and penstock
- Discharge test on regulating and flow controlling gates/spillway gates
- Working condition of flow meter
- Start machine at low speed and check for any sound or interference in the machine
- Unit rotation starts at low speed and then increase gradually from 25% and 30% to 100% speed
- Online field test of generator
- Bearing temperature stabilization, etc.
- All protective devices, lock out relays and emergency stop systems are checked at low speed for functionality

4.11.4.2.2 No Load Test
The following are the no load activities:

- Check start-up and shut down sequence including mechanical braking
- Conduct initial mechanical run including over speed test
- Conduct mechanical run for bearing temperature
- Measure shaft run-out (Eccentricity)
- Conduct functional test for generator protection relay
- Conduct functional test for excitation system during no load running
- Measure short circuit and no load curve

- Determine generator reactance and time constant
- Synchronization with grid

4.11.4.2.3 *Load Test*

- Conduct load rejection test on one generating unit to check response of turbine governor or excitation system under step up load condition from 25%, 50%, 75%, 100% to 110% of design load
- Conduct reactive capability test
- Conduct power system stability test
- Measure V-curve characteristics
- Check temperature rise and power output
- Time taken by machine in stopping after application of brake
- Measure vibration

4.12 Liquidation of Punch List

The punch list furnished shall be liquidated before conducting a performance test.

4.13 Performance Test

A field performance test is carried out to verify the efficiency of the turbine and generator. A field acceptance test is one that is done to verify the efficiency of the turbine as per IFC 600041. Individual losses are measured to find out the efficiency of the governor.

After satisfactory commissioning, trial operations shall commence. The duration of the trial run depends on the size of the project.

Smaller Project: 3–10 days for continuous run and major project for 30 days per unit.

4.14 Project Close Out Report

The project close out report is an informative document pertaining to a project which is prepared by the planning group after completion of a project in the commissioning stage. It is the final status of the project and is prepared by the planning group who collect necessary information during execution of the project. This report should contain the following:

- Detailed data about the project such as name, location, project authority, PMC, contractors, starting date, budget, duration, etc.
- Final status of the plant and status of acceptance of the plant by the project authority
- Status of acceptance checklist which do not affect commissioning but need to be liquidated within timeframes given by the operation and maintenance group
- Resources used during execution vis-à-vis planned
- List of constraints faced and corrective measures adopted to accomplish the work
- Lessons learned during execution and how will it be tackled in the next project in similar situation to avoid any slippage in time schedule
- Vivid report on standard health and safety
- QA/QC management for delivering a quality product which ultimately save time for speedy completion

Bibliography

AHEC-IITR, "3.12 Electro Mechanical: Erection, testing and commissioning", standard/manual/guideline with support from Ministry of New and Renewable Energy, Roorkee, December 2012.

AHEC-IITR, "1.8 General– Project Management of Small Hydroelectric Projects", standard/manual/guideline with support from Ministry of New and Renewable Energy, Roorkee, December 2012.

5

Site Mobilization and Kick-Off

5.1 Introduction

Construction management consists of many phases starting from mobilization of manpower and machinery, building of infrastructures, execution and finally handing over for operation. Construction is a unique faculty of project management which does not match others. Construction is generally a short term activity and very temporal, yet the relationship developed amongst the construction group within the short duration is everlasting. Many groups with a different mindset come from different parts of the country to build the facility. They come, construct, go and leave behind a footprint on a place they don't belong where they developed a relationship, fellow feelings and fraternity in their mind forever.

Construction is not rocket science. It is only 30% knowledge and 70% common sense. A construction manager with these abilities can move effectively with the above skilled people and guide them to work as a team to accomplish the work with a common goal, objective and strategy.

The construction phase is very important and vital for a project activity. Quality of work depends on workmanship and management. Hydropower projects are constructed in remote areas where infrastructure facilities are limited. Access and approaches are difficult. Working conditions are tough. Health care facilities are a scarce commodity. A dam site is an exceptional and problematic working environment which should be tackled in a mature way showing all sort of tolerances.

The project authority has some vital responsibilities for timely completion of the project by developing required infrastructures so that business associates of various units can start the work without delay immediately after the kick-off meeting.

The following infrastructures shall be in place before the contractor for the construction of the main component of a hydropower project reports for taking up the work.

5.2 Infrastructure Development by Project Authority

Infrastructures provide support systems to the construction of main project. Infrastructures such as roads, power, water, communication facilities, temporary accommodation, warehouses, etc., are made ready before taking up the main construction effort.

5.2.1 Construction of Road Network, Bridges, Culverts, Cause Ways, and Others, in Project Area

A main road shall be made across the project area and it should be connected to the nearest rail head or highways for transportation of construction materials from different parts of the country. This road will be the lifeline of the project and it will feed all branch roads leading to various project sites like dam, intake, tunnel, power house, etc. This main road shall be made focusing on the following requirements to facilitate movement of heavy equipment during the construction phase:

- Crust thickness of the road
- Upward and downward slope conforming to ruling gradient
- Turning radius; suitable curve shall be provided so that trailer with multiple wheels can negotiate the curve smoothly
- Provision of super elevation shall be reviewed for negotiation of heavy trailer

Keeping in mind the magnitude of the project, bridges and culverts shall be made considering HFL of the area so that roads remain during rainy seasons. Some causeway shall also be made per requirements. These bridges and culverts are made for facilitating smooth construction activities that include transportation of materials, manpower, machinery, etc.

5.2.2 Construction of Haul Roads

Various haul roads shall be made to expedite the construction activities such as transportation of stone, earth for filling and disposal of surplus material, etc.

- Stone quarry site to work sites
- Borrow area to work sites
- Crushing plant site to work sites
- Disposal area for surplus material, etc.

5.2.3 Service Road

Service roads shall be constructed to connect residential areas, offices, contractors areas, labor camps and utility services like the medical facility, school, police station, etc.

5.2.4 Project Townships and Utility

A central project township shall be constructed to accommodate employees of various cadres and central offices of all department like dam, dyke, tunnel, power house, etc., shall be constructed in a temporary fashion and permanent township shall be constructed gradually during the construction phase as required.

5.2.5 Other Utility Accommodation and Facility Buildings

Some accommodation shall be constructed for employees posted at sites, including utility facilities like a hospital, school, guest house, police station, bank, post office, commercial center, central store and warehouses, laboratory, and central workshop.

5.2.6 Water Supply

A water supply network shall be laid considering peak load during construction.

5.2.7 Construction Power

Total power requirement can be assessed, and power shall accordingly be taken from the nearest grid while backup diesel generating (DG) shall be made available for emergency and construction power.

5.3 Role of Construction Manager

The construction manager has a tremendous role to play and responsibility to create minimum facilities in all fields of human requirements, build requisite infrastructures and mobilize the work force to complete the work within budget and timeline.

5.3.1 Phases of Construction Management

Construction management is generally in the following stages:

- Mobilization and development of infrastructures
- Construction stage
- Completion and handing over to client

5.3.2 Assumptions of Construction Management

The following are the assumptions of construction management that are taken as belief to be true in future.

- Scope
- Cost
- Schedule

5.3.3 Constraints of Construction Management

The following are the constraints of construction management. All assumptions become true if no constraint is encountered during execution.

- Scope
- Time
- Cost
- Quality
- Safety
- Tolerance

5.4 Planning for Mobilization and Kick-Off Meeting

After contract is awarded with the Zero date, a first meeting is held between the client, PMC and contractor to understand project background, working procedures, formality, etc., and the contractor is to furnish their plan like mobilization of manpower, machinery etc., as described below.

5.4.1 Kick Off Meeting

The kick off meeting is the first meeting held between the client, PMC and contractor where all personnel come to know each other, discuss various aspects of the project starting from planning, scheduling, mobilization of manpower, and machinery. This meeting develops understanding, reliability, and trust amongst the groups so as to take the project in the right direction.

5.4.2 Convener of the Meeting

The meeting is generally convened by the contract manager of the client after issuing the Letter of Intent to the contractor with the Zero S date of the contract, and the Zero D date of the contract is reckoned from the date of receipt of LOI, with one copy of LOI returned to client duly signed as a token of acceptance. The kickoff meeting is generally held in the project conference room of the client.

Participants of the meeting:

- Construction manager
- Engineering manager
- Contract manager
- Planning manager
- Quality assurance and quality control manager
- Material & store
- Safety manager
- Manager (HR)
- Accounts and finance manager
- Authorized representatives of contractor

5.4.3 Documents to Be Furnished before Site Mobilization

Contractor will furnish the following for fulfilling contractual obligations:

- Power of attorney issued in favor of site in charge
- Mobilization plan of manpower and machinery with starting date of work
- Organogram with date of mobilization of personnel
- Credentials of each member
- Company's quality manual
- Company's safety manual

5.4.4 Documents Required by Contractor before Site Mobilization

Contractor will ask concerned department of client to provide the following and enable them to start the work.

5.4.4.1 By Contract Department

The contract department should furnish the following documents to contractor:

- Original contract document along with specification, BOQ and SOR, etc., with agreement
- Format for bank guarantee for SD and mobilization advance

5.4.4.2 By Engineering Department

The engineering department should furnish the following documents to the contractor:

- Status of drawing index and IFC drawing
- Status of design accomplishment
- Soil investigation of dam site
- Status of design approval
- Status of material approval

5.4.4.3 By QA/QC Department

The QA/QC department should furnish the following documents to the contractor:

- Guide line for quality plan
- Job procedure and ITP, etc.

5.4.4.4 By HSE Department

The HSE department should furnish the following documents to the contractor:

- Guidelines for HSE
- Procedure for getting licenses for blasting operations including transportation, storage of explosive and blaster's license, etc.

5.4.4.5 By Planning Department

The planning department should furnish the following documents to the contractor:

- Overall schedule
- IFC drawings and index of drawings
- Format for daily progress report
- Format for weekly progress report
- Format for manpower deployment report
- Format for machinery deployment report

5.4.4.6 By HR Department

The HR department should furnish the following documents to the contractor:

- Documents related to labor license

5.4.4.7 By Account and Finance

The Account and Finance department should furnish the following documents to the contractor:

- Information and procedure of billing, etc.

5.4.4.8 By Construction Department

The construction department should furnish the following documents to the contractor:

- *Information regarding survey*
    - Detailed survey drawings with master control pillar
    - Contour of the area
    - Status of diversion work
- *Area*
    - Offices
    - Store
    - Labor camp
    - Bar binding area
    - Shuttering making area
    - Storage for scraps
    - Testing laboratory
    - Maintenance workshop
    - Batching plant
- *Various connectivity*
    - Power connection to office, construction site, etc.
    - Construction water connection
    - Drinking water connection
    - Internet and telephone connection

Minutes of meetings shall be prepared with the date of commitment of all activities which will be signed off and issued to groups for taking necessary action per MOM.

5.5 Site Mobilization

After the kickoff meeting the contractor should start the following activities to start work.

5.5.1 Site Infrastructures Development

After allocation of areas as per kickoff meeting, the contractor should mobilize the following immediately to start the work:

- *Porta cabin*
  - CM's office
  - Engineer's office
  - Administrative office
  - Office and testing lab
  - Office and store
  - Dining area
  - Toilet as per HSE norm
  - Creche
  - First aid room
- *Office furniture and fixtures as per requirement*
  - Chair, table, computer table, cabinets, AC, fan, light, etc.
  - Water cooler and water dispenser, etc.
- *Office equipment*
  - Computers
  - Server
  - Printer, color and B&W
  - Copier A3/A4
  - Heavy duty scanner
  - Drawing plotter
  - Drawing copier
  - Camera (still) and video
  - LCD TV and projector
  - First aid box
  - Fire extinguisher as per HSE
- *Software*
  - Primavera version latest
  - MS office
  - Anti-virus
- *Survey equipment*
  - Total station
  - Leveling instrument

5.5.2 Mobilization of Machinery and Construction Equipment

The contractor should mobilize the following equipment per the mobilization plan furnished at the kickoff meeting:

- *Earth moving equipment*
  - Power shovel like Poclain, JCB
  - Loader
  - Dozer
  - Dumper/tripper
  - Roller
- *Rock drilling equipment*
  - Pneumatic jack hammer
  - Wagon drill
  - Core drilling machine
  - Breaker with chisel
  - Crawl mounted JCB fitted with breaker
  - Air compressor
- *Equipment for concreting*
  - Batching plant of requisite capacity along with all materials necessary for making bins, etc.
  - Drag line
  - Transit mixers
  - Boom placer
  - Bucket of capacity of 1.5 cum for carrying concrete
  - Concrete pump
  - Needle vibrator (petrol, diesel)
  - Plate vibrator
  - Table vibrator
  - Concrete mixer
- *Equipment for consolidation grouting*
  - Grout pumps with all accessories
  - Multi-staged packers
  - Measuring tank
  - Agitator
  - Flow meter, etc.

- *Lifting equipment*
  - Cranes of various capacity per the mobilization plan
  - Cable way
  - Hydra
- *Bar bending yard*
  - Bending machine
  - Cutting machine
  - Coupler machine
- *Others*
  - Water tanker
  - Truck mounted water tanker
  - Cars
  - Mini truck
  - Pickup van
  - Laboratory testing machines, etc.
  - Welding machine

5.5.3 Manpower Mobilization

Mobilization of manpower shall be done progressively per requirements, but the following shall be mobilized immediately:

- Engineers, surveyors, supervisors, technician store personnel and office staff
- Subcontractors
- Suppliers

After the development of initial infrastructure, the following concurrent activities shall be taken up aggressively in order to start main construction activities at the site.

5.6 Initial Site Activities

The contractor is to initiate preliminary work as indicated below:

- Surveying of dam site
- Surveying of coffer dam sites

- Surveying for location of installation of cable way for carrying of concrete in dam block
- Survey for locations for crane installation and for transportation of concrete in other blocks not reachable to cable way
- Survey for fixing passages for direct unloading of concrete by transit mixer/truck on blocks
- Considering the above survey, block wise excavation plans can be drawn to facilitate progress of work
- Finalization of stone quarry and sand quarry
- Installation of batching plant
- Installation of testing lab

Other Administrative Activities

- Submission of bond/BG for SD and mobilization advance
- Application for labor license, etc.
- Application for explosive licenses and blaster license

5.6.1 Surveying of Dam Site

The axis of the dam shall be marked. Both upstream and downstream shall be marked as per layout drawings. The area shall be divided into 5 meter grids and the level of the entire area shall be taken and levels taken shall be matched with contour of the dam site furnished by the client. Any discrepancy in level shall be brought to the notice of authority before taking up the removal of overburden.

A detailed construction program of each block of dam shall be drawn based on the permissible limit of height difference between the first and last block and permissible limit of height difference between two adjacent blocks. So, raising of the height of all blocks shall be done simultaneously, conforming to permissible height differences. Accordingly, all activities like excavation of hard rock, drilling holes and grouting shall be planned.

5.6.2 Stone Quarry

A quarry is an extractive rock surface where rocks are extracted by blasting or by any other means and broken into the required size for feeding into the crusher to get coarse aggregate of various sizes varying from 80 mm to 10 mm for construction work.

Before finalization of the quarry, some statutory along with technical requirements are to be fulfilled as stated below.

5.6.2.1 Selection of Stone Quarry

The site shall be selected carefully so as to avoid any future problems.

- Quality of rock should conform to specifications like basalt, granite, etc.
- Should be hard, durable, strong and free from deleterious material & coating
- Quantum of deposit should be per requirement
- Away from resident and sensitive areas
- No damage to flora and fauna
- Easy access
- Overburden shall be minimum
- Face can be generated for effective production by blasting
- Within striking distance from construction site

5.6.2.2 Operational Requirement

Quarrying is a hazardous activity which invites some statutory requirements to be fulfilled before the start of quarrying activity, and the matters shall be taken up with the client who is the principal employer:

- Application for development of quarry
- Environmental Health Act
- Occupational Health Safety Act
- Proof of ownership
- Environmental impact assessment
- Quarry Environmental Management Plan
- Hours of working
- Noise control, etc.

5.6.2.3 Technical Requirement

Samples are to be collected from the proposed quarry and shall be tested as indicated below for acceptance:

- Specific gravity
- Bulk density
- Aggregate crushing value
- Aggregate impact value
- Aggregate abrasion value

- Soundness of aggregate
- Limit of deleterious materials

5.6.2.4 Alternative Proposal

If any running quarry is available in close proximity the possibility shall be explored. The capacity of the quarry and capacity of supplying of various sizes of aggregates shall be inquired and judged to meet the peak requirement.

Suppliers should furnish the following test results of the product:

- Specific gravity
- Bulk density
- Moisture content
- Absorption value
- Aggregate crushing value
- Aggregate impact value
- Aggregate abrasion value
- Flakiness index
- Elongation index
- Deleterious materials
- Soundness
- Classification of rock

In addition, a sample of 150, 80, 40, 20 and 10 mm size aggregates shall be collected and tested in an approved laboratory.

5.6.3 Batching Plant

A concrete batching plant consists of a material batching system, mixing and control system up to being able to print out the product slip.

Selection of a batching plant shall be done very carefully considering that the requirement of concrete is huge in a gravity dam. The capacity of a batching plant shall be chosen accordingly. Capacity shall be decided based on the monthly requirement of concrete according to the preliminary schedule. The capacity of batching plant is expressed in M^3 per hour. Various capacities of batching plants from 20, 30, 40 and so on M^3/per hour are available on the market.

5.6.3.1 Location of Batching Plant

Location of the batching plant shall be closed to the dam site because temperature of concrete shall be controlled and is to be maintained below

18°C at the time of pouring. It is preferable to keep low to avoid a thermal effect on concrete.

5.6.3.2 Selection of Batching Plant

The batching plant can be selected per the following parameters:

- Per hour requirement of concrete
- Type and number of grades shall be mixed
- Type of aggregates to be used
- Type of additives to be used
- Consistency
- Type and batch size of mixer (filling and compacted volume)
- Mixing time
- Aggregate storage bin with discharge gate facility
- Aggregate weighing hopper with electromagnetic loading cell
- Cement/fly ash hopper with electromagnetic loading cell
- Water weighing hopper and admixture weighing container, etc.
- Height of platform
- Height of pouring, etc.

If the in-line silo type is chosen, the following shall be considered:

- Number of compartments with bin capacity
- Type and number of binders
- Water operating pressure
- Control system

5.6.3.3 Installation System of Batching Plant

Installation of the batching plant is done in various stages.

- Preparation of foundation. Foundation shall be cast as per foundation drawings supplied by the manufacturers.
- Preparation of the storage bin shall be made per different sizes of aggregates and sands to be used in mixes.
- Partitions between the bins shall be made strongly with proper support systems so that various size of aggregates and sand do not get mixed together.

- A separations wall between the plant and bins which will be up to the height of platform of BP shall be designed, anchored and supported by steel plates and structural steel so that it should retain the horizontal pressure exerted by aggregates.
- Width of each compartment of aggregates should be such that loading equipment can work easily.
- Size of stack yard shall be large enough to meet the requirement of the project.

All above activities shall be done by the site construction group. Installation of the batching plant shall be done by the installation group of manufacturers.

Installation is divided into many parts like mixing host unit, cement silo, batching equipment and others.

5.6.3.4 Calibration

All ingredients are measured by weight with scales and are put into the batching plant. These scales are to be calibrated before operation and during operation, and all scales are to be calibrated periodically to maintain its accuracy. A calibration chart shall be displayed in the batching plant indicating the last date of calibration and the due date of the next calibration.

Calibration of the batching plant shall be conducted through an efficient weight and measurement group in the presence commissioning group of manufacturers to ensure safety.

Calibration is a simple process to check the scales by predetermined weights. So before checking, prepare enough standard weights. Zero setting shall be done. Weighing tests from zero to max in order and unloads the weight to zero in the same order (Table 5.1).

Average variation = Sum total of percentage variation/No. of reading
= 0.08% (Permissible limit) ± 3% for aggregates

Permissible tolerance:

- Cement and mineral admixture—±2%
- Aggregates, chemical admixture and water—±3%

All weighing and measuring equipment shall be tested and calibrated over its full working range in specified intervals:

- Mechanical system—Once in 2 months
- Electrical/load cell system—Once in every 3 months

TABLE 5.1

Calibration of Batching Plant

	Loading			Unloading			
Sl No.	Standard Weight in Kg (A)	Actual Weight in Kg (B)	Variation in % $\frac{B-A}{A} \times 100$	Standard Weight in Kg (C)	Actual Weight in Kg (D)	Variation in % $\frac{D-C}{C} \times 100$	Remarks
1	0	0	0	150	150.5	0.33	
2	10	10	0	140	140	0	
3	20	20	0	130	130.5	0.38	
4	30	30	0	120	120	0	
5	40	39.5	−1.25	110	110.1	0.09	
6	50	50	0	100	100	0	
7	60	60	0	90	91	1.11	
8	70	70	0	80	79.5	−0.62	
9	80	80	0	70	70	0	
10	90	90	0	60	60	0	
11	100	101	1	50	50	0	
12	110	110	0	40	40	0	
13	120	120.5	0.41	30	30	0	
14	130	130	0	20	20	0	
15	140	140	0	10	10	0	
16	150	150.5	0.33	0	0	0	

5.6.3.5 Routine Maintenance Check

Some routine maintenance checks are to be taken up for safe working of the plant per standard codes of practice. The codes prescribe some daily, weekly and monthly check ups for maintaining the health of the plant.

- *Daily routine*
  - Clean weigh dial
  - Weighing hopper shall be cleaned properly
  - Wash mixer pan
- *Weekly routine*
  - Hoppers and doors shall be in clean condition
  - Central mixer blades and peddles shall be checked for tightness and shall be adjusted if necessary
  - Remove any concrete if build up from mixer
  - Check oil level in air line lubricator
  - Check drain water tap on air line

- Check pipe work for leaks
- Check wire line and electrical apparatus
- Routine check and servicing of mixer
- *Monthly routine*
  - Check all calibration of weighing scales
  - Check calibration of water meter

5.7 Blasting Operations in Stone Quarry

Blasting operations are part and parcel of construction of a gravity dam. A gravity dam is constructed in a hilly area that requires hard rock excavation by blasting and also huge quantity of aggregates to be manufactured from rock obtained by a blasting operation.

A blasting operation shall be carried out as per the Explosive Act of the country. Each country has its own Explosive Act and it is essential to obtain the following licenses before conducting any blasting operation.

Blasting operation includes possession, handling, transportation, storage, use of explosive.

- License for storage of explosive
- License for handling and transportation
- License for conducting blasting operation, i.e., Blaster's License

5.7.1 Blasting Activity

Blasting activity is divided into five steps:

- Storing, handling and transporting explosives in blasting area
- Drilling of blast holes in blasting area in specific patterns
- Loading and stemming the blast holes with explosives
- Fire the loaded holes by electro-mechanical blasting machine
- Clearing the blasted rock and make ready for next operation

5.7.2 Explosives

Explosives consist of three components:

- Gelatin
- Detonator (ordinary/electrical/millisecond)
- Fuse coil/detonating cord

5.7.2.1 How It Works

A vertical small hole is made through the center of the gelatin cartridge and a detonator in conjunction with detonating cord is inserted into the hole of the cartridge. This primed explosive is then inserted into the drilled hole and stemmed tightly so as to develop adequate pressure in the gas released due to detonation within the hole. This release of heat and high pressure gases expand with sufficient force to develop cracks near holes, and under this tremendous pressure the rock is fragmented, displaced and thrown away. Stemming plays a vital role in this operation because adequate pressure cannot be developed if it leaks resulting in poor production of rocks fragmentation.

5.7.2.2 Efficiency of Blasting

Efficiency of blasting depends on several factors:

- Breaking, fracturing and shattering of rock
- Creation of face to have free thrust on rock
- Spacing of blast hole
- Type of spacing, that is, regular or staggered
- Depth and diameter of blast holes
- Type of blast hole, that is, vertical and inclined or combination
- Type of detonator, that is, ordinary or electrical/millisecond
- Explosive charges
- Proper stemming of holes

5.7.2.3 Creation of Face of Rock

Free rock face is required to be created to have the maximum effect of blasting for better efficiency. The theory of blasting operation is that compressive forces are created after detonation. Rock, like concrete, is strong in compression and weak in tension. When pressure generated by blasting exceeds the strength of rock medium, a crack is developed in the rock. A face is created where over burden is small and high vertical free face which will allow pressure generated by blasting to dislodge and move away to have better production as shown in Figure 5.1.

5.7.2.4 Pattern and Spacing of Blast Holes

The arrangement of blast holes are done in a grid pattern with regular spacing of holes. Spacing of blast holes is one of the key factors for efficiency

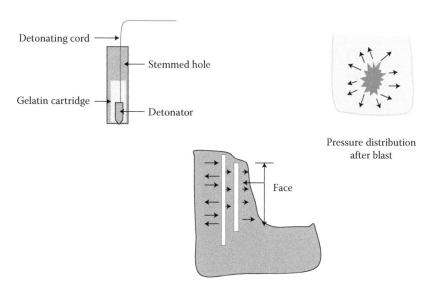

FIGURE 5.1
Description of blast hole.

of blasting; spacing is the horizontal distance from center to center between the holes. It is observed that staggered spacing of blast holes will have better and better efficiency than regular spacing.

Spacing of holes depends on the depth of holes and burden of rock face; the burden of rock is the distance from the center of a hole to the external open face of rock. The spacing shall be kept less than half of the depth of the hole or 1.1 times the burden, whichever is less. Blasting produces a radius of influence over the area. The area remaining unaffected by blasting operations in regular spacing is much more than the blasting conducted in staggered spacing of holes. Staggered spacing of blast holes is preferred for better efficiency (Figures 5.2 and 5.3).

5.7.2.5 Diameter of Blast Hole

The minimum diameter of blast holes is decided on the size of explosive cartridge. The size of the gelatin stick available varies from 20 mm to 40 mm in various countries, and the diameter of a drilling bit of pneumatic jack hammer is about 35 mm. Generally, the diameter of a blast hole is kept 35 mm. Larger diameter of holes can be drilled about 100 mm depending on the quantum of work.

The bottom diameter of a hole should be able to accommodate an explosive cartridge. The gap between the inner wall of a hole and the cartridge should be minimal, as entrapped air in the gap reduces the effect of blasting. Care should be taken during the drilling process about the diameter of the drilling

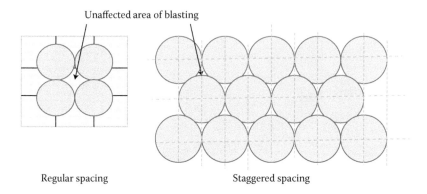

FIGURE 5.2
Spacing of blast hole.

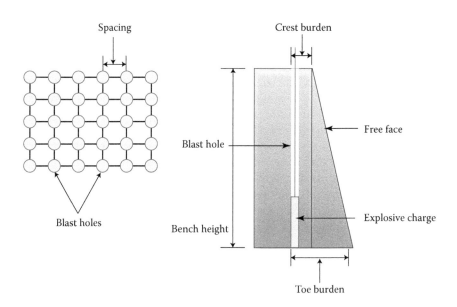

FIGURE 5.3
Grid pattern of blast holes.

rod which becomes worn due to wear and tear. It becomes difficult to insert explosive in the holes.

It is needless to explain that all other dimensions of holes are interrelated to each other. Other features are depth, spacing, burden, bench height, etc.

5.7.2.6 Bench Height

Bench height shall be fixed of the blasting operation with respect to the size of the hole.

$$\text{Bench height} = \frac{\text{Diameter of hole}}{15} = \frac{35}{15} = 2.3 \text{ m}$$

5.7.2.7 Depth of Hole

The depth of the hole shall be equal to bench height or a little deeper, about 10% more than bench height.

5.7.2.8 Burden

It is the distance from center of the hole to the free face of rock. Rock face cannot be developed exactly vertical. So the burden at the top and bottom of a hole may vary. Burden is dependent on the diameter of hole. The burden shall be equal to 30 times the diameter of the hole.

5.7.3 Type of Explosive

There are many types of explosive such as gun powder, nitrate mixture, nitro compound, chlorate mixture, liquid oxygen, etc.

Gelatin, which comes under nitro compound, is generally used in construction activities along with Ammonium Nitrate Fuel Oil Mixture (ANFO).

High efficiency can be achieved by adding ANFO in the right proportion with the nitro compound.

Detonator: The detonator is a capsule filled with a high power explosive like lead azide as a base charge and diazodinitrophenon as a primary charge. It detonates and controls a huge charge of explosive. It is initiated by electrical energy or any other means.

Detonators of many kinds that are used in construction, namely ordinary detonators and electrical detonators, and electrical detonators are divided in to three categories: normal electrical detonator, millisecond detonator and second detonator. Millisecond and second detonators are called delay detonators.

Blast holes are fired using an instantaneous electric detonator and short delay electric detonator with a delay in steps of 25, 30, 40, etc., milliseconds. These detonators can be used in sequence to have a better production of blasting operation.

Safety Fuse: Safety fuse is used to initiate an ordinary detonator in blasting activity.

5.7.4 Design of Explosive Charge

Blast holes are filled with a continuous column. The effectiveness of blasting depends on the charge and size of stemming. Stemming enhances the effectiveness of blasting. Design of the charge depends on some factors like burden, stemming and charge length. Burden can be calculated and taken as equal to 30 times the diameter of the blast hole. Stemming length can be taken as a maximum of 1.2 times burden and length of charge can be calculated by deducting stemming length from length of the hole.

Clay and sand shall be mixed in proportion 1:4 and shall be used as stemming and shall be compacted properly. Clay used in stemming will act as a binder and sand will absorb the impact of the blast.

Water can also be adopted as a stemming material using polyethylene tube of appropriated length. Water stemming reduces dust in the air and absorbs toxic gases.

5.7.5 Equipment Required for Drilling Blast Holes

Drilling equipment as mentioned below are required for making blast holes in rock.

- Rock drilling machine like pneumatic jack hammer drilling machine
- Wagon drilling machine
- Air compressor

Procedure: Over burden on the surface of rock shall be removed. Free face shall be created by doing shallow depth blasting in order to open the face. A required bench height can be established. Required spacing of holes and length of hole can also be obtained.

Drilling of holes shall be done in grid with staggered spacing and holes in the front two rows shall be supported by inclined hole in the 3rd row to have better efficiency in blasting operations. A horizontal hole will have two components, vertical and horizontal due to inclination where the horizontal component will push the rock toward the free end. A millisecond detonator of various categories like M0, M1, M2 etc., shall be used as indicated in the drawing. This arrangement of delay detonators shall be done row to row. Delay M0 in Row 1 will blast first and develop pressure all around. Pressures developed towards R2 and the bottom shall be counteracted by the huge compact rock. Pressure developed towards the free face is greater than counter force and will dislodge some chunks and throw them away from the free face of bench cut which will create a new free face for row R2 and subsequently M1 and M2 will blast serially one after an other having a free face created by R1 and R2 and dislodge and push every part of the rock to produce better output of the operation (Figure 5.4).

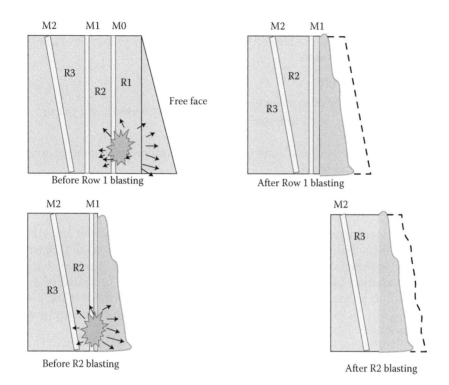

FIGURE 5.4
Blasting sequence of millisecond detonator.

What will happen if delay detonators are not used serially? Let us take an example, if M2 delay detonator is placed in R-1, M0 in R-2 and M1 in R3. As per this system, M0 in R-1 will blast first and the pressure created due to this blast will be counteracted from all three sides. No free face is available and confined from three sides except the top. M1 will blast next without any effect as it is confined in all three sides and the last one to blast is M2 in R-1 will have some effect due to the availability of the free face and efficiency of blasting will be reduced.

There are no rules available about spacing, depth and design of charges because it differs from rock to rock, their joints, etc. It is suggested to do some trial blasts with different parameters to assess the power factor. The most effective depth, spacing, charges, etc., should be finalized for the particular site.

Before conducting blasting operations, levels shall be taken in a suitable grid in the proposed blasting area and the area of free face shall also be measured in M2 and note the quantity of explosive used in in each hole. After blasting, all blasted materials shall be removed and levels shall again be taken on the blasted bottom plane of the same area where initial levels were taken before blasting. The volume of blasted rock shall be calculated. Also assess wastage of rock that occurred during blasting. Power factors can be calculated by dividing the quantum of blasted rock by the quantity of explosive used for this particular operation.

Power factor (PF)

$$= \frac{\text{Quantity of explosive}}{\text{Quantity of rockblasted}} \quad \text{and will be expressed in Kg per m}^3$$

If PF of 3 number of trials are 0.3, 0.35 or 0.4, all parameters used in trial blast shall be considered.

5.7.6 Controlled Blasting

Blasting operations shall be controlled due to close proximity of any village or human habitation or any structure which shall be protected. Blasting operations shall be conducted in this situation by controlling the charges, and kentiledges like steel plates shall be placed over the proposed blasting area and an adequate number of sand bags shall be placed over the plates to arrest fly rocks so as to avoid any possible damage occurring due to flying rocks.

5.7.6.1 Method of Blasting in Dam Foundation

Hard rock excavation by blasting in dam foundations shall be conducted carefully. Care should be taken in such a way that blasting should not damage the rock structures and blasting operations shall be done in precision. Depth of holes, spacing and charges of explosive shall be optimally used.

Depth of hole should be terminated at least 20 cm above the decided founding level of foundation to avoid cracking on the finished level of the foundation. Generally, radial cracks are found on the surface level where the blast hole terminated. The balance 20 cm of depth shall be removed by chiseling and by breaker, and the surface should be free from any cracks that occurred during blasting operation. This operation is done by the breaker duly fitted with chisels.

5.7.7 Storage of Explosive

Explosives shall be stored in the premises approved by the competent authority under the Explosive Act of the country. A location plan duly indicating all necessary dimensions such as location of magazine from nearest habitation, necessary arrangement of installation of magazine as per the Explosive Act and proper lightning conductor, etc.

The explosive magazine shall be selected with respect to the type of explosive to be stored in the magazine. Magazine type 1 is for high explosive, type 2 and 3 portable or moving is also for high explosive with some laid down conditions. Type 4 is for blasting agents.

5.7.8 Transportation of Explosive

Explosives shall be transported by licensed explosive road vans permitted under the Explosive Act of the country with a restriction of transport of explosive by other vehicles.

Explosive shall be loaded and unloaded in the road van and transported with care.

- Loading and unloading of explosive shall be carried out when the explosive van is stopped and all wheels are chocked
- Loaded explosives should not move in the van during transportation
- Explosives shall be protected against friction and bumping
- Vehicle shall be driven very cautiously and it should not stop on the way in many places
- Vehicle should carry working fire extinguisher
- All electrical connections shall be checked
- Fuel tank shall be checked for leakage
- Van lighting shall be on when it stops at night
- Permission copy shall be carried by the driver during transportation of explosive

5.7.9 Precaution

Blasting operations shall be conducted in accordance with the appropriate Explosive Act. Precaution as mentioned below shall be taken while strictly adhering to rules in order to avoid uncalled for accident:

- A competent person will be in charge of the blasting operation; he should have a license under country's prevailing Explosive Act
- Blasting operations shall be carried out by a person having a Blaster's License under the prevailing Explosive Act
- Blasting of one class can be converted into another class
- Explosives shall be brought to the site from approved storage in original and unpack conditions
- Explosive shall not be taken to site if site is not ready for charging
- Detonator shall be carried to site in different container
- No person will carry or handle explosives from sunset to sunrise; Restrictions shall be imposed of handling explosives after sunset

- No person will smoke, light a fire, or keep inflammatory substances like petrol, diesel, acid or any other hazardous articles at any time within 15 meter distance from where explosive is stored
- Blasting operations shall be conducted in the specified time as approved by the competent authority under the Explosive Act of the country
- When a hole is charged it shall be marked
- Stemming shall be done without applying any force by a rod made of non-sparking material
- The supervisor shall inspect the area and ensure that everything is alright as per safety norms
- Time of blasting shall be displayed in the area
- Warning siren shall be played one hour before blasting operation to caution people
- Entire area shall be cordoned off by guard with red flags
- No person or vehicles shall be allowed to enter the area where blasting shall be carried out
- All clear sirens shall be played after blasting is conducted
- After blasting is conducted, area shall be inspected and shall be marked if any hole is found unblasted

5.7.10 Blast Log

A blast log shall be maintained for utilization of explosive with following information:

- Location
- Date and time of blast
- Blaster name with license number
- Hole details such as hole number, diameter of hole, depth of hole, spacing, etc.
- Length of stemming and material used
- Explosive type and quantity charged
- Type of detonator used
- Any kentiledge is used
- Safety measures taken
- Any misfires occurred and taken
- Any damage or incident

6

River Diversion

6.1 Introduction

A concrete gravity dam is constructed across the river and construction activities are mainly carried out in the river bed. Construction engineers face difficulties constructing the dam due to the flow of water of the stream. A dam cannot be constructed on the river bed if the river is not diverted effectively from its main course to make the working area completely dry or partially dry. So, it is mandatory to divert the stream before taking up any construction activities, which will ensure safety during the construction period.

An effective scheme of river diversion shall be designed that will facilitate taking up construction activities on the river bed. Prior to the start of the construction activities, a detailed survey shall be conducted about the geological and hydrological characteristics of the site and river, respectively, and the effective scheme of river diversion shall be designed accordingly.

6.2 Diversion Requirement

River diversion is essential to fulfill the following reasons:

- To provide a safe, secure and partially or fairly dry working area in a flowing river bed for the construction of dams, bridges, etc.
- To divert water from the main river to another manmade or artificial channel for distribution of water for irrigation, domestic and industrial purposes.
- To divert water from the main river to another channel in order to control flooding and save all downstream components from the fury of high floods.

6.3 Method of Diversion

There are many methods of diversion of a river:

- River diversion through open channel
- River diversion through tunnel
- River diversion through conduit
- River diversion through conduit provided below the block of the dam
- River diversion through temporary sluice

6.4 Various River Diversion

6.4.1 River Diversion: Open Channel on Edge of the River

A prime requirement for the construction of a dam across the river is to have a fairly or partially dry area to take up construction work on the river bed. This condition can be achieved by constructing a temporary barrier, known as a coffer dam, and water can be diverted from the deep portion of the stream to another part of the wider river valley to make the working area fairly dry. This method is adopted due to geological constraints of the site where construction of a diversion tunnel is not possible. In this case, water is allowed to flow over the partially completed blocks during flooding when construction is not in progress. Capacity of a diversion channel can be fixed by collecting data on floods during the monsoon period for the last 25 years. A quantity of water shall be allowed to pass through a temporary sluice and balanced water shall be diverted through a diversion channel. But these are overall guidelines to control water flow. Action shall be taken on the prevailing site condition. After making all these arrangements, the working area in the river bed will be subjected to great seepage, which is unavoidable. Huge dewatering arrangements are to be made. Many high capacity dewatering pumps shall be installed to pump out accumulated water from the working area where block concreting is planned to take place. The lowest block shall be taken up, and a number of pipes of requisite diameter, as per requirements, shall be laid to allow seepage water to flow through these pipes and to be buried under blinding concrete of blocks. This arrangement shall be used to pass on any extra seepage water during construction. These pipes and temporary sluice shall be sealed after the construction is over (Figure 6.1).

Precaution: The channel shall be designed with proper slope so as to protect channel side slope from collapsing.

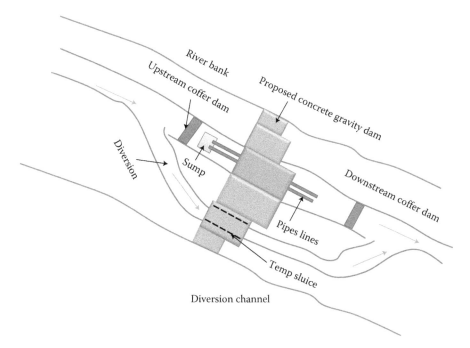

FIGURE 6.1
River diversion.

6.4.2 River Diversion: Diversion Channel and Coffer Dam

Diversion work consists of a diversion channel along with an upstream and downstream coffer dam. Depending on site conditions, a diversion scheme can either be applied on the right bank or on the left bank development, and the river will pass through the diversion channel. The coffer dam will contain a sluice, which will discharge construction water to the river. A suitable slope of the diversion channel shall be maintained based on the bed level of the river at inlet and outlet.

The diversion channel shall be designed considering the maximum flood recorded in the last 25 years. The wall of the diversion channel shall be designed as a cantilever RCC wall with suitable height based on discharge, and suitable bed width shall be provided to pass the water smoothly.

The coffer dam on upstream shall be made with suitable construction materials as per requirement and design. A stable section of the coffer dam of designed height is constructed with plumb concrete or PCC of 1:2:4 proportion. The section will have adequate top width and the downstream slope shall be made providing adequate tread and rise (Figure 6.2).

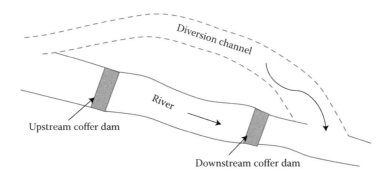

FIGURE 6.2
Diversion channel and coffer dam.

6.4.2.1 Construction of Upstream Coffer Dam

Construction of the upstream coffer dam shall be taken up from both banks simultaneously with designed top width and height, stability of dam is provided with horizontal tread 1.5 m and rise 1.5 m in the downstream face of the coffer dam and concrete of 1:2:4 or CSG (cement, sand, gravel) shall be used. Concreting shall be done with a transit mixer, and an insert plate shall be provided for the erection of the sluice. Consolidation grouting shall be taken up in the upstream coffer dam as required, and the erection of the sluice gate shall be taken up as the last activity.

- *Material requirement*: Material requirements for the upstream coffer dam shall be calculated.
  - Earth work in excavation
  - Rock excavation
  - Consolidation/curtain grouting
  - Concrete
  - Insert plate
  - Sluice gate
  - Disposal of muck

6.4.2.2 Construction of Downstream Coffer Dam

The downstream coffer dam shall be made with muck in combination with impervious material. The dumped rock shall be provided with a proper slope of 1.7 H:1 V, with some thickness of stone pitching. The construction of the coffer can be taken up in two phases. The downstream coffer dam prevents backwater and seepage.

- *Material requirement*
  - Earth work in excavation
  - Rock excavation
  - Rock fill material
  - Filler material
  - Insert plate for fixing sluice
  - Sluice gate material

6.4.2.3 Construction of Diversion Channel

Construction of a diversion channel shall be carried out from both ends. Excavation work shall be carried out by excavator where normal soil is encountered. The drilling and blasting operation shall be carried out where hard rock is encountered. Suitable mucking arrangements shall be made to transport surplus excavated material. Surplus rock material can be used in the construction of the downstream coffer dam.

The slope of the channel bed shall be checked with the level of the river bed at inlet and at outlet. The excavated bed shall be leveled and compacted properly. Blinding concrete in 1:3:6 shall be carried out before taking up floor concrete, and when floor concrete is over the cantilever wall shall be concreted.

- *Material requirement*
  - Earth work in excavation
  - Hard rock excavation
  - PCC
  - RCC in cantilever wall
  - Disposal of surplus material
- *Equipment requirement*
  - Excavator
  - Loader
  - Dumper
  - Wagon drill/jack hammer
  - Air compressor
  - Transit mixer
  - Needle compactor
  - Crane for erection of sluice gate
  - Welding machine

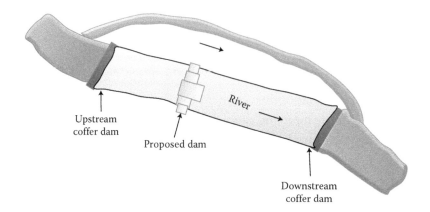

FIGURE 6.3
Diversion tunnel.

6.4.3 River Diversion: Diversion Tunnel and Coffer Dams

Excavation of the diversion channel becomes difficult in hilly terrain where a diversion tunnel is the only option for diverting the river during construction of the dam on the river bed. It should be decided whether one diversion tunnel is sufficient; otherwise, a second diversion tunnel can be carried out in the other bank of the river (Figure 6.3).

Under this scenario, a river diversion system consists of:

- Diversion tunnel of required length, diameter and shape with proper slope based on river bed level at inlet and at outlet. Service gate, emergency and stop log gate may be provided for regulating the flow of water.
- Rock fill upstream coffer dam of designed length, height, width and distance of upstream coffer dam from the axis of the dam upstream. The upstream coffer dam keeps the working area fairly dry so as to facilitate the starting of excavation work.
- Rock fill downstream coffer dam of height, length, width and distance of downstream coffer dam from the axis of main dam. The downstream coffer dam prevents back water and seepage.

6.4.3.1 Construction of Diversion Tunnel

Construction of the diversion tunnel can be carried out from both faces simultaneously. An adit can also be considered from the river bank to the midsection of the tunnel. The excavated diameter of the tunnel shall be fixed considering the thickness of shotcreting and tunnel lining.

Quantity of excavation and concreting shall be calculated.

- Quantity of rock excavated
- Additional quantity of rock for over break (max 20% of drawing quantity)
- Quantity of travelling form for tunnel
- Quantity of steel
- Rail
- Quantity of lining
- Quantity of shotcrete

6.4.3.2 Construction Methodology

- Excavation shall be carried out from both faces simultaneously.
- Hard rock excavation shall be carried out by drilling and blasting in two stages through heading and benching method. Heading and benching method is adopted when quality of rock is found to be unsatisfactory. Heading shall be carried out for the top, followed by benching in remaining part in the bottom of tunnel section.
- Horizontal drilling holes of designed length shall be carried out in the top half of the section and to be blasted to make room for drilling of vertical holes in the bottom portion. This method is very effective because the top is done before the bottom portion which makes a clear open face on top and side to produce better results with minimum utilization of explosive (Figure 6.4).
- Concreting in the diversion tunnel is carried out in three stages: in kerb, crown and invert.

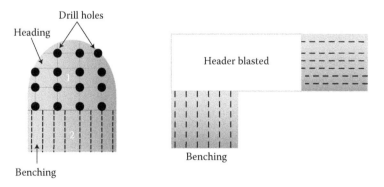

FIGURE 6.4
System of diversion tunnelling.

- The first activity is kerb, where the shutter is placed and concreted.
- Rail shall be placed on the kerb for the movement of travelling form required for crown concreting, which is the second stage of operation.
- The third stage of concreting is invert, which shall also be done by the help of a travelling steel form.
- Disposal of muck mixed with rock and soil shall be done in the identified dumping area. Dumped muck shall be used in the construction of upstream and downstream coffer dams.

6.4.3.3 *Planning and Scheduling of Material, Equipment, Machinery, and Others*

Construction of the diversion tunnel is a critical activity for which material, equipment, and machinery planning shall be done with a fixed timeline.

- Construction schedule considering all activities
  - Excavation of tunnel
  - Excavation of adit
  - Concreting in tunnel
  - Concreting in adit
  - Disposal of excavated muck
- Equipment and machinery requirement for planning
  - Excavator
  - 2 boom jumbo drill
  - Wagon drill
  - Jack hammer
  - Air compressor
  - Dozer
  - Loader
  - Tripper/dumper
  - Concrete pump
  - Transit mixer
  - Grout pump
  - Water pump

6.4.3.4 *Construction of Upstream Coffer Dam*

The upstream coffer dam is a rock fill dam which is composed of fragmented rock with an impervious core.

- *Selection of coffer dam*: Rock fill coffer dam is selected due to availability of plenty of rock fill material and impervious core material to make the dam water tight. Space is also not a constraint and a rock fill dam is preferable for any foundation condition and flow.

- *Preliminary dimensions*
  - *Free board*: Free board is provided to stop the overflow of water due to wave and wind action. Generally, 1.5 m free board is provided in a coffer dam.
  - *Top width*: Top width depends on the stability of the dam, and top width allows safe movement of construction equipment over it.
  - *Side slope*: Generally, 1.5:1 slope is provided on the upstream side and 1.25:1 slope is provided on the downstream side.
- *Forces acting on coffer dam*
  - Dead load
  - Hydrostatic pressure
  - Uplift pressure
  - Earth and silt pressure
  - Reaction force of foundation
- *Requirement of materials*
  - Quantity of excavation
  - Quantity of rock excavation
  - Quantity of coarse filler material
  - Quantity of fine filler material (sand)
  - Quantity of consolidation and curtain grouting
  - Quantity of impervious core
- *Methodology*
  - Construction shall start from both ends.
  - Excavation work of dam should start after completion of DT.
  - Diversion of river water shall be done through the diversion tunnel by constructing a temporary dyke to start placement of fill materials.
  - Material obtained from excavation shall be used as core material.
  - Material obtained from excavation of bed shall be used as filler material after screening test, etc.
  - Borrowed material shall be used as shell material.
  - The dam shall be constructed in two stages. The first stage is to be built by using excavated material up to a certain level, and the second stage shall be constructed by standard method without using any water. The rock fill zone shall be compacted

in layers from 300 to 600 mm, and compaction shall be done using a heavy vibro roller.

- Drilling holes for consolidation grouting shall be carried out in designed spacing and depth, and pressure grouting shall be conducted.
- Curtain grouting for greater depth shall be conducted in upstream of dam.

- *Planning of construction*: A detailed schedule for the construction of coffer dam shall be prepared along with equipment planning and requirement of equipment and machinery. For detailed planning, the following information is required:
  - Total number of activities involved
  - Quantity of each activity
  - Number of work fronts available
  - Time required for completion
  - Number of shifts
  - Output of each activity per shift
  - Identification of sequential and concurrent activities
- *Requirement of equipment and machinery*: Exact numbers of equipment and machinery can be evaluated considering efficiency of each piece of equipment and quantum of work of each activity. Here are the type of equipment required for the work:
  - Excavator for ordinary earthwork
  - Loader
  - Dozer
  - Dumper/tripper
  - Demag for loading blasted rock
  - Crawler drilling machine
  - Wagon drilling machine
  - Jack hammer
  - Air compressor
  - Arrangement for flushing grout holes
  - Grouting machine with packers
  - Water pump
  - Vibratory roller
  - Vibratory compactor
  - Water tanker

6.4.3.5 Construction of Downstream Coffer Dam

Construction of the downstream coffer dam will follow the same procedure of construction except consolidation and curtain grouting activity.

- *Requirement of materials*
  - Quantity of excavation
  - Quantity of rock excavation
  - Quantity of coarse filter material
  - Quantity of fine filter material (sand)
  - Quantity of impervious core
- *Methodology*: Methodology of construction of the downstream coffer dam shall follow all activities except the following activities:
  - Drilling holes for consolidation grouting shall be carried out in designed spacing and depth and pressure grouting shall be conducted.
  - Curtain grouting for greater depth shall be conducted in upstream of dam.
- *Equipment Requirements*
  - Excavator for ordinary earthwork
  - Loader
  - Dozer
  - Dumper/tripper
  - Demag for loading blasted rock
  - Jack hammer
  - Air compressor
  - Water pump
  - Vibratory roller
  - Vibratory compactor
  - Water tanker

6.4.4 River Diversion: Diversion Tunnel/Channel and Overtopped Coffer Dam

Construction of the river diversion with a diversion channel/tunnel and coffer dam is time consuming and expensive. The diversion system is designed considering data of high floods for the last 25 years. This flow of high flood water does not continue for a long time. It may happen at a maximum of 20% of the year and balance 80% of time, the flow of the river is much less than the design flow of the river. In a hilly area, an overtopped coffer dam with a diversion channel/tunnel is adopted. When this system of river diversion is adopted expenditures and time requirements are reduced due to the reduced volume of the coffer dam. Nature is full of uncertainties and data of highest flood for last 25 years which may not happen again and again, the economic benefit will be huge if maximum flooding does not occur during construction and may be called a risk taking benefit.

The flow of the river is diverted through a diversion tunnel/channel in the period of low discharge and allowed to flow through the overtopped dam during high discharge. The construction of the main dam shall be stopped when water is allowed to flow over the overtopped coffer dam.

This system of river diversion sorts out financial problems and also timelines, but it invites some technical problems of scouring of the dam foundation and erosion due to flooding. It causes huge economic loss if the proper energy dissipation system is not in place.

6.4.5 Discussion

Q1: Why do you build a coffer dam in a river valley project?

Ans: A dam is constructed in a river valley project across the river. It is mandatory to make the river bed dry prior to taking up construction work. The coffer dam is built across the river to divert the river flow and make the working area fairly dry.

Q2: Name some types of coffer dams?

Ans: Types of coffer dams are as follows:
- Concrete coffer dam
- Rock fill coffer dam
- Earthen coffer dam
- Cement sand gravel (CSG) coffer dam

Q3: What is an overtopped coffer dam?

Ans: When flood water is allowed to pass over top of the coffer dam it is called an overtopped coffer dam.

Q4: What is impervious core and where is it used?

Ans: Impervious core material is used in the core of a rock fill coffer dam to make the dam a water tight coffer.

Q5: What kind of soil condition is required for making a rock fill coffer dam?

Ans: A rock fill coffer dam is suitable for all kinds of foundations.

Q6: What is shotcreting?

Ans: Shotcrete is a concrete that is pneumatically projected with a high velocity through a gun on the surface to achieve high strength and low permeability.

Q7: What are the processes of shotcrete?

Ans: Dry-mix and wet-mix process.

Q8: What are the factors that should be considered for selecting a diversion system?

Ans:

- High flood data of the river
- Geotechnical features of the site
- Availability of construction material
- Type of coffer dam

Q9: Why is lining provided in a diversion tunnel?

Ans: Smooth surfaces provide less frictional force and increases discharge through the tunnel.

Q10: Why do you consider diversion through conduit in place of diversion through a tunnel?

Ans: River diversion through conduit is used when rock strata is not suitable for tunnelling.

Bibliography

EIA & EMP Executive Summery, Heo Hydro Electric Project.

7

Reinforcement Cutting and Bending Yard

7.1 Purpose

Concrete is weak in tension and strong in compression, but steel is equally strong in both; a composite structure made with both steel and concrete is the order of the day. A structure can be constructed considering its strong points which will have constructability, buildability and serviceability. To compensate for the weakness of concrete, steel is introduced in concrete to reinforce the structure which is known as reinforcement. Reinforcement can withstand both tensile force and shear force. Reinforcement is used as a stirrup in the beam and column and reinforced cement structures (RCC) have become very popular in the engineering world.

The contribution of steel in the construction world is enormous. It is about 15%–20% of the total cost of concrete work. Producing reinforcement of the required size and shape is a time consuming activity and fixing is also a time consuming activity. It has become a challenge and constraint for the achievement of the desired target, so all activities pertaining to reinforcement shall be controlled and monitored in a structured way.

These controlling measures include planning, quantification, indenting, procurement, storage, utilization, measurement, reconciliation, wastage and scrap management.

7.2 Type of Steel Used in Civil Engineering Work

Types of reinforcements used in work:

- Carbon steel plain and deformed bar
- Ribbed tor-steel
- Hot rolling/cold working high yield strength plain and deformed bar
- Stainless steel bar
- Galvanized carbon steel

- Epoxy coated carbon steel
- Fiber reinforced polymer bar

The following strength grade of deformed rebar are used in reinforced concrete structures:

- Fe 415, Fe 415 D, Fe 415 S
- Fe 500, Fe 500 D, Fe 500 S
- Fe 550, Fe 550 D
- Fe 600

Fe indicates the specified minimum 0.2% proof stress or yield stress, in N/mm². The letters D and S following the strength grade indicates the categories with same specified minimum 0.2% proof stress/yield stress, but with enhanced and additional requirements.

7.2.1 Chemical Properties

Chemical properties of all constituents of all categories of steel varies and should be tested.

Carbon content shall be within 0.25–0.30, sulphur content from 0.45 to 0.06, and phosphorous varies from 0.06% to 0.04%.

7.2.2 Mechanical Properties

Mechanical properties of rebar include proof/yield stress which should vary from 415 to 650 N/mm² (0.2%), elongation percentage on particular gauge length of definite cross sectional area for various category like Fe 415, Fe 500, Fe 600, etc. (Table 7.1).

7.2.3 Nominal Size of Rebar

Normal sizes of rebar are 6, 8, 10, 12, 16, 20, 25, 28, 32, 36, and 40 mm (Table 7.2).

TABLE 7.1

Mechanical Properties

Properties	Fe 415	Fe 415 D	Fe 415 S	Fe 500	Fe 500 D	Fe 500 S	Fe 550	Fe 550 D	Fe 600
0.2% proof stress/yield stress, min, N/mm²	415	415	415	500	500	500	550	550	600
Elongation, percent, min. on A, where A is √gauge length 5.65 the cross-sectional area of the test piece	15.5	20	20.0	15.0	18.0	20.0	12.0	16	12.0

TABLE 7.2

Nominal Size of Rebar

Sl No.	Nominal Size (mm)	Cross Sectional Area (mm²)	Weight per Metre (Kg)
1	6	28.0	0.22
2	8	50.0	0.40
3	10	78.5	0.62
4	12	113	0.88
5	16	201	1.60
6	18	254.0	2.0
7	20	314.0	2.46
8	25	491.0	3.85
9	28	616	4.83
10	32	804.0	6.32
11	36	1018.0	7.99
12	40	1257.0	9.86

7.2.4 Inspection and Various Checking on Rebar

Rebar is available in the market in straight form in all diameters except 6 mm diameter, which is available in the market in coil form. Available length varies from 10 to 12 m.

- *The following checks are required before direct procurement*
  - Vendor shall be approved by the project authority
  - Grade of rebar shall be per specification and approved brand
  - Vendor shall furnish the material testing report
  - If it is a new vendor, sample shall be collected of approved brand to be tested
- *If materials are supplied by the project authority, the following checks are to be conducted*
  - Indent for material shall be raised with grade, size and quantity per specification, etc.
  - Material shall be checked and accepted as conforming to specification
  - Test of material shall be conducted and frequency of tests shall be done per terms and condition of contract
  - If test results of any lot does not conform to specification it shall be brought in to the notice of project authority for rejection
- *Precautions to be taken on receipt of material*
  - Stacking of material shall be done diameter-wise on wooden sleepers to avoid any contact with soil if surface is not hard faced
  - Check the material whether it conforms to specification or not

- Identify the stack with tag number
- Sample shall be collected for necessary laboratory test
- *Precaution to be taken during bending of rod*: If any bar breaks or cracks during bending, testing of sample to be carried out for carbon content of the particular lot. Special care should be taken for 25 mm diameter and above.

  Particular lots shall be marked and used in work after clearance of test result.
- *Testing of samples*
  - Chemical and mechanical tests are to be conducted as per code requirements.
  - Checking for rolling margin at site is essential because material is issued from store by weight and material used in work shall be measured by linear and multiplied by weight to get quantity. In a big project where a large quantity of steel is involved, it creates a huge gap between material issued and material used during the reconciliation of material. The percentage of this gap becomes even more than the allowable percentage of waste indicated in the contract. Generally, it occurs in a larger diameter of rebar.

    To check it, a sample of various diameter of rebar shall be collected and weighed.

    Revised mass can be calculated.

    $$\text{Revised Weight} = \frac{\text{Weight of the sample}}{\text{Actual length of sample}}.$$

    This revised weight cannot be taken for payment, but it shall be considered for the reconciliation of material. Allowable waste as per contract is 3%:

    2½% is accountable and ½% is unaccountable waste.
  - *Check for reconciliation*: Reconciliation of material shall be done at the end of the work when all measurements are finalized. Reconciliation is done with material used in work plus allowable waste against material issued.

Example:

1. Total material issued = 2500 MT
2. Total material used in the work per measurement = 2400 MT
3. Allowable wastage @3% = 72 MT
4. 2½% accountable wastage = 60 MT (to be returned as unserviceable material with specified length as per contract)

5. ½% unaccountable = 12 MT (can be returned in the form of scrap for which credit shall be given)
6. Material to be returned to store = 2500 − 2400 − 72 = 28 MT
7. If contractor fails to return 60 MT to store they shall be penalized

So, it is essential to plan, organize and prepare BBS per normal processes, and cutting, and bending of steel shall be done to control extra wastage so as to avoid recovery for not returning the material.

7.3 Planning and Setting Up Bar Bending Yard

In mega projects a centralized bar bending yard is required to be set up for effective management of large quantities of rebar in a controlled manner so as to produce quality cut bar with low waste.

7.3.1 Requirement

Before establishing the bar bending yard, a detailed study of the contract document, technical specification and drawings of the project is necessary to know about quantity and item rate, measurement system and allowable waste, dealing with generation of scrap, laps and chairs, and diameter-wise quantification of steel (Figure 7.7).

A per month requirement of steel shall be calculated from the project schedule.

7.3.2 Preparation of Bar Bending Schedule

A Bar Bending Schedule (BBS) is a technical document that is prepared in accordance with IFC drawings reading in conjunction with specification, standards and codes to facilitate proper cutting and bending of reinforcement. Cut length, shape and bending criteria of reinforcement are adopted per code ACI315-99. The BBS will provide all details like location, mark, type, length, shape, size of each category of rebar along with abstract of quantity of rebar.

BBS is a tool by which reinforcement can be organized, controlled and waste can be minimized. Steel required for site work are cut and bent into the required shape and size per the BBS.

An indicative format of the bar bending schedule is drawn with detailed descriptions in Table 7.3, with the Bill of Material of each category of rebar.

7.3.3 Format of Bar Bending Schedule

Refer to Table 7.3 in Section 7.3.2 for BBS format.

TABLE 7.3

Bar Bending Schedule

Sl No.	Location	Bar Mark	Size and Type	No of Set	No per Set	Total Number	Length	Shape	Total Length	Unit Weight/M	Total Weight
1	Training Wall	T-1 SR1	TMT 25ø	5	6	30	4.5 M	Straight	135 M	@3.85 Kg	519.75 Kgs
2	Do	T-1 SR2	TMT	4	4	16	3.8 M	L Shape		1.58 Kgs	96.06 Kgs

7.3.4 Guidelines for Preparation of BBS

The engineer who prepares the BBS should have the following information:

- Reinforcement detail drawings of structural elements
- Location of structural elements
- Bar mark
- Diameter of bar
- Concrete cover
- Length of bar
- Cut length of bar
- Spacing of bar
- Number of bars
- Shape of bar

BBS shall be prepared keeping focus on fixing and placement of reinforcement of structural elements:

- Try to use the full length of the bar as far as practicable to minimize waste and reduce lapping. It will save material and time which will reduce the cost
- Avoid providing lapping in one place and provide laps in a staggered manner
- Avoid providing lap in tension zones, especially in the bottom rebar of beam in mid span and use mechanical couplers
- Do not provide lap in beam and column joints
- Provide laps in closer spacing as specified in lapping zone
- Provide mechanical couplers in higher diameter bar instead of conventional lap and avoid congestion in lap zone to facilitate proper compaction of concrete.

All of the above details can be obtained from R/F detail drawings except cut length, number of bars and shape of bar.

Bar length: BBS must show all bar dimensions with bar length including bends and hooks.

Hooks and bends: Hooks and bends are specified in code of practice to standardize the preparation of BBS.

Rebar under bending due to the ductility property of rebar gets elongated, so a deduction factor shall be applied to as per code of practice to get the cut length of the rebar.

- K factor with respect to type of steel shall be applied per IS2502 to find out cut length
- R factor shall be applied per BS 8666 to find cut length

7.3.4.1 Calculation

Let us consider Figure 7.1, which represents a drawing of a bucket of a dam consisting of 20 mm dia reinforcement at 150 c/c in main reinforcement. The transverse measurement of bucket is 8 m and cover is 50 mm.

$$\text{No of main rebar} = 8000/150 + 1 = 53.3 = 54 \text{ nos}$$

$$\text{Cut length} = \text{Horizontal length} + \text{Vertical length}$$
$$- 2 \times \text{Clear cover} - 4 \times \text{Diameter}$$
$$= 10000 + 1500 - 2 \times 50 - 4 \times 20$$
$$= 11320 \text{ mm} = 11.32 \text{ m}$$

Refer to Figure 7.2 for calculation of cut length of stirrup.

7.3.4.2 Recommended End Hook for All Grades

Standard Hooks and Bends and Recommended End Hook for all grades are per ACI 315-99. Refer to Figure 7.3 and Tables 7.4 through 7.6 for hooks with regards to finished bend diameter D.

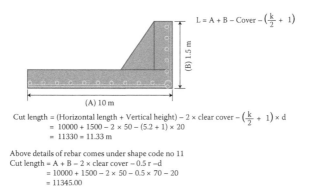

$$L = A + B - \text{Cover} - \left(\frac{k}{2} + 1\right)$$

(B) 1.5 m

(A) 10 m

Cut length = (Horizontal length + Vertical height) − 2 × clear cover − $\left(\frac{k}{2} + 1\right)$ × d
= 10000 + 1500 − 2 × 50 − (5.2 + 1) × 20
= 11330 = 11.33 m

Above details of rebar comes under shape code no 11
Cut length = A + B − 2 × clear cover − 0.5 r −d
= 10000 + 1500 − 2 × 50 − 0.5 × 70 − 20
= 11345.00

FIGURE 7.1
Shape of rebar in bucket.

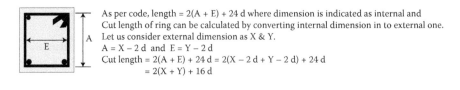

As per code, length = 2(A + E) + 24 d where dimension is indicated as internal and
Cut length of ring can be calculated by converting internal dimension in to external one.
Let us consider external dimension as X & Y.
A = X – 2 d and E = Y – 2 d
Cut length = 2(A + E) + 24 d = 2(X – 2 d + Y – 2 d) + 24 d
          = 2(X + Y) + 16 d

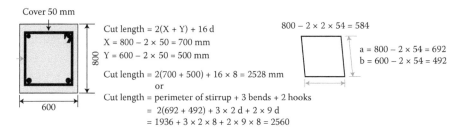

Cut length = 2(X + Y) + 16 d
X = 800 – 2 × 50 = 700 mm
Y = 600 – 2 × 50 = 500 mm

Cut length = 2(700 + 500) + 16 × 8 = 2528 mm
                or
Cut length = perimeter of stirrup + 3 bends + 2 hooks
          = 2(692 + 492) + 3 × 2 d + 2 × 9 d
          = 1936 + 3 × 2 × 8 + 2 × 9 × 8 = 2560

800 – 2 × 2 × 54 = 584

a = 800 – 2 × 54 = 692
b = 600 – 2 × 54 = 492

FIGURE 7.2
Cut length of stirrups.

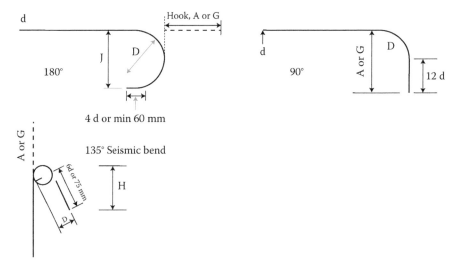

FIGURE 7.3
Standard hooks and bends as per ACI315-99.

Figure 7.4 represents pier of concrete gravity dam and reinforcement details of pier. Sections of pier indicate reinforcement arrangements with 25 mm dia main bar at 160 mm c/c with stirrups with 8 mm bar.

Arrangement of stirrups profile has been done per ACI 315 when the main bar spacing is more than 150 mm. Four types of stirrups are used as shown in Figure 7.2. U stirrups and cross ties are used as marked in Figure 7.2.

TABLE 7.4

Recommended End Hooks for All Grades of Concrete

| Bar Size No. | D ft (mm) | 180° Hook | | 90° Hook |
		A or G in ft (mm)	J in ft (mm)	A or G in ft (mm)
3(10)	60	5(125)	3(80)	6(155)
4(13)	3(80)	6(155)	4(105)	8(200)
5(16)	3¾(95)	7(180)	5(130)	10(250)
6(19)	4½(115)	8(205)	6(155)	1–0(300)
7(22)	5¼(135)	10(250)	7(175)	1–2(375)
8(25)	6(155)	11(275)	8(205)	1–4(425)
9(29)	9½(240)	1–3(375)	11¾(300)	1–7(475)
10(32)	10¾(275)	1–5(425)	1–1¼(335)	1–10(550)
11(36)	12(305)	1–7(475)	1–2¾(375)	2–0(600)
4(43)	18¼(465)	2–3(675)	1–9¾(550)	2–7(775)
18(57)	24(616)	3–0(925)	2–4½(725)	3–5(1050)

TABLE 7.5

Recommended Hooks for Resistance to Earthquake

| Bar Size No. | D ft (mm) | 135° Hook | |
		Hook A or G	H
3(10)	1½(40)	4¼(110)	3(80)
4(13)	2(50)	4½(115)	3(80)
5(16)	2½(65)	5½(140)	3¾(95)
6(19)	4½(115)	8(205)	4½(115)
7(23)	5¼(135)	9(230)	5¼(135)
8(25)	6(155)	10½(270)	6(155)

Source: Adopted from Table1 of ACI315-99.

TABLE 7.6

Recommended Stirrup and Tie Hook for All Grades

| Bar Size No. | D ft (mm) | 90° Hook | 135° Hook | |
		Hook A or G ft (mm)	Hook A or G ft (mm)	H ft (mm)
3(10)	1–½(40)	4(105)	4(105)	2–½(65)
4(13)	2(50)	4–½(115)	4–½(115)	3(80)
5(16)	2–½(65)	6(155)	5–½(140)	3–¾(95)
6(19)	4–½(115)	1–0(305)	8(205)	4–½(115)
7(22)	5–¼(135)	1–2(355)	9(230)	5–¼(135)
8(25)	6(155)	1–4(410)	10–½(270)	6(155)

Source: Adopted from Table1 of ACI315-99.

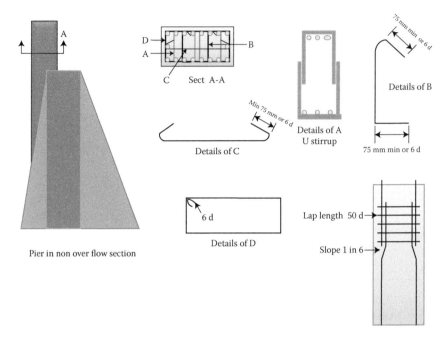

FIGURE 7.4
Rebar details of pier of dam.

7.4 Splicing and Its Types

Splicing of reinforcement is a crucial activity in construction projects. Design and performance of splicing depends on engineers.

7.4.1 Conventional Lapping

Verticals are lapped and staggered. At least 50% of bars shall be spliced in one point. The slope of the inclined portion of the main bar in lap length that provides some offset shall not be more than 1 in 6. Lap length shall be provided 50 d. Additional ties shall be provided in lap length when the main bars are bent to provide offset. The spacing of ties shall be d/4 or greater than 100 mm but less than 150 mm, and it should be placed from the point of the bend. Reinforcement in the lap portion becomes congested due to the splice which poses difficulty in concrete compaction.

7.4.2 Mechanical Splicing/Bar Coupler

Mechanical splicing is also preferable where couplers are used to splice. Rebar shall be threaded up to required length, and couplers shall be tightened. It does not congest the coupling area (Figure 7.5).

Coupling reduces the usage of rebar. It accelerates the pace of the work and avoids congestion in the splice zone of the steel area.

Lapping of rebar is done to transfer the load, and this transfer of load depends on concrete. The system will fail if concrete does not conform to the specification. Mechanical splice will sustain the load without the help of concrete in which it is positioned.

Couplers are of two kinds: Threaded and nonthreaded, and are available in various sizes from 10, 12, 14, 16, 20, 22, 25, 28, 32, 36, 40 and 50.

7.4.2.1 Standard Tapered Threaded Coupler

The standard tapered threaded coupler consists of an internally threaded sleeve with two right handed, threaded taperings mid length of coupler. The end of the bar is saw cut, and threading is provided on the taper end by threading equipment. Initially it is tightened manually into the sleeve, and finally the sleeve is tightened by torque wrench.

This type of coupler is most suitable for the joining of rebar of the same diameter.

7.4.2.2 Parallel Threaded Coupler

This type of coupler is similar like the tapered threaded coupler except its ends are not tapered. Some couplers are available for joining different diameters of bar, etc. (Table 7.7).

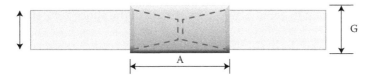

FIGURE 7.5
Mechanical splicing.

TABLE 7.7

Dimension of Coupler

Dimension		16 mm	20 mm	25 mm	28 mm	32 mm	36 mm	40 mm	50 mm
Length	A	70	74	90	100	112		138	170
Diameter		25	30	36	42	46		55	70
Weight		0.17	0.24	0.41	0.66	0.85		1.90	2.22
Torque		110	165	265	275	285		330	350

7.4.3 Installation of Coupler

The coupler is fixed to rebar and placed in concrete and cast. After concrete is cast and the next phase of work is to start, remove the plastic cap from the coupler and position the continuation bar inside the sleeve and tighten by torque wrench.

7.4.4 Test for Coupler

Couplers shall be tested as follows:

- Static tensile test
- Slip test
- Cyclic tensile test
- Fatigue test

7.4.5 Splicing by Welding

Welded splice is permitted by all codes of practice. The strength of the welded joint shall be taken equal to or more than 100% of the strength of the bar. Generally, 85–90% of the strength of the bar is considered for the bar in tension and 100–125% for the bar in compression. Welding shall be avoided on the bend of the bar. Splicing by welding shall be done in staggered condition. Special care shall be taken when the subject structure is under dynamic loading.

IS 9417 permits welding splice in a deformed bar where parent material should be weldable quality per IS 1786, and electrode used shall be per IS 814.

Column or pier in dam, etc., where the percentage of steel as per design is greater than and between 4% and 6%, mechanical coupler or welded splice shall be taken up to avoid congestion of steel in the splice zone (Figure 7.6).

7.4.5.1 Electrodes

Electrode size shall be as follows (Table 7.8).

7.4.6 Welding Method

Welding procedures shall be adopted either by flush butt weld or by metal arc welding. The butt joint may be carried out by the flush butt welding process and lap joint shall be carried out by the metal arc welding process.

7.4.6.1 Butt Welding

The bar end shall be shaped into a bevel face by shearing and machining by using a power saw and grinder. The bevel faces should be finished by hand

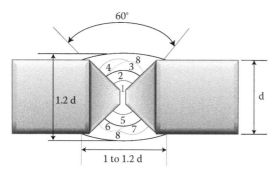

Sequence of welding beads

V butt joint

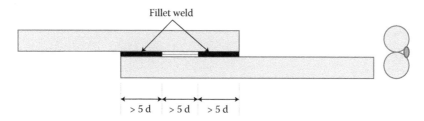

FIGURE 7.6
Welded splicing.

file. The bars should be aligned and set in their axis in one line by using clamp or guide so as to avoid any rotation. The joint should not be out of alignment more than 25% thickness of the thinner material (Figure 7.7).

7.4.6.2 Sequence of Welding

Butt welding is a crucial activity. This welding is carried out in sequence. A small gap of 2–3 mm is provided between two bevel faces for root run as shown in Figure 7.6. 1 to 4 beads are made first. A small stop shall be given after every second bead. The bead shall be cleaned and rebar shall be

TABLE 7.8

Electrode Size (mm)

Bar Size ø mm	Up to 10 mm	10–18 mm	18–28 mm	28–49 mm
Electric Arc weld	Root run 2–2.5			
	Other Run	3.15	4.0	5.0
Lap weld with longitudinal beads	2.5	3.5	5	5

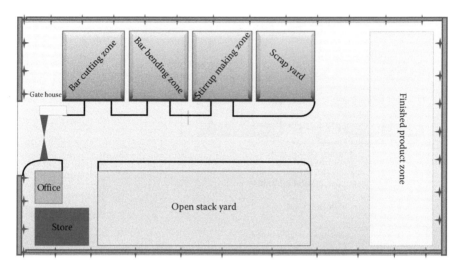

FIGURE 7.7
Bar bending yard.

allowed to cool down. The temperature during welding in each bead shall be monitored and it should not be more than 300°C. The next bead shall be carried out when the temperature is about 250°C.

After completion of beads, the bar shall be turned 180° and bead 5, 6 and 7 shall be completed. The bar is rotated full circle after completion of the final beads. Diameter of the weld shall be 1.2 times the diameter of bar, d. The bevel angle should not be more than 60°.

7.4.6.3 Lap Weld

Figure 7.7 shows lap weld with a horizontal bead. The length of the weld bead shall be 10ø. Max bead length shall be limited to 5ø. Longer longitudinal beads shall be @5ø with a gap in between the beads. This arrangement of welding is on one side only. If welding is done on two sides of the rebar, the length of the bead shall be 5ø each on both side equivalent to 10ø.

7.5 Setting up Bar Bending Yard

The bar bending (BB) yard shall be developed in accordance with the quantum of rebar required for the project. The bar bending yard can be located at site if adequate space is available; otherwise, it shall be located outside the site.

The layout of the yard shall be formulated in such a way that loading and unloading of materials can be done easily and adequate space shall be provided for smooth negotiation of trucks and loading equipment, etc., with proper entry and exit as shown in Figure 7.7.

Yard should contain the following:

- Storage for straight bar
- Storage for coil
- Office
- Store
- Area for cutting, bending machine
- Area for stirrup making machine
- Area for finished product with proper loading arrangement
- Space for movement of truck, crane and hydra, etc.
- Area for scrap storage
- Only one entry and exit with security arrangements
- Gate house to maintain entry and exit record with registration number, etc.
- Area to be fenced

7.5.1 Equipment/Machines Required

After the BB yard is set up, the following equipment and machines shall be installed in desired location.

- Bar cutting machine
- Bar bending machine
- Stirrup making machine
- Bar de-coiling machine
- Overhead travelling crane for movement of materials
- Hydra and small capacity cranes

7.5.2 Guide Lines for Selection of Machines

Guidelines for selection of equipment and machines for BB yard are as follows:

- *Bar cutting machine*
  - Overall dimensions
  - Overall weight
  - Max cutting diameter

- Cutting stroke per min
- Power of motor
- *Bar bending machine*
  - Automatic/semiautomatic/digital
  - Overall weight
  - Overall dimension
  - Max bending diameter
  - Rotational speed of working disc
  - Power of motor, speed of motor
  - Bending control system
  - Angle setting system
  - Bending radius with electrical braking system
  - Limit switch in both in front and back
- *Stirrup making machine*
  - Single strand/double strand
  - Pulling speed
  - HS version
  - Bending speed
  - Ave power consumption
  - Daily output
  - Adjustment of bending angle
  - Various shapes, etc.

7.5.3 Work as per BBS

After installation of the bar bending, bar cutting and stirrup making machines at the BB yard, the engineer in charge should fix the priority of cutting and bending as a monthly program. He should identify the structures that are considered in the program. He should take out the IFC drawing of the structures and corresponding BBS prepared for the particular IFC drawing and start the work accordingly. Fabricated rebar made per BBS shall be bundled and stacked in the finished product zone with tag for future identification for dispatching to work site.

The engineer can assess the quantity exactly fabricated and percentage of waste generated. He can identify unserviceable materials and find out the possibility of using these unserviceable material in chairs and other ancillary work to avoid waste for effective reconciliation of material as per contract.

7.5.4 Advantage of BBS

BB yard utilized with proper BBS will add the following advantages:

- Working with proper BBS will reduce material waste
- Cut length and shape of each bar made in BBS with finer details in accordance with code and IFC drawing shall save costly items like steel
- BBS provide quality control of the work
- BBS provide accurate quantity of steel which helps do proper estimation of the work
- BBS helps to control the stock of material
- Rebar fabricated per BBS can be easily laid in position within form work which accelerates the pace of work
- Work can be accelerated without any placement constraint and time can be saved

7.5.5 Safety Norms to Be Adopted at Bar Bending Yard

The workforce engaged in the BB yard shall strictly use the following personal protection equipment (PPE) as per safety norms:

- Helmet
- Safety shoes
- Hand gloves
- Goggles
- Ear plugs

Crane and hydra operators should have a valid operator's certificate and shall be equipped with helpers. All moving equipment should have a backing horn to avoid accidents during backing.

7.5.6 Safety Hazard Identification in Bar Bending Yard

The safety engineer is to identify the safety hazards in the yard. He should ensure that everyone working in the yard adheres to safety norms and adopt safe working practices.

- Improper housekeeping results in injury
- Cuts and bruises in palm due to improper handling and without usage of gloves

- Toppling of hydra and cranes
- Hit by trucks and tailors
- Improper earthling system
- Failure of crane hook or guy

7.5.7 Advantage of BBS

Bar Bending Schedule is required to be made to improve the quality of work, as well as reduce cost and time as described below:

- Working with proper BBS will reduce material waste
- Cut length and shape of each bar made in BBS with finer details in accordance with code and IFC drawing shall save costly items like steel
- BBS provide quality control of the work
- BBS provide accurate quantity of steel which helps do proper work estimates
- BBS helps to control the stock of material
- Rebar fabricated per BBS can be easily laid in position within form work which accelerates the pace of work
- Work can be accelerated without any placement constraint and time can be saved

Bibliography

ACI315-99. Details and Detailing of Concrete Reinforcement. Reported by ACI Committee 315.

8

Investigation and Exploration
of Dam and Reservoir Site

8.1 Fundamentals

A dam is a national property that is constructed for the national economy with a huge investment of national and other resources. The safety of a dam is of prime importance for safeguarding national investment and benefits derived by the nation from this project. An unsafe dam causes hazards to human life and property downstream. Any unsafe design of a dam may lead to national disaster.

The world has already experienced many failures of dams. These failures have forced technical groups to think of new kinds of technical development that can provide safety to the dams and prevent dams from collapsing. Subsequently, geotechnical investigation has gained importance.

A dam is an impervious barrier made across a river impounding water upstream in order to create the requisite water head. A dam is a water retaining structure and shall be designed to be safe against forces acting on the dam. One of these forces is water. It may be below the foundation or on and through the body of the dam, so safety of the dam rests on the nature or the degree of permeability duly attained in the dam foundation and also in the body of the dam. The permeability through the foundation of the dam is nothing but uplift force being generated due to built up hydrostatic head upstream of the dam. Permissible limits of permeability of concrete depend on the grade of concrete adopted and degree of compaction achieved during construction. Permeability through concrete can be controlled or can be kept within the permissible limit if concrete is done as per specifications and guidelines laid down in the Quality Control Manual. A dam becomes vulnerable if permeability of the dam exceeds the permissible limit. Our point of discussion in this chapter is permeability through the foundation on which the dam rests and the water tightness of the reservoir area where water is impounded.

8.2 Dam Site Investigation

A dam is constructed in a hilly terrain having various type of rocks. It is very important to have an adequate assessment of the geotechnical condition of the dam site for the evolution of the safety of the dam. The primary aims and objectives of an investigation are to carry out a geological exploration to evaluate the geotechnical and geo-mechanical characteristics of the dam foundation. The foundation exploration may be carried out for some extraordinary geologic conditions such as fault zone, shear zone, clay seam, fissures, cleavages, karst, buried channel, etc., if some abnormal characteristics are observed during preliminary investigation. These adverse geological conditions shall be assessed and understood correctly for appropriate treatment of the foundation to make the foundation suitable for safe dam construction.

All significant data obtained during investigation that affect the feasibility of the foundation site and require treatment should be clearly envisaged before design and construction so that the chances of geotechnical surprises are minimized and avoided during construction.

The author takes the liberty to advise students that the complexity and uncertainty of the natural geological environment that make the dam site's geological condition is unique. They should be careful while dealing with hydro structures in the future and pay due respect to the geological condition of the proposed site.

Before going through the details of the investigation process, it is essential to clarify and define some special and extraordinary geological features to the readers, especially those who are students, to have a clear understanding of the adverse geological constraints/problems. These features, if remained unidentified and untreated, may pose a threat to the stability of the dam.

- *Joints*: Fractures along which particularly no displacement of rocks has occurred. A crack produced in the rock under the action of internal forces during its cooling and drying.
- *Fault*: Fractures along which the rocks on one side have been displaced relative to those on the other side. A fault is a fracture surface along which rocks have been relatively displaced in both directions vertically and horizontally. An earthquake may occur due to the sudden movement of a big fault.
- *Fault breccia*: Angular or sub angular rock fragments produced by fracture and grinding during faulting and distributed within or adjacent to the fault plane that will have very low bearing capacity.
- *Cleavage*: Fissures that develop in rocks under the action of external tectonic factors.

- *Fissures*: A long, narrow opening that has occurred in the rock by cracking or dislocation.
- Seam is a thin layer or strata of rock, coal or mineral, etc.
- *Bedding*: Some rocks such as sedimentary and metamorphic rocks occur in the form of layers or strata bounded by parallel surface.
- Foliation
- Cavities
- Dyke
- Sinkhole

Water is the main factor in the chemical weathering of rocks, and if water seeps or percolates through fissures, joints and faults, etc., it makes the foundation vulnerable and unsafe.

8.3 Geological Exploration

Geological exploration of the dam site is conducted to study and investigate the type, nature, feature and property of geological structures, both on and below surface, and exploration shall be conducted in an adequate area around the dam.

The purpose of exploration is to find out the character and integrity of the geological and structural features of the foundation material available at the proposed site. All typical and critical geological conditions that may exert influence on the stability of dam structure shall be investigated, observed and tested. Their possible remedial measures shall be envisaged before starting the design of the structure.

Dam site geology, in hills or plains, is treacherous as well as mysterious due to the flowing of water for eons. The vulnerability of a dam due to some adverse geologic condition shall be investigated, planned and programmed in a structured way to determine the following information concerning the competency of the foundation material.

- Type of rocks, their nature, characteristics and properties, scaling, extent of weathering, and its depth, formation of soil, debris and silt carried by river, etc.
- Structural features like strikes, dips, bedding, cleavages, faults/shear zones, folds, seams, joints, cavities, fissures, etc.
- Type of valleys, hills, course of river, nature of slopes and type of probable landslide, bed rocks and its depth

- Thickness of overburden and its properties
- Tectonic features and presence of any active fault in and around dam
- Location of proposed dam
- Ground water table conditions, springs, seepage, wells, cavities, etc.

It is clear that the extent of investigation depends on the size and magnitude of the project. The group assigned for investigation shall study and interpret the available information about the proposed project site before taking up further investigation. The sources of this available information are topographical maps, tectonic maps, hydro-geological maps, geological maps, available literature and various geological reports made by different and relevant approved organizations of the country.

The content of each category of map is described for the benefit of readers like students or new engineers for their benefit.

- *Topographical maps*: It is an important geographic tool which is prepared by the national mapping organization of each country. The Survey of India have drawn this map for India. This map serves the purpose of a base map and is used to draw all other maps. This map shows the important natural and cultural features such as relief, vegetation, water bodies, cultivated land, transportation, etc.
- *Tectonic maps*: The tectonic map of India was produced by the Geological Survey of India & ONGC based on the geological map and topographical map of India. The tectonic map of India shows international boundaries, water features and populated places. Relief is shown by contours. Depth is shown by bathymetric tints.
- *Hydro-geological map*: A hydro-geological map shows the unconsolidated/semi consolidated and hilly formations, and also indicates the ground water potential for each type.
- *Geological map*: A geological map is a map which shows geological features like strikes, dips, faults, folds, foliage, etc.

8.4 Stages of Geological Exploration

Geological exploration is taken up in stages as mentioned below:

- Reconnaissance or pre feasibility stage
- Preliminary investigation or feasibility stage
- Detailed investigation or DPR stage
- Construction stage

8.4.1 Reconnaissance or Pre Feasibility Stage

A survey group comprised of engineering geologists and civil engineers should do some homework before they make a visit to the site for a reconnaissance survey. The survey group is to find out and study the available data such as geological, tectonic and hydro-geological maps, etc., so as to have a comprehensive idea about the sites before the inspection. Accordingly, they visit to assess the overall geology of the site and critically observe the foundation conditions. The selection of the site shall be done on the basis of geology, topographical expression and anticipated depth and quality of rock.

The inspection group is to record their observations and findings as general geological, tectonic condition of the area, potential seismic hazard, presence of faults, fissures, seams, cracks and type of bed rocks, overburden, etc.

Inspection groups examine the suitability of the preferred site by studying the information obtained in the reconnaissance survey in conjunction with previously available data and suggest the tentative layout of the dam, general geological map and its section.

8.4.2 Preliminary Investigation or Feasibility Stage

In this stage, general geological data are collected to formulate the project layout. Exploration coverage shall be made in such a way that it will be suitable for assessing the techno-economic feasibility of the dam site.

8.4.2.1 Method of Preliminary Investigation

The area preferred for the dam site shall be extensively explored geologically in a largely adequate area so as to confirm the site is geologically fit for supporting the construction of a dam; exploration is comprised the following steps:

- Surface geotechnical mapping
- Exploration by test pits and trenches
- Exploration by drilling
- Exploration by seismic survey, GPR
- Determination of water tables and field permeability by percolation test
- Standard penetration test and field density test on overburden
- Surface geological mapping

Geological mapping is a study of geologic features like type of rock, age, pattern, etc., of a proposed site of the dam. Mapping shall be done in a specific scale of 1 in 1,000 with contour interval of 1–5 m as approved by the authority. Area coverage of mapping shall consider the area equal to twice the height of the proposed dam. The criteria shall be followed both upstream and downstream. The survey shall be preferably extended up to 100 m above the dam for an immature area and 25 m for mature topography.

The first step of surface geological mapping is to do the reconnaissance survey regarding accessibility to the proposed site, area for camping, ideas about terrain and how to start the work.

The second step is to find out the equipment required for field work, that is, base map, campus with clinometers, GPS, air photos, hand lens, camera, pencil and note book, geologic hammer, sample bag, personal protective equipment like safety clothes, safety shoe, sun glass and first aid kit, etc.

The author feels that students and new engineers should be briefed about the utility and application of equipment required for geological mapping and a brief description of each is indicated below for the benefit of the students and freshers.

- *Hammer and chisel*: Geologists working in the field need a hammer to break the rock for specimen collection. Sometime hammering does not break the rocks by itself, and a chisel is required for cutting the rock so as to collect specimens.
- *Compass and clinometers*: A compass is used to take bearings in the field.
- *Hand lens*: Geologists should always carry a hand lens with him in the field and should have the capacity of magnification between 7 and 10 times.
- *Tapes*: 3 m steel tape is very handy and useful to carry and anything from grain size to bed thickness can be measured.
- *Map cases*: Geologists work in the field in sun, rain and mist. A map case is essential for protection from heat and cold, rain and mist. A map case consists of a rigid base so that geologists can plot and write easily and it will protect the map.
- *Field notebook*: A field book should be of good quality, with a strong hard cover, rain proof and a good binding. A hard cover is necessary for a good surface for writing and should be kept in the pocket. Size should be adequate enough so that it can be kept on the palm for note taking.
- *Scales*: Geologists should keep a scale of at least 15 cm long.
- *Pencil and eraser*: Pencils are required for mapping in the field. A hard pencil is required to plot bearings, a soft pencil is required to plot strikes and write notes. A third soft pencil is required for writing in the notebook.
- *Acid bottles*: A 10% concentration of HCL is required to work on limestone if necessary.
- *Global positioning system (GPS)*: When geologists are in the field, it becomes difficult for them locate their position on the map due to lack

of proper details. That's why geologists use GPS to locating themselves in the map as well as a navigational system. It is a small hand held instrument that picks up signal from orbiting satellites constantly transmitting time and their position. The time delay between transmission and reception allows the satellite—receiver distance to be calculated. By simultaneously using signals from several satellites, the position of the receiver on the ground can be determined.

- *Personal protection equipment:* Geologists must wear proper clothes suiting to the atmosphere in the field. They should have proper personal protection as indicated below:
  - Gloves
  - Safety shoes
  - Sunglasses
  - Hat
  - Waist coat
  - Rain coat

The third phase is starting the mapping activity to collect field data. Field data are photographs, notes, measurements, physical samples, etc.

Two numbers of topographic base maps are required for plotting geological observations made by the geologist. The geologist will plot the observation on the first map in the field in a specific scale and the requisite grid pattern, and a fair copy will be made by interpreting the observation for final submission. Rocks and other features observed at the site shall be marked on the map. The following are to be mapped at least on the geological map:

- Type of rock and contact
- Shape of rock, age, porosity and permeability
- Type of weathering and its pattern
- Geologic structure
- Strike, dip, fold, fault, joint, bedding, cleavages, etc.
- Geothermal manifestation
- Ground water characteristics and location of springs, their temperature, etc.

8.4.2.2 Sub Surface Mapping

Sub surface mapping shall be done in the guidelines of surface mapping. The entire river bed and dam area will be filled by debris and other materials carried by the river. A geophysical survey shall be conducted to establish the depth, nature of bed rock and ground water table, etc.

As per code, for a dam of up to 30 meter height, exploration by trial pits, bore holes and trenches are sufficient. A dam of height above 30 meter requires additional exploration.

Sub surface exploration shall establish the following:

- Depth of bed rock, thickness of overburden and weathering patterns
- Litho logical character of various rocks such as color, texture, grain size, composition, etc.
- Details of joints, major or minor fault zones, etc.
- Depth of water tables
- Permeability strata

8.4.2.3 Feasibility Report

After completion of the preliminary investigation or feasibility stage investigation of the dam site, all geological data shall be analyzed and a feasibility report shall be made to include the following as specified in the code:

- Geological plan of the dam site along with location of the dam
- Geological section along the axis of the dam
- Geological section of test pits and trenches
- Bore logs and photographs of cores along with recovery
- Field and lab test result
- Geological evaluation of dam foundation

8.4.3 Detailed Investigation or DPR Stage

Detailed investigation is taken up after the feasibility report is made and the geologic condition of the dam site is well understood during the feasibility stage. In addition to it, a detailed investigation of the dam site's surface and sub surface is required in the DPR stage. This exploration shall be done via close interaction between the geologist and design engineer.

8.4.3.1 Detailed Sub Surface Exploration

Exploratory drilling is the best method for the exploration of rocks. It is the most practical and accurate method by which reliable sub surface data can be obtained and a practical grouting program can be made with the data. Core drilling is the process of obtaining a cylindrical rock sample by rotating a hollow core barrel with the attachment of a bit. It is either diamond, tungsten carbide or polycrystalline. Tungsten or polycrystalline can be used for soft rocks. The most important aspect of this drilling is the recovery of the material drilled. If recovery is less than 50%, then the drilling is to be abandoned in that location. Diameter of the core and depth of bore hole is to be decided by the geologist. Larger diameter bits give better core recovery and less disturbed core samples. The location and spacing of bore holes shall be fixed by the geologist per site conditions. Generally, bore holes are preferred to be done along the axis of the dam and some additional bore holes, if the geologist feels, are to be taken in doubtful patches within the dam area like any gorge, abutment area, or any portion with suspected color texture or any other unpredictable geological feature. It is advisable to drill 2 or 3 bore holes in the toe area of the dam. At least one bore hole shall be drilled in the energy dissipation silting basin area, and 3 bore holes shall be done in the selected location in alignment with the diversion tunnel depending on the geology and topography of the area. A permeability test shall be conducted to find the permeability of the foundation material.

The location of each bore hole shall be marked on the rock and numbered in red. The level and coordinate of each hole is to be recorded before the start of the drilling procedure. The drilling machine operator should be experienced to ensure smooth operations. Rock structure is not homogeneous and it will have some unseen seams, joints or any weak zone below which may pose problems in drilling operation, resulting in the drill bit becoming stuck in the joint or fractured part of the rock. It becomes difficult to get the drill bit freed from this hazard and delays caused to the operation, but experienced drillers are capable of assessing this situation well in advance by observing drilling performances and sludge return during drilling, and may take up necessary precautionary measures in order to avoid an undesired situation. It is also important to ensure that boulders of cemented soil are not mistaken as bed rock.

Undisturbed samples shall be collected as far as practicable and shall be wrapped or waxed to retain the moisture. Samples collected shall be kept in a wooden box as per length of the samples. The box should contain the bore number, location, reduced level, depth of hole, and date, and it shall be graduated on the top of the box so as to identify the depth wise data for the preserved sample. The length of a collected sample may not match with the length of the hole drilled because the recovered sample may be smaller due to the presence of a joint/fracture in the rock. Once the entire sample is collected, the whole bore will be logged. The logging of each sample shall be recorded in the prescribed format. The sample collected in the box in a systematic manner shall be used for laboratory tests to determine engineering properties, and shall be preserved for future reference during the construction stage. The format shall contain adequate information of the bore hole such as bore hole number, location, level, date of starting and completion, depth, type of bit, type of rock, percentage of core recovery and any other information like water table if encountered.

When it becomes difficult to understand the geological condition in river bed by vertical drilling, it is advisable to take up inclined drilling from both abutments with bore holes crossing each other in the river bed. It may go deeper in the bed to detect any fault/shear zone that exists below.

Additional criteria are prescribed by relevant IS Code for a dam having a height of up to 50 meters. One drift shall be excavated in each bank at or near the mid height of the dam. Two drifts with cross cuts shall be excavated on each bank at or near 1/3 or 2/3s height of the dam. Three or more drifts are recommended when dam height is more than 100 meters. The length of the drift shall be decided by the geologist. The grout ability of foundation shall be tested in this stage.

8.4.3.2 Geotechnical Property of Foundation Material

The geotechnical properties of the foundation material shall be evaluated by testing collected samples in a laboratory as well as field in-situ tests as indicated below:

- Laboratory test
  - Unconfined compressive test
  - Modulus of elasticity
  - Specific gravity and water absorption
  - Slake durability index
  - Shear and compressional ultrasonic velocity
  - Shear strength of infilling material
  - Swelling
  - Joint stiffness test
- In-situ test
  - Shear strength of rock mass
  - Modulus of deformation
  - Seismic wave velocity
  - Bearing capacity
  - Geo-hydrological characteristic of foundation shall be defined
  - Geo-technical evaluation of foundation/abutment treatment should be made for finalization

8.4.3.3 Detailed Project Report

A detailed project report shall be made after exploration and testing is completed and project layout shall be done as the final step.

8.4.4 Construction Stage

Overburden is removed and the rock surface of the foundation is exposed where detailed exploration shall be taken up to appraise construction and design engineers about geological and geotechnical features of the exposed foundation like fault zone, clay seam, any other weak strata which may require grouting, dental treatment, etc., to strengthen the foundation and make it water tight.

The foundation surface shall be cleaned by the application of air and water jet and shall be handed over to the geologist to make the final foundation grade geological mapping of the foundation. This mapping shall be done with a scale of 1 in 100 with 1 meter contour interval in grid of 2 m × 2 m by total station.

8.4.4.1 Exploration by Pit

Exploration by pit is a low cost method. Location of pits are selected by examining the site condition based on the advice of the geologist. It is very effective in shallow depth. Pits are excavated manually or mechanically in all types of soil depending on site conditions. Manual excavation is beneficial for taking an undisturbed sample of soil. The size of the pit can be fixed with respect to the depth of pit and also type of soil (soft, medium, hard or murrum). The depth of the pit should preferably be taken below the water table. The width of the bottom of the pit shall be fixed while considering adequate working space for workforce and also for placement of ladder so that the technician can collect samples for laboratory tests and to perform any in-situ test at the bottom. However, the bottom width should be less than 1.5 m × 1.5 m. The bottom shall be made level. The advantage of pit exploration is that all strata are exposed on the wall of the pit and can be logged at each and every layer. As required, SPT can be performed at prescribed intervals at the change of strata for the determination of liquefaction of material and modeling of deformation.

Liquefaction is a state of soil in saturated, cohesion less conditions when shear strength of soil is reduced to almost zero due to pore pressure caused by the vibration of earthquakes, and soil tends to behave like a soil mass in this condition.

8.4.4.2 Safety Precaution

Excavation shall be done with a suitable side slope of 1:1.5 or 1:2 depending on the nature of soil; it can also be protected by bench cutting. Excavated material shall be stacked at least 1.5 m away from the pit. Hard barricading shall be done around the pit for safety. A pipe shall be provided in the case of a deeper pit to allow fresh air to come inside the pit and a length of pipe shall be taken at least 1 meter above the mouth of the pit. The pit shall be filled immediately after inspection and sampling are completed and records are generated. When a water table is encountered in the pit, a suitable dewatering arrangement shall be made taking proper care for safety.

8.4.4.3 Collection of Undisturbed Samples

Stratum wise undisturbed samples can be collected from the pit for laboratory tests. A pillar 40 cm × 40 cm shall be left undisturbed in the middle of the pit for the collection of samples from each layer. Special care shall be taken to protect the moisture content of the sample. The sample collected shall be wrapped to avoid loss of moisture.

8.4.4.4 Exploration by Trench

The exploration by excavating a trench is quite similar to exploration by pit. The difference is that the length of the trench provides continuous exposure of the condition of the profile. The exploration by trench is most suitable for a slope. All other procedures are the same as in explorations by pit, including safety procedures.

8.4.4.5 Exploration by Core Drilling

Exploratory drilling is an important activity of sub surface exploration. Exploration by core drilling shall be taken up in various locations as advised by the geologist and shall be recorded as described in Section 8.4.3.1.

8.4.4.6 Geologic Interpretation of Geological Mapping

Geological mapping indicates strikes, dips, folds, faults, strati-graphic boundary, lithological type, drainage, etc.

8.5 Permeability Assessment of Foundation

It is essential to clearly differentiate between permeability, percolation and seepage before coming to an assessment of permeability.

- When water passes through a zone of non-saturation to the underlying zone it is called percolation.
- Seepage.
- A phenomenon when water passes through a rock or soil freely is known as permeability. It is the capacity of a rock to allow water or

other fluid to pass through it. It is measured in rate of flow. The rate of movement of water depends on permeability of rock through which it flows and its hydraulic gradient.

Assessment of permeability of foundation can be done by various methods, but the most popular and effective method is the Lugeon Test which is an in-situ test. The test is conducted on site. A bore hole shall be drilled as suggested by the geologist up to the designated depth. The test is done through a packer connected with the air and water through a pump and air compressor. This is a multi-staged test. The packer shall be placed in the hole. It means that the packer will isolate the depth where the test is conducted from the balance depth of the hole to find the permeability in each isolated portion of the hole because porosity is not uniform throughout the depth of the hole. This procedure will indicate almost the exact performance of the hole in each section by measuring the intake of water under the specified pressure up to the specified refusal of intake. The Lugeon Test is conducted to measure the quantum of water injected to a section of hole under specified pressure. The value is indicated as loss of water in liters per min per meter run of the hole.

8.6 Definition of Grout and Determination Grout Holes

Bed rock is not homogeneous and isotropic. It consists of many defects such as joints, fissures, cavities, seams, interstices, any other discontinuities, etc., that permit water to percolate through these defects and make the dam unstable, vulnerable and weak.

The process by which these defects of rock are sealed in order to reduce the permeability of the foundation rock mass is known as grouting. Grouting is done under application of pressure to fill the interstices, fractures, fissures, seams and joints, etc., to strengthen the foundation. The pressure shall be sufficient enough to help the grout flow through interstices and voids without dislodging the rock main structure. The grouting material may be cementitious, colloidal or any chemical solution.

The grouting procedure becomes successful if all interstices, seams, and discontinuities are filled with proper grout materials, but becomes ineffective in the long run if clay or any ungroutable material or any loose materials remain in the fissures, seams or interstices, etc. So, it is pertinent that all loose materials, clay or any other materials present in the rock mass shall be cleansed by flushing the holes with controlled air and water being injected in the holes by packers. Detailed procedures will be narrated in the next chapter. This cleansing procedure shall be taken up thoroughly before taking up grouting activity.

Engineering geologists have a vital role to play about the foundation treatment of any dam site. Before designing any treatment process, it is essential to clearly understand the geologic and hydrologic condition of the proposed dam site. The initial geologic and hydrologic idea can be had from a pre investigation report, and knowledge can further be upgraded during the process of construction and a grouting method can be improvised accordingly. Moreover, it is to be examined during the process by the geologist whether any special corrective measure shall have to be adopted for strengthening the foundation in addition to grouting. Accordingly, the engineering geologist will design the pattern, size, and depth of the holes based on the height of the dam and characteristic of rocks encountered in the dam site so as to have a safe and economic dam.

As per the geologic and hydrologic condition of the dam site, different types of grouting are prescribed by the geologist, but our discussion in this chapter shall be limited to the methodology of grouting as indicated below for the benefit of the readers.

- Consolidation grouting/area grouting
- Blanket grouting
- Curtain grouting

Consolidation grouting is done to strengthen the rock beneath the dam. Consolidation grout holes are generally made of a shallow depth up to 10 meters. The spacing of grout holes depends on the characteristic of rock present in the proposed site. The primary holes are drilled and water tested assess the permeability, and are then grouted.

Overburden shall be removed from the foundation area and rock surface shall be exposed and shall be offered to the geologist for geological mapping of the area. The geologist inspects the exposed rock profile and its various features like strike, dip, fold, fault, etc., and will incorporate it in the mapping process. This geological mapping of the entire footprint of the foundation shall be used to select a special foundation treatment. A geological section drawing shall be made duly indicating possible fissures, cracks, seam, etc., along with vertical and inclined grout holes. The location of grout holes shall be marked on the drawing in grid pattern duly indicating vertical and inclined holes.

Bibliography

AHEC-IITR, "1.4 General: Report Preparation: Reconnaissance, Prefeasibility, Feasibility/Detailed Project Report and As Built Report", standard /manual/ guideline with support from Ministry of New and Renewable Energy, Roorkee, August 2013.

9

Construction Methodology of Dam Foundation and Technology of Its Foundation Treatment

9.1 Foundation

A dam is a water retaining structure that fails on many occasions, resulting in loss of life and causing economic and environmental losses. Construction of a dam foundation is a critical and careful activity to ensure the safe life of the dam. It is seen that a dam does not fail due to overflowing from a sudden flash flood, but that the main reason of failure of a dam is due to the failure of the foundation.

Dam site geology is a complex phenomenon, and it is extremely difficult to envisage its complexity. The foundation rock may not be as solid and strong as it looks. It may consist of clay seams, fissures or a fault zone/shear zone beneath the rock where water, after impounding, may percolate or seep through the rock structure and make the foundation vulnerable which may lead to a disaster. It is very important to do the correct assessment and investigation of foundational material during construction.

An experienced geologist should be assigned to associate with the construction group during the construction period to assess the risk of the foundation due to the presence of any fault/shear zones, clay seams, fissures or any other geotechnical constraints, and advises the construction group to take up necessary treatment of the foundation and make the foundation as safe as practical.

9.2 Layout of Foundation

The construction group should have the following documents before starting construction activities.

- Pre-construction investigation report
- Pre-construction geological mapping
- General arrangement drawing (GA) and Foundation Layout Drawing Issued For Construction (IFC)
- Specifications and relevant standards pertaining to dam construction
- Contract document of the work which includes General Condition of Contract (GCC), Special Condition of Contract (SCC), Schedule of Rate (SOR), etc.
- Quality Manual
- Safety Manual

9.3 Identification and Removal of Over Burden

The foundation, as marked, shall be cleaned by stripping vegetation, removing debris and felling trees to make the area suitable for establishing the survey requirement, as well as for marking the center line the dam and establishing the depth of overburden that shall be removed.

- Reference pillars consisting of RL and coordinates shall be established in suitable locations so that these pillars are not disturbed during construction and are protected for future reference.

- A contour survey shall be taken up at the agreed upon interval and a contour drawing shall be prepared for calculation of quantum of earthwork and rock excavation done during execution.

- The axis of the dam foundation shall be marked, and the level on the axis shall be taken so the profile of the river (cross section) can be prepared.

- Overburden, if any, shall be removed by deploying JCB/Poclain to expose rock surface.

- A suitable carting arrangement shall be made to transport out excavated material to a demarked dumping area and a dozer shall be deployed to level the dumping area as required. The distance from the place of working to the dumping area shall be measured for the payment of carting.

- The entire area of the dam foundation is divided in blocks as required, and block wise levels are taken in specified intervals and shall be handed over to the resident geologist for geological mapping of each block.

9.4 Geological Mapping

After removal of overburden, a joint inspection along with the geologist is to be carried out for the inspection of exposed rock surface. The engineer and geologist are to observe the following:

- Type of exposed rock, classification and its properties
- Inclination of the rock surface
- Presence of any shear/fault zone
- Level of exposed rock surface whether it needs further excavation

Excavation of hard rock/soft rock shall be taken up with blasting or chiseling to reach a satisfactory foundation level. The blasting operation shall be taken up in a controlled fashion so that it does not crack the rock in the deeper strata due to impact of blasting. It advisable to avoid a blasting operation if the depth of excavation of hard rock by blasting is less than 50 cm. Removal of such thickness shall be done by chiseling. The downward slope of the rock surface shall be kept from downstream to upstream, which will provide better resistance against the sliding of foundation – a paramount requirement of stability analysis of the dam. Bench cuts in many occasions are provided to act as a key in the foundation.

After excavation is completed, the entire area shall be cleaned by air and water jet and the site is to be released to the resident geologist for geological mapping.

The geologist is to advise and conduct core drilling up to a depth of at least 10 m to assess the geological feature of the rock foundation. This exploratory drilling construction period shall be taken up by a core drilling machine with a diamond bit and a sample of core shall be inspected and logged by the geologist. The core of the drilling shall be preserved in the graduated core box so that the length of the recovered core shall be measured.

Geological mapping of the specific block shall be made and furnished by the geologist, duly indicating the following information:

- Dip and strike
- Type of rock encountered
- Fold
- Fault
- Clay seam, fissures, cleavage, etc.
- Location and depth of grout holes including vertical and inclined holes
- Section to indicate location of probable clay seam that are intercepted by vertical and inclined drill holes.

9.5 Foundation Treatment

Bedrock that is encountered at a site upon which a dam is to be built should be strong enough to carry the load of the dam. Dam site geology is unique and unpredictable as bedrock may be fractured, weathered, vulnerable and susceptible to hazardous seepage through foundation materials that may lead to instability of the dam foundation. It is evident that water seepage influences the stability of a dam. Water percolating through cracks, joints, creases and fissures within the rock structure creates a certain hydraulic head in different points. Any fault/shear zone that exists adjacent to the dam foundation may become active to make the dam foundation unstable and unsafe.

The technique used to treat the above cited weaknesses of bedrock, flaws and geotechnical constraints is known as foundation treatment.

The most popular methodology of drilling and pressure grouting of cracks, joints, seams and fissures within the rock is by cementitious materials, which helps to reduce seepage and consolidate the structure to enhance the load bearing capacity of the rock.

There are different techniques of foundation treatment that consolidate the foundation.

- Consolidation grouting
- Blanket grouting
- Dental treatment
- Curtain grouting

9.5.1 Drilling of Holes in Grid Pattern per Geologist

The geologist will furnish a drawing consisting of locations of drill holes marked in a grid pattern along with spacing, size and depth. The depth and spacing of grout holes are selected per geological characteristics of the bedrock and hydrostatic pressure. The directions of holes are determined in such a way that the maximum number of holes intersect the seams within the rock structure. The drawing is made in three stages. The number of holes present in the first stage are called primary holes, second stage holes are called secondary holes, and third stage holes are known as tertiary holes (Figure 9.1).

The drilling system adopted must be efficient to be able to permit continuous and straight penetration in the rock in order to avoid getting the drilling rod stuck within the hole. It becomes very difficult as well as time consuming to get the stuck rod out of the hole. An experienced driller and competent supervisor should be engaged who will ensure the effective production of holes of constant diameters and full depth of the holes, and see that debris being accumulated within the holes during drilling are removed

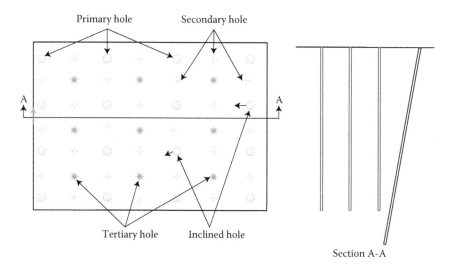

FIGURE 9.1
Grid pattern of grout holes.

by blowing air inside the holes. This is a vital requirement of foundation grouting.

There many methods of rock drilling to have the requisite grout holes made to match the desired size and depth such as:

- Rotary with different rotational speed, torque and speed
- Percussion drilling

Grout holes with a 38 mm diameter are drilled by jack hammer being run by an air compressor with a capacity of 350 cfm. Vertical holes are drilled by jack hammer and inclined holes are drilled with a wagon drill. These drilling machines are of the percussion type, and the most economical form of a drilling jack hammer is only suitable for drilling vertical holes, while inclined holes are drilled by the help of a wagon drill. Expert drillers can envisage and give ideas about the characteristic of rock or presence of any seams filled with clay beneath the rock surface by observing the smoothness or impact of the drilling operation and judging the rock fragments obtained during the drilling operation. Sometimes, drill bits get stuck inside due to drilling through a cavity, or on some occasions the drilling bit runs free inside without obstruction, and it provides information of the presence of a seam, joint, etc., within the rock structure. Construction engineers should interact and do close monitoring with the drillers in order to obtain all this small but vital information to make the foundation safe and stable.

Holes must be used for their intended purpose and should not be left alone for a long time to avoid possible collapse of the wall of the holes. The mouth of completed holes should be plugged immediately after the holes are cleaned.

9.5.2 Safety Compliance during Drilling Operation

Supervisors must ensure that the method satisfies the safety requirement of the project authority and should adhere to safety norms for protection against noise, vibration, the arrest of dust being produced during operation and other environmental restraints.

The supervisor is to ensure that the driller uses ear plugs, masks, hand gloves, safety goggles and safety shoes during the drilling operation as required.

9.5.3 Establish Connectivity between the Holes by Flushing for the Removal of Clay from the Clay Seam Existing within the Rock

After completion of the drilling of all primary holes, it is essential to wash these holes in order to take out clay, dirt and other foreign material, if any, within the drilled holes or seam, fissures or in intercepted seam/fissures, etc.

The following are the purposes of washing the drilled holes in the dam foundation:

- Wash out all dirt, clay or any other deleterious materials from seam through interconnected holes.
- Establish the connectivity between the holes to formulate the grouting sequence.

9.5.4 Washing Procedure

When holes are the drilled through weathered rock or rock made of a clay seam they require pressure washing to remove the caked clay seam under controlled air while water is being applied alternatively. This activity is conducted with the help of a packer. The packer is made of a GI pipe of diameter which should be at least 60–70% of the diameter of the hole. A combination of air and water under controlled pressure shall be applied alternatively through the holes at desired level.

The following equipment is required for the washing system:

- Air compressor
- Water pump
- Packer

A packer is made of a GI pipe equivalent to the depth of the hole. It is connected with two lines from the air compressor and water pump for the supply of air and water. The flushing of the holes shall be done alternatively with air and water.

It will be easier to understand if the procedure of flushing is narrated with figures consisting of numbered primary holes.

Set the packer above the zone to be washed in the first hole and blow air and water alternately into the hole and observe the intake of water inside the hole. Any refusal of water by the hole will indicate that the hole is not connected to any other holes through any seam. It is very difficult to envisage the zone of clay within the holes, so washing is to be accomplished by setting the packer at multiple levels in the depths of the holes which is known as multi-staged flushing.

Now, draw the packer out of the first hole and insert it into the second hole and start pressurizing the hole with air and water alternately. The muddy water found coming out from hole number 7, establishes the connectivity between the holes. Continue with flushing until fresh and clean water comes out from hole number 7 which will indicate that all dirt and clay has been flushed out from the clay seam and proves that hole number 7 is connected through a clay seam with hole number 2. Now, plug hole number 7 tightly so that no water can come out and again start pressurizing hole number 2 with air and water which will develop pressure within the rock. Muddy water which was found to be oozing out from adjacent hole number 6 establishes that hole number 2 and 6 are interconnected. Flushing with alternate air and water shall be continued until fresh and clean water comes out. Now, plug hole number 6 and start pressurizing hole number 2 in the same manner as described earlier and continue until fresh and clean water comes out from the hole number 5. Continue pressurizing for some time hole number 2 and stop the operation if hole number 2 refuses to further intake water, which concludes one cycle of flushing and establishes that hole number 2 is connected to holes 5, 6, and 7. Now, plug hole number 2 and insert the packer into hole number 5. Hole number 5 is pressurized with air and water to find that no water oozes out from either hole number 6 or 7 to prove that these holes are not interconnected. Hole numbers 6 and 7 shall be tested in the same manner to find no connectivity between the hole 6 and 7. So, hole numbers 5, 6, and 7 are not interconnected but connected to hole number 2 only with a clay seam running from the West-East direction as shown in Figure 9.2.

In the same procedure, connection between the holes numbered 3, 4, 8, 11, 12, 15 and 16 and hole numbers 9, 10, 13 and 14 are established. It is evident that three sets of interconnected holes are connected with three sets of independent clay seams existing within the rock structure.

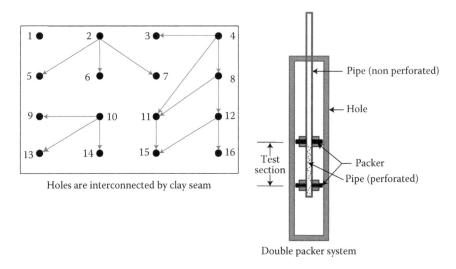

Double packer system

FIGURE 9.2
Washing of grout holes.

9.6 Consolidation Grouting in Foundation Area

Bedrock upon which the dam is to be constructed is found weak and incompetent due to the presence of seams, fissures, joints, etc., of various thickness, and these geological challenges make the foundation vulnerable. A dam is a water retaining structure that creates a reservoir in the upstream side after impounding, and water will tend to percolate through the geological defects which will make the dam unstable and unsafe.

To make the rock consolidated, compact and strong so that it can't collapse under the worst conditions, a procedure of dam foundation grouting is formulated by which geological defects are sealed by injecting cementitious material through grout holes, and is known as consolidation grouting.

The purpose of consolidation grouting is to strengthen the rock and reduce the seepage by filling the gap of the seams, fissures and joints in order to make a strong monolithic rock foundation to sustain the loading, and also to make an underground curtain to prevent the flow of water resulting in a reduction of uplift pressure under the foundation.

After the washing of clay seam/fissures/joints as specified in the Section 9.7.1, which create voids or hollow spaces inside the cleaned seams, these said spaces are to be filled up by consolidation grouting to strengthen the rock beneath the dam foundation.

Consolidation grout holes are of relatively shallow depth varying from 5 to 12 m, and holes are drilled in a grid pattern and are grouted in single or multi stages as per the geological condition of the rock and the discretion of the geologist. Spacing of grout holes are specified depending on geological condition of the rock; sometimes it may be at regular intervals or split intervals as per decision of the geologist, and should be decided based on the result of the water test and test holes for grouting. Drilling of grout holes shall be conducted perpendicular to the surface and sometimes, as required, some inclined holes are prescribed by the geologist to intercept the clay seam beneath the foundation. A dam site is never flat everywhere, but it consists of a sloped surface starting from the bank of the river to its bed. So in the sloped of ground surface, holes shall be drilled perpendicular to average slope of the ground surface.

As per site experience of grouting in medium rocks where holes were grouted at a pressure of 0.25 kg/m² per meter length of hole, if the situation demands, higher pressure may be applied carefully so that the rock structure is not fractured or to take care about any upheaving of surface of rock due to uplift being created by this pressure. Continuous observation of the surface of ground heave is required, and grouting shall be stopped immediately. An upheaval gauge for measuring heave is available on the market. So, the controlling of high pressure is a must in order to avoid the occurrence of any additional defect in the bed rock. The pressure must be sufficient enough to help the grout mix flow into the interstices smoothly and to ensure that the optimal quantity of grout mix is injected into the rock without causing any harm like fracture or displacement to the original rock structure. The smallest opening of 2 mm size can be grouted by cement grout, which gives permeability of about 50 Lugeon and can be taken as a criteria for groutability.

Various tests were done at sites and they found a rough correlation of Lugeon value with co-groutability in granite bedrock, and it may vary from site to site as each condition is unique (Table 9.1).

The report of washing the grout holes reveals that about three clusters of grout holes are interconnected. These clusters consist of hole numbers 2, 5, 6, and 7; 3, 4, 8, 11, 15, and 16; and 9, 10, 13, 14. Let us consider the first cluster of holes 2, 5, 6, and 7, where holes 5, 6, and 7 as per the report are connected to hole number 2. Grouting then commences with hole number 2 for multi-staged grouting. Multi-staged grouting is conducted to effectively seal all

TABLE 9.1

Co-Relation between Lugeon Value and Groutability

Sl No.	Lugeon Value	Classification	Rock Condition
1	From 0 to 1	Very Low	Ungroutable
2	1–5	Low	Poor groutable
3	5–20	Medium	Groutable
4	20–50	High	Groutable with many fissure
5	Above 50	Very high	Highly groutable

the seams present in various levels in the entire depth of the hole where the packer is placed in multiple levels in the hole. The depth of hole number 1 is 10 m and it is decided to do grouting in three stages by placing a packer in each level. Place the packer at the top of lowest grouting stage, that is, at the depth of 7 m from the mouth of grout hole. The mouths of hole numbers 5, 6, and 7 shall be tightly capped suitably so that any grout material does not ooze out. Any leakage of this kind from any grout hole during grouting operations will reduce specified grout pressure. The supervisor shall be vigilant enough to observe or locate any such leakage of grouting material from any other portion of rock surface. If any leakage is noticed, then and there it should be blocked in the following procedure:

- Identify the location of the leakage
- Stop the grouting operation immediately
- Take out all loose rock fragments from the point of leakage
- Insert a GI nipple inside the opening and grout with rich cement concrete or mortar depending on the size of opening and resume grouting after it is set.

Grouting procedures starts with the thinnest grout mix (for example 1:8) so that the narrowest interstices can be sealed by the injection of the grout mix, otherwise the narrowest interstices cannot be grouted if the grouting operation starts with a thicker mix; so it is desirable to start always with the thinnest grouting mix. After observing and considering the grout intake, the proportion of grout mix shall be changed to a richer proportion such as 1:6, 1:4 and so on; refusal of grout intake provides a fair idea about grouting in the bottom most level is accomplished.

Refusal of grout intake is considered when the grout flow of intake is less than 2 liters per minute sustained for 10 minutes for grout pressure at 2 kgs/cm^2.

Now, lift the packer and place it for the second stage of grouting at 4 m from the mouth of hole number 2. The procedure shall be followed in the same fashion as narrated in the case of the first stage of operation and followed by the third stage of operation.

After completion of multi-staged grouting in cluster number 1, the mouths of all holes shall be properly capped with a marking of the hole number and date of grouting. The entire process of grouting of cluster number 1 shall be documented in the prescribed fashion of the grout register as per the QA/QC manual.

Grouting of cluster numbers 2 and 3 shall also be accomplished in the same procedure. It is mandatory that the grouting operation shall be carried out under observation of the resident geologist for effective and transparent execution.

The geologist, after the completion of grouting operation, will advise on the location of requisite test holes' water and grouting so as to confirm the effectiveness of completed grouting in the particular area of the foundation.

After completion of the water test and grout test, the geologist will advise to go for drilling of the secondary and tertiary holes to take up further grouting work to make the foundation compact and consolidated.

9.6.1 Grouting Equipment

Management of a hydropower project is very critical and cumbersome due to its remoteness. A manger has to think far ahead of starting any critical activity so that the progress of the work is not hampered due to want of requisite material, machine and manpower.

A grout plant consists of a mixer, agitator, grout pump and supply fittings for effective execution of the grouting technique in the dam foundation. The plant is established to mix cement, water and other admixture, if any, in designed proportions for a grout mix to be used for grouting the dam foundation. Water and cement is mixed in a mixer. After mixing, the grout is then pumped into the agitator. The grout mixture is agitated slowly in the agitator and finally put into the grout hole through packers for application of the grout in the dam foundation.

- *Grout mixer*: It is a high speed vortex mixer which mixes the mix quickly and evenly. Water, cement or bentonite is mixed to prepare a homogeneous slurry and supply to the agitator. It is available in various capacities like 300, 400, 800 L with output indicated as 4.5, 15 m³/hr etc. with specific water-cement ratios.

- *Agitator*: The cement grout mix is pumped to the agitator where cement starts setting after mixing in the mixer. The grout mix shall be agitated continuously in order to prevent the grout mix from setting and also to help keep the desired level of uniformity before entering the pump. The agitator is usually a tank having an agitating mechanism. Agitators are available in various capacities from 300, 800, 1200 L, etc., with various rated revolutions such as 36, 54 rpm, etc.

- *Grout pump*: A grout pump is used to draw grout from the agitator. It has an arrangement of controlling pressure and flow. Sometimes, the grout pumped is coupled with a geared unit for enhancing pressure and flow. Pumps are available in single drum and twin drum arrangements and used depending on the size of project for continuity of flow of grout. The circulation line starts from the pump and leads to the mouth of the hole and sends back unused grout mix to the agitator. The grout pump has various capacities indicating pressure and output as expressed in 5 MPa and 200 L/min, 75 L/m and 5 MPa high speed, and 38 L/m and 10 MPa low speed.

9.6.2 Role of Ingredients of Grout Mix

The following ingredients are required to prepare a grout mix:

- *Water:* Water is the most commonly used component of grout mix. It should be portable and free from any organic and deleterious salt such as sulphate. Water should have pH values not more than 8. The water shall be tested as per available code of practice so that suspended matter, total dissolved salt, sulphides, chlorides, etc., are within the permissible limit.

Water plays several roles in grout mix. Water takes part in the hydration process of cement and also enhances the fluidity of the grout mix so that it can be pumped and percolate through the fissure/seam and joints in the rock structure effectively. Viscosity and cohesion of grout mix is reduced by the help of water. It facilitates smooth penetration of grout mix into thin fissures. The theory of grouting is to start grouting with a thinner mix so as to penetrate the grout mix into the thinnest fissures and then switch on to a thicker mix. The basic idea is to start the grouting with water of very low yield stress and to increase cohesion step by step using liquid of higher and higher viscosity until a rich mix is completed.

But, it is a fact that thin mix is not a stable mix because it shrinks more than thick mix. Rich or thick mix is called stable mix with limited bleeding.

The thick mix has the following advantages over thin mix:

- It fills the void completely because no bubble is present in the mix due to usage of less quantity of water
- It provides high strength
- It reduces shrinkage

A disadvantage of thick mix from the start of grouting is that it cannot penetrate into thin fissures and it gets deposited in a thick fissure blocking a thin fissure which remains ungrouted and makes the structure vulnerable.

- *Cement:* Ordinary Portland cement is used for grouting and no other specific cement is required. All properties shall conform to relevant code of practice. The shape and grain size distribution of cement influences several properties of grout mix. Fine grained cement is preferred to coarse grained cement for better hydration process due to a large specific area. It also influences the porosity, and larger porosity reduces the bleeding of grout. Penetrability of grout also depends on grain size of the cement.

During the grouting operation, the construction engineer and resident geologist coordinate with the operator in the following observations:

- Intake of grout under constant pressure is to be measured, and if the intake of grout is found to be lower than requirement the grouting shall be stopped.
- Intake of grout increases under constant pressure. If intake is increased due to opening of joint/seam/crack, grouting shall be stopped and restarted after some interval.
- If intake of grout and pressure is constant grouting may be stopped.
- Pressure and intake of grout that rises to some value and falls rapidly means the opening of new joints/cracks and after some time; if pressure remains constant, then stop grouting after observing permissible intake in specific allowable time period.

9.7 Discussion on Procedure of Grouting, Grouting Material, and Equipment

Foundation grouting is a significant activity in construction of dam whose objective is to seal the voids, cracks, fissures, porosity, etc., with grouting materials. It is essential to have an idea and knowledge of materials and their properties so that the foundation can be made impervious and safe. The following factors that affect the flow of grout are discussed at length.

9.7.1 Factors Affecting the Flow of Grout

Certain properties of materials used in the grout mix that have influence on the effective performance of the grout mix will be discussed based on some fundamentals:

- *Rheology*: Rheology is a science of deformation and flow of matter. Flow is the main criteria of grout which draws a relationship between stress, strain and time. Viscosity, cohesion and internal friction are the parameters that are related to rheology of the grout mix. Rheology is important because it characterizes fresh cement paste or grout and it gives an idea of how it would behave and perform in practical application. Viscosity controls the flow rate in fissures, cracks, and seams under equal pressure. Cohesion controls the internal pressure required to start the flow of grout. Cohesion also controls the minimum distance of the flow of grout. A grout mix having high internal friction is not suitable for injection because excessive pressure will be required to pump the grout mix within the seams/cracks/fissures, etc.

- *Viscosity*: Viscosity is one of the critical factors to be considered for grout mix during the grouting operation because the grout material flows and penetrates into the fissures. Viscosity indicates the resistance of grout to flow due to internal friction. It decreases the internal friction of moving grout mix. A grout mix with large viscosity resists motion and a grout mix with low viscosity flows easily and smoothly because its molecular makeup results in very little friction when it is in motion. Grout flow is inversely proportional to the viscosity of grout, so a grout mix is added with some super plasticizer which reduces viscosity resulting in an increase in the flow of grout mix.

- *Cohesion*: Cohesion is a phenomenon through which all molecules stick to each other or unite together. It is nothing but intermolecular attraction between molecules. It happens due to cohesive attractive forces between like molecules. Surface tension is caused by cohesion which is resistance of a surface to rupture when under stress. Grout pressure shall be designed in such a way that the pressure must be sufficient enough to overcome cohesion, otherwise grout mix won't flow inside the fissures/cracks/seams.

- *Internal friction*: A fluid is a material with no fixed shape and deforms easily when external forces influence its nature. Fluid friction is the force that resists motion within fluid itself. There is internal friction, which is a result of the interaction between molecules, and there is external friction, which indicates to how a fluid reacts with other matter. Molecules within the fluid need to move relative to each other to deform by squeezing past or displacing each other. This is internal friction that prevents a fluid from moving. Grout mix that has high internal friction is not suitable for injection because excessive pressure will be required to pump the grout mix within the seams/cracks/fissures etc.

- *Specific gravity*: Specific gravity has a vital role to play in designing the grout mix. The effectiveness of grout depends on the density of the material. The most effective and stable grout mix can be designed with low density particles. A grout mix being prepared with high density particles is an unstable mix due to the fact that solid particles in the grout mix settle fast. Grout mixes made with low density particles are more stable than the grout mixes prepared with sand. Grout mixes made with water and cement at the right water-cement ratio can be pumped effectively through cracks/fissures/seams.

- *Bleeding*: Bleeding of grout is an intrinsic and deep rooted problem. There are many factors that are responsible for causing bleeding of grout, which can be defined as separation of water from the mix which floats on the top surface of it due to its density being lower than cement, creating a void within fissures or fractures in the rock. The effectiveness of grouting is measured by its penetration and grout intake until refusal of grout up to the permissible limit.

Factors that play a vital role and govern bleeding are sedimentation, consolidation, hydration and flocculation of grout mix. Sedimentation is a process where particles settle at the bottom due gravity and all particles compress one another until they establish an equilibrium.

Consolidation is a nice space occupying effort between the particles already settled and some newly settled particles by their weight. This effort continues until they establish equilibrium and squeeze out pore water from the earlier settled particles. Expelled pore water moves upto the surface of the grout mix, causing bleeding of the grout mix.

Flocculation is the process of separation of solid and liquid which removes suspended and colloidal materials. This process influences both bleeding and penetrability.

9.7.2 Grain Size Distribution of Cement

Cement grout used in consolidation grout of the dam foundation is influenced by shape and grain size distribution of the cement. The hydration process occurs between water and cement that provides strength to the grout mix. The rate of hydration depends on the surface area of the ingredient. The rate of hydration with fine grained cement is much more than that of coarse grained cement. It is due to the fact that small grained size cement provides a more specific area.

Rate of hydration that depends on grain size is very much important in regard to bleeding and penetration because the grout could stay in the agitator more than half an hour before grouting, which helps to perform the grout mix effectively.

- *Effect of temperature*: As in the case of concrete, temperature has a huge role to play to achieve strength. The rate of chemical reaction increases with the temperature of the ingredients. Temperature of water that increases during mixing of grout mix will increase grout yield stress. Yield stress is proportional to temperature, which can be kept near 25°C.

- *Pressure filtration*: The grout hole is interconnected by fissures, cracks and seams at various levels in the entire depth of the grout hole. During the grouting operation, there is a tendency of the formation of filler cake in the walls of the grout hole, which blocks the passage of the grout mix to the fissure crack/seam or seam radiating out from the wall of hole. This caked wall is formed due to increased grout pressure. Usage of a grout mix with a high water content ratio may reduce the tendency of formation of this cake.

- *Porosity*: Water from the grout mix is absorbed during the grouting operation in porous and dry rock, which will increase viscosity and cohesion and will affect the penetrability of the grout mix. It is advisable to do water tests before taking up the grouting operations, or grouting in such type of rock shall be conducted with higher water content ratios.

- *Connectivity between the holes*: The main purpose of conducting grouting is to fill the interstices/fissures/seams within the rock structure so as to increase the strength and to make a barrier for percolation of water through the rock structure, resulting in a reduction of uplift. Most of the seams/fissures contain clay within them. It is essential to remove this infill clay from the fissures/seams before taking up grouting operations, otherwise grouting without removal of clay material won't serve the purpose.

9.7.3 Percolation Test

Percolation or water test is conducted on the rock bed of the foundation of a gravity dam to check the permeability of the rock bed so as to find the presence of joints/cracks, fissures or seams within the rock structure.

The water test is generally conducted in three stages. The first stage of the test is conducted during design of the project, the second stage is during the individual grouting operation, and the last one is done for confirmation. This test provides ideas about how to plan the grouting program for the foundation of the dam, and also helps to design the depth of curtain grouting.

A vertical hole shall be drilled not less than 35 mm diamer by pneumatic jack hammer up to a designed depth, which is preferably equivalent to the depth of hole varying 5–10 m of consolidation grouting.

Clean water shall be pumped into the bore hole through a mechanical or pneumatic packer. The packer may be a single or double system, which can be decided considering the nature of rock strata encountered and test section is considered accordingly. A single packer system is adopted for conducting a percolation test in soft rock where the wall of a hole can't stand and tends to collapse in full depth. So, the entire depth of the hole is not drilled in this method but in a double packer system. The entire hole is drilled at one time because this system is implemented in hard rock strata.

The water is strictly to be cleaned and free from any clay or salt. The presence of impurities will provide the wrong result because it will clog the wall of the test hole section, resulting in delivering an unsuitably low permeability.

9.7.4 Equipment Required

The following details of equipment are required for the work:

- Drilling machine, compressor, drilling rod, casing and casing driver, etc.
- Water meter of standard make
- Pressure gauge of requisite range
- Centrifugal pump is preferred of min capacity of 500 L/min with pressure 30 kgs/cm^2
- Water pipe, perforated pipe, packer, etc.

9.8 Curtain Grouting

Curtain grouting is a high pressure grouting that is conducted in the upstream side of the dam for the following purposes:

- To build a curtain wall upstream of the dam to control the seepage in order to reduce uplift force beneath the core of dam.
- To safeguard the foundation against erodibility hazards.

Curtain grouting is done on the axis of the dam and in practice, holes are drilled in the drainage gallery of the dam and in the tunnel for grouting in an abutment. In a low gravity dam where a drainage gallery is not provided, curtain holes are to be drilled in the upstream fillet of dam (USBR). Holes are drilled vertically as well as inclined per design requirements. The incline of the holes shall be maintained as far as practicable within 45° to the verticality. The angle of hole inclination shall be fixed with respect to distance of the drainage gallery from the upstream edge. Holes may be inclined a maximum of 15° upstream from the plane of axis, and if the gallery is very closed to the upstream face, then grout holes are to be drilled vertically. The curtain grouting operation shall be allowed to be taken up when a minimum of 40% of the work of the entire dam is completed. A 0.6 m length of pipe shall be inserted in the floor of the drainage gallery to facilitate drilling work. It is a fact that depth, spacing and orientation of holes are designed per geological features. Inclined holes are preferred when permeability obtained by the percolation test is mainly due to a vertical joint system. However, it is very difficult to envisage the type of joint system existing within the rock structure, so it is suggested to adopt a mixed set of hole patterns to seal each type of discontinuity.

Design of curtain grouting is formulated based on data available during investigation and additional information being observed or obtained during the consolidation grouting operation in the foundation. Many codes of practice advise limiting Lugeon value in various geological conditions to which curtain grouting shall be taken up. The thumb rule prevails at sites that have curtain grouting to reduce seepage with the following conditions:

- For dams exceeding 30 m height, curtain grout should be carried out when water absorption exceeds 2 Lugeon.
- For dams under 30 m height, curtain grout should be carried out when water absorption exceeds 4 Lugeon.

Curtain grouting is done in a single row and if the situation demands, an additional row is also taken up. The pattern of holes of a curtain is dependent on the geological feature. The pattern can be single line or multiple line curtains.

9.8.1 Single Line Curtain

A single line curtain can be adopted if the encountered rock contains a uniform size of openings such as fissures, cracks, seams, etc. In a single line system, holes are generally drilled in wider spacing which are known as primary holes, and after grouting of primary holes, a second group of holes are drilled in smaller spacing in between the primary holes which are called secondary holes. A third group is drilled, and are known as tertiary holes. The spacing of primary holes shall be kept between 6 and 12 m. Secondary and tertiary holes can be drilled accordingly. Percolation tests and grout observation shall be studied keenly and shall be compared with the previous set of holes so as to decide whether secondary or tertiary holes are to be drilled or not. However, sometimes it is advised to take up another line of holes with a different orientation and inclination. If spacing of the primary holes is less than 6 m, a percolation test is to be carried out in some holes to decide the forward path of further grouting.

9.8.2 Multiple Line Curtain

If the encountered rock consists of bigger sized openings, cracks, fissures and seams which are non-uniformly distributed, as per methodology, grouting should start with a thinner mix in order to fill the smaller interstices so the defects of this kind consume much more grout mix during injection under high pressure. If the proportion of grout mix in this situation is changed to make the mix thicker in order to control consumption of grout which may block the wall of the hole, it will prevent penetration of grout in smaller fissures, cracks and seams. This sort of situation demands the adoption of a multiple lines curtain. In this case, the outer line shall be grouted initially with a thicker mix, and if consumption of grout is observed to be higher and there no sign of refusal of grout it is advised to stop the grouting in order to tackle the situation; this action can be taken based on experience, knowledge and vital information being acquired from the consolidation grouting process.

9.8.3 Depth of Hole

The depth of hole is designed based on the hydrostatic head of the dam and percolation test being performed on the rock which project the characteristic of rock bed at the dam site. Depth of the hole can be calculated from an empirical formula advised by various relevant codes of practice.

$$d = 0.33\,h + c \quad \text{as per USBR}$$

where
d = Depth of hole
h = Height of dam in meter
c = Constant varies from 8 to 25

Depth of hole shall be 30–40% of hydrostatic head as per the thumb rule where rock is dense.

Depth of hole shall be 70% hydrostatic head where rock encountered is poor.

Grout pressure is dependent primarily on interstices of the rock structure, depth of hole and over burden.

The thumb rule prevails at the site if grout pressure varies from 0.75 to 1 kg/cm² per vertical depth of hole.

It is found that toughness and density of rock is increased with the depth of rock. Grout pressure that is suitable for deeper strata may dislodge rock in upper levels. In order to avoid this sort of situation, the type of grout process has been developed and is described below:

- *Descending stage*: In this process holes are drilled to a limited depth or up to a certain seam. Holes are to be cleaned and grouted up to that depth. Grout shall be allowed to initially set and then clean out the grout and drill further and grout in the next stage by placing the packer in the previous grout level, which will prevent back flow of grout mix. This process is repeated until the final depth of hole is reached.

- *Ascending stage*: In this process the grout holes are to be drilled up to full depth, and a section of grout shall be selected before the starting of the grouting operation. The packer shall be placed on the top of the selected section being measured from bottom of hole. The packer will resist the upward flow of grout which will develop pressure and force the grout mix to enter the seam/crack/fissures. After completion of this stage of grouting, the pipe is raised and the packer is placed on the top of the next stage and is grouted with lower pressure. The process will be repeated until the hole is completed from bottom to top.

9.8.4 Drainage System

Drainage holes are drilled in the drainage gallery of the dam. These holes are provided to control seepage in order to reduce uplift force at the bottom of the dam. The shape of a drainage gallery may be rectangular or horse shoe. Curtain holes are drilled on the floor towards upstream and holes for drainage are drilled on the floor downstream from the high pressure curtain. The diameter of drainage holes are 75 mm. Spacing and depth is designed based on geological conditions of the site.

9.9 Dental Treatment of Fault Zone, Shear Zone

A fault is a fracture surface along which the rocks have been relatively displaced. An earthquake occurs from sudden movement along faults, but the displacements are rarely more than a meter. Structures like great dams and lengthy bridges should not be constructed across a fault that is still active. But in some inevitable cases, the fault shall be deactivated to make the foundation safe and stable.

Geologists and construction engineers must assess the geological condition of foundation bedrock and envisage the consequences and risks involved due to the presence of a fault zone. The fault shall be inspected by conducting an extra investigation and exploration in the following line:

- Length of the fault
- Width of the fault
- Depth of the fault by localized excavation or by drilling
- Type and category of fault filling, also known as breccias
- Throw of the fault
- Direction of the fault

The area identified shall be excavated to take out all objectionable materials that are nothing but angular fragmented rocks comprising the fault limbs, cemented by some secondary material such as calcite, iron, oxides, silicones, etc., known as fault breccias. Fault breccias has a very low bearing capacity, and is unsuitable as foundation material. Excavation shall be done, if possible, up to solid rock bottom, or the depth and width of excavation shall be done with respect to the height of the dam along with the cut-off shaft in the upstream as per guidelines given in the relevant code of practice of various countries. An arrangement of a drain system shall be designed to drain seepage water that is likely to occur. Consolidation grouting work shall be taken up as required in the excavated area to enhance the strength and also to make a barrier for seepage, and the entire area of the weak zone shall be concreted with a suitable grade of concrete with adequate reinforcement. This entire system is known as dental treatment for a fault zone of a dam foundation.

It is preferred to remove the entire suspected area until hard and competent rock is encountered, but on many occasions it is not possible to remove the entire depth of the weak zone where the depth is evaluated as per USBR guidelines as indicated below:

$$d = 0.0066 \, bh + 1.5 \quad \text{for } h < 46 \text{ m}$$

$$d = 0.30 \, b + 1.5 \qquad \text{for } h > 46 \text{ m}$$

where
  h = Height of dam above foundation in meters
  b = Width of weak zone in meters
  d = Depth of excavation of weak zone

9.9.1 Protection against Piping

A fault zone detected in the foundation area is treated with the process known as dental treatment, and some grout holes were also grouted along the length of the fault. Despite that, the dam may not be safe against possible piping action. The fault may consist of some materials which may initiate piping action beneath the foundation, so appropriate action is to be taken against piping action. The cutoff trench both upstream and downstream shall be excavated, and the dimensions of the shaft shall be equal to the width of the weak zone plus 0.3 m on both sides. The depth of the cut off trench shall be equal to $W = h - d$ (USBR) where W = Width of cutoff trench, h = Reservoir head above ground, and d = Depth of cutoff trench. The cutoff trench shall be filled by concrete of the required grade.

9.10 Case Study

A descriptive and exploratory analysis of some critical events that occurred during the construction process are described below.

9.10.1 Case Study Number 1

How a fault was detected during construction of dam and remedial measure adopted through dental treatment?

Foundation preparation work was in progress in block number 3 of the Umrang Dam in the Kopili hydroelectric project in Assam, India.

Foundation preparation means cleaning the area, removal of debris and dirt, stripping off and scaling, and finally applying water jets to make the area ready so that it can be offered to the geologist for inspection to facilitate binding course in the foundation.

During the process, a pink patch measuring 4 m length was discovered in block number 3 on the left bank. The spot was inspected and observed carefully and they decided to take up some exploratory work to ensure the integrity and competency of bed rock.

Initially, further excavation work in the suspected area was taken up with the help of a pneumatic jack hammer fitted with a chisel run by an air compressor (350 cfm). During excavation, a soft weathered and disintegrated rock strata was encountered at a 1 m depth from the surface.

A decision was taken to go ahead with detailed exploration activities in a structured way in order to assess the nature of discontinuities such as fault/shear zone in the following guidelines:

- 4 bore holes of a diameter of 100 mm were done by core drilling from upstream to downstream along the length of block, and core samples were collected. Recovery of the core was significantly low up to a 5 m depth due to weathered and disintegrated/fractured rock and below, while at an 8 m depth a strong rock classified as condolite was encountered.

- 4 more holes of the same diameter up to an 8 meter depth were done in the upstream side in alignment with the suspected fault line, and also the same number of bore holes were done in the downstream to explore the possibility of extension of the fault line; fortunately, the fault was limited to the foundation area and terminated a few meters from the upstream face, which made it clear that this part won't be a passage for seepage in the future after impoundment of the reservoir.

- The entire suspected area was decided to be ripped open up to a 5 meter depth or until hard and competent rock is encountered.

- Accordingly, all objectionable materials such as fragmented rock particles and other soft particles known as fault breccias were removed up to a 5 m depth on average and hard condolite was encountered.

The identified fault is a localized fault started from 3 meters from the upstream face of block on 3 and stretched downstream covering block number 2 obliquely. The configuration and layout of the fault is sketched in the footprint of the blocks (Figure 9.3).

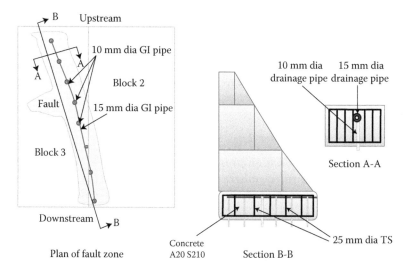

FIGURE 9.3
Dental treatment of dam foundation.

The treatment methodology was designed and accomplished as indicated below:

- The entire length of the suspected area was excavated from upstream to downstream up to a depth of 5 meters where hard rock was encountered.
- Eight grout holes were drilled along the length of the fault up to a depth of 5 meters from the bottom of excavation. These holes were washed and grouted under 2.5 kgs/cm^2 pressure.
- Three holes were drilled along the length of the fault up to the depth of 5 meters, and water tests were conducted and the results observed to be within the limit of 50 Lugeon.
- As the direction of the fault was from upstream to downstream, an arrangement for draining out any possible seepage likely to occur after impoundment was made to arrest this seepage and discharge this seepage downstream of the dam. It was designed to release uplift force likely to be developed due to seepage. Figure 9.3 indicates the arrangement as follows:
  - 10 cm diameter GI pipes were put vertically at 3 meters along the length of the fault.
  - All vertical pipes were connected to a horizontal pipe of 150 mm diameter sloping down from upstream of the dam to drain out seepage water to the downstream of the dam.
- The trench was concreted with concrete grade A20S 210 with adequate reinforcement as shown in Figure 9.3.

- Blocks 3 and 2 were modified to accommodate the fault in one block so as to avoid construction of two blocks having a construction joint on the fault line.
- No cutoff trench was made on the upstream side. The cut off trench was constructed in downstream and was filled with concrete of A40 S185.

9.10.2 Case Study Number 2

Anchor foundation of a high pressure tunnel (penstock) was intercepted by a sink hole in a project where 50% of storage area was occupied with limestone. How was this geological defect sorted out and the work accomplished?

The Kopili hydroelectric project is situated at Garampani, Assam, India. The river Kopili is a tributary of the mighty river Brahmaputra. The entire region where this project is situated consists of three natural basins named Kopili, Umrang and Longlai, which is the lowest basin topographically. The Umrang storage area is 100 m lower than the Kopili storage area. These three natural basins are situated on a treacherous and mysterious limestone bed. The entire area is in tertiary, sandstone, limestone and shale which overlie gneiss and granite. Nearly 50% of the area of Kopili and Umrang is occupied by limestone which is comprised of many karsts/sinkholes and surface solution channels. These natural phenomena occurred due to the action of water with limestone as the area is a heavy rainfall area. About 500 sinkholes were identified with various diameters during geological mapping.

The karstic condition that threatened water tightness and competency of reservoir – a nightmare of the hydropower project.

Water is the prime factor that initiates the chemical weathering of rock. When sub surface and vadose water percolates into depths and accomplish complex geological work, the solution of limestone and rainwater becomes charged with carbon dioxide. In a limestone infested belt, water readily works its way down through joints and along bedding planes until it reaches an impervious layer, which may be within the limestone formation or beneath it. The water then flows in the natural drainage directions until it finds an exit, perhaps many kilometers away from the intake. Once a through drainage is established, but not until then, dissolved material is carried away and a fresh supply of water coming in from above continues the work of carrying the solution primarily along the joints and bedding planes, and a cave is dissolved out of limestone.

The surface openings become gradually enlarged in places where the contours of the ground favor the special concentration of the flow-off, and funnel shaped holes known as sink holes are developed. Sink holes are a form of karst topography that develop on the surface of soluble rocks. They are of various sizes varying from a few meters to some that are deeper.

- Karsted crevices are formed by the dissolving action of sub surface water.
- Karst caves and channels are of great concern and attract many researchers because these are distinguished by many special features.

This sinkhole or cave originates near the upper boundary of the saturation zone where underground water flows through the crevices being created by the karstic action, which is the most favorable formation condition of underground caves and channels. These types of caves are comprised of a system of horizontal channels, and also water is found to be dripping vertically from the top. They have many irregular branches, are interconnected by narrow passages, and suddenly widen out to great area. These are not straight but zigzag and are serpentine, being created due to a complex system of joint dissecting rocks. The limestone was found to be compact and dense. The project authority was wary about secondary permeability which has developed in the rock due to the presence of divisional planes such as cavities, fissures, faults, shear zones and joints. Solution along this fracture plane has further widened the passage, giving rise to the leakage possibility of impounded water under the reservoir condition. There is possibility of escape of impounded water from the Khandong reservoir to the lower Umrang reservoir, and further escape to the lowest Langlai valley through various interconnected sinkholes making the working reservoir dry. Water tightness and competency of the reservoir was a great concern owing to the presence of numerous sinkholes in the valley.

Geo-hydrological investigation was proposed to carry out tests to find whether the cavernous limestone is competent enough to hold water under reservoir conditions.

The following geo-hydrological tests were conducted to find out the condition of ground water and to establish whether the sinkholes are interconnected.

1. Measurement of piezo-metric head
2. Tracer testing
3. Water balance study
4. Temperature log

9.10.2.1 Effect of This Particular Geologic Condition on Construction

Penstock foundation intercepted by buried channel: Construction of the foundation of a high pressure tunnel (penstock) was in progress in this project with the rare geological distinction of prized karsts/sinkholes. Penstocks 4.6 m in diameter and made of steel were under fabrication in the shop. These isolated foundations were being constructed at fixed intervals. Excavation work of one of such foundation measuring 8 m by 6 m with a depth of 3 m was completed. Excavated material from the bottom of the excavation was removed, and the area was made ready for necessary scheduled exploration before taking up concreting work.

When the construction group turned up at the site in the next morning to take up the scheduled exploration work, they were surprised to see that the excavated foundation pit collapsed and had sunk by at least 3.5 m, forming a deep cavity and the rippling sound of flowing water was heard coming from inside the cavity. Effort was made to remove the collapsed materials to assess the exact condition of the newly formed cavity, and the materials could be removed to some extent. Water was seen flowing through the remaining debris which confirmed the existence of a sink hole in that area, which was a matter of concern for the construction group.

The direction of the incoming flow was known as it was descending from upland, but the source was not known which was the only priority at that time.

A search group was formed and started searching for the source, which was a difficult task as the area was a deep forest mainly infested with Silver Oak and Gooseberry trees (Phyllanthus Emilica) due to the presence of limestone.

A small stream 3 m wide was located about 1.5 km upstream from the excavated pit which could possibly be the source of this flow of water.

A series of bore holes were drilled to intercept the buried stream and tracer tests were conducted by using chemicals and fluorescence dyes that gave positive results to prove interconnection with the sub surface solution channel.

However, a divergent point of the buried solution channel from the surface stream could not be envisaged. An anticipated length of the stream was chosen for conducting V notch tests in the anticipated length of the stream. This length of 500 meters were divided into 10 equal parts to conduct tests to measure discharge in each section, as shown in Figure 9.4.

Discharge Q1, Q2, Q3, Q4, Q5, Q6, etc., were measured with an idea to find out any difference in discharge between Q1 and Q2, Q2 and Q3, Q3 and Q4, Q4 and Q5, etc. Any insignificant variation in quantity of discharge between Q1 and Q2 will indicate no escape of water in this section of stream through

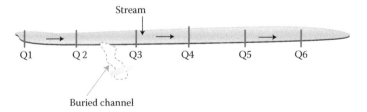

FIGURE 9.4
Location of buried channel.

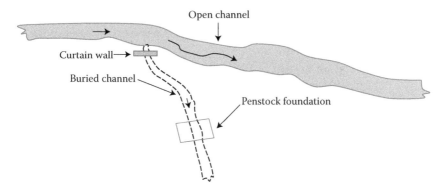

FIGURE 9.5
Penstock FDN encountered with sink hole.

any buried channel. Any significant difference in discharge quantity in any of the section will confirm the escape of water from any point in this section. It was found that discharge quantity Q3 was much less than the discharge quantity Q2, which confirmed the escape of water from any point in this particular section. Further, the same was exercised making a smaller section between Q2 and Q3, and the point of diversion was located where the buried channel was originated as shown in Figure 9.5.

The flow of water was stopped by making a curtain wall a few meters away and downstream, along with some curtain holes that were grouted with high pressure. The channel in question was not running parallel to the alignment of the penstock foundation but was crossing the alignment. So, the foundation was shifted and the collapsed pit was filled with concrete followed by grouting.

9.11 FAQ on Grouting Grout Pressure, Multistage Grouting, Permeability Test, and Others

The following is a frequently asked questions (FAQ) section to help readers quickly understand grouting and its related activities.

- Why do you require consolidation grouting in the foundation of dam?
  - To enhance the strength of rock foundation of the dam
  - To reduce the permeability and control the uplift force
  - To reduce deformability of fractured rock
  - To make the rock bed compact, consolidated and competent

- What is Lugeon value?
  - Lugeon is a limit of permeability used in a water test and grouting operation in a dam.
  - 1 Lugeon unit means 1 liter of water consumed per meter of test length of hole per minute, at 10 bar pressure.
- What is the suggested limiting Lugeon value for void filling in the dam foundation?
  - 3–5 Lugeon value
- What should be the diameter of a hole of consolidation grouting?
  - It is 38 mm in diameter
- What should be the spacing and depth of grout holes for consolidation grouting?
  - Depth and spacing of holes depend on geological conditions, but generally, depth of holes varies between 5 and 10 meters.
  - Spacing of grout holes are determined per geological condition of the site. Spacing is considered in regular interval or sometimes in split interval based on the result of water tests and test holes of grouting. Generally, spacing of holes are kept between 0. 9 m to 1.2 m.
- What additives are generally used in grout mix?
  - Fly ash and calcined shale
- What is the purpose of washing grout holes?
  - Removal of fragmented material deposited during drilling operations.
  - Establish connectivity between the holes and remove clay and deleterious material from seams/cracks/fissures.
- When do you conduct a water test?
  - A water test is conducted before and after completion of the grouting operation.
- What should be the diameter and depth of a water test hole?
  - The depth of the hole should not be less than 5 times the diameter of the hole.
  - The diameter should be a minimum of 32 mm, but a hole of larger diameter is preferred so as to minimize frictional loss.
- What should be the test pressure for hard igneous and metamorphic rocks?
  - It is 0.23 $kg/cm^2/m$.
- What is the type of pump to be used in a water test?
  - A centrifugal pump of minimum capacity of 500 liters per minute with a pressure up to 30 kgs/cm^2.

- When do you adopt a single packer system for percolation?
  - A single packer system is adopted for conducting percolation tests in soft rock where wall of hole can't stand and tends to collapse in full depth.
- Why do you require clean water free from any dirt and clay for the percolation test?
  - The presence of clay/dirt/impurities will provide a wrong result because these impurities will clog the wall of the test hole section, resulting in low permeability.
- What is the penetration length of consolidation grout?
  - The maximum path that grout mix can travel through opening, fissure, joint and crack from a bore hole under designed grout pressure. Penetration of grout in the fissure/discontinuities/ seams depend on grout pressure, water pressure, yield of grout and aperture of the fracture. Penetration is a function of accumulated friction between grout and the fracture wall where aperture is assumed to be constant.

9.12 Technical Discussion on Karst or Sinkholes

Earth's atmosphere contains different constituents of gases like nitrogen, oxygen, carbon dioxide and other gases. The percentage of nitrogen and oxygen is very high in comparison with other gases. Our focus point on this discussion is the presence carbon dioxide, which is about 0.04–0.05%, but it plays a major role to form a very treacherous and mystic topographic feature on Earth's surface and beneath the Earth, which is called karsts topography.

9.12.1 How Is It Formed

When it rains, the rain water gets mixed with carbon dioxide present in the atmosphere and rain water becomes acidic from the reaction with carbon dioxide and turned into an acid known as carbonic acid. Soil on Earth also contains some percentage of carbon dioxide. It intensifies the acidic behavior of rain water further when it touches the ground. Water becomes one of the main factors for the chemical weathering of rock. All natural waters are slightly dissociated into H^+ and $(OH)^-$ ions. The acidity of water is measured by the concentration of $(OH)^-$ ions which is called pH values. Water is acidic when the pH value is less than 7, and it is alkaline when it is more than 7. Chemical weathering gets accelerated due to an increase of acidic behavior of water. The pH value of rain water lies between 4 and 7. This acidic water seeps through soil and rock by means of gravity and comes in touch with soluble rock like limestone, dolomite, marl, rock salt, gypsum, etc.

Limestone is a sedimentary rock having calcium, and rain water reacts with this carbonate rock.

$$CO2 + H_2O = H + (HCO_3)$$

Limestone does not react with pure water and CO_2 which is present in the $CaCO_3$. Limestone gets slowly dissolved and removed as calcium bi carbonate $Ca(HCO_3)_2$, and it turns slowly into a furrowed-like surface. This is called karsting, which is a process of leaching out of soluble fractured rock by moving underground and vedose water, resulting in the formation of caves and channels deeper down.

A sinkhole is a form of karst topography that develops in soluble rock. This is a cavity in limestone and is formed due to action of water with limestone as described earlier. The scouring action of running water also plays a part in the development of this particular type of topography. As the rain water seeps through the joints, water gradually dissolves, corrodes and carries away particles of limestone, thus widening the joints. It also penetrates minute fissures, and water slowly converts these minute fissures/cracks into big hollows. It may be of any measurement having a large cave entrance and a length ranging from many meters to kilometers. These type of caves consist of a complex underground drainage system. Inside the cave there may be a flow of water diverging on both sides from the main cave. There will be deposits of many interesting shapes in the form of stalactites, stalagmites and other interesting shapes. Sometimes water drips down the wall which creates a flowstone or rimstone deposit on the floor, but sometimes the floor of sink holes are found to be filled with water at least up to ankle height.

9.13 Sink Hole/Karst Topography Is a Matter of Concern for a Hydel Project

Karsts topography/sinkholes are a great concern for a hydropower project, especially in a storage scheme where a reservoir is created as storage. A valley of limestone will have karsts topography/sinkholes that have complex underground drainage in the reservoir area, which will pose a threat to the water tightness of the reservoir. Water from the upper valley will be transferred to the lower valley through these buried channels resulting in emptying the reservoir. Before making any hydropower project the entire proposed reservoir area is to be inspected and sinkholes are to be identified and explored; the extent of activities and water tightness is to be established by various tests. The most fascinating experiences a fresh civil engineer will have are surprises to find a disappearing stream on this topography. The disappearing streams are those streams which suddenly terminate their flow and vanish into the ground, and also a sudden appearance of a stream through a wall of the hill from a source unknown, which indicates the presence of an underground buried channel. A hydropower engineer who works on hydel projects should find out the source of the suddenly appearing stream as well as the destination of the disappearing stream; otherwise, the buried channel present in the reservoir area will have no integrity of tightness of reservoir.

9.13.1 Caves and Sinkholes and Their Characteristics

Karsts topography consists of caves and sinkholes with unique characteristics which educate civil engineers, fascinate geologists and attract tourists who make this place a venue for recreation.

Caves are splendid for their crystal formations on their ceilings, walls and floors, which are known as speleothems. The excess calcium gets deposited on the ceiling, floor and wall of the cave, creating a unique formation known as speleothems. It is the crystallized form of the mineral calcite. The acidic water seeps through fissures/cracks and dissolves some limestone, and when it reaches the cave it loses some carbon dioxide to cave air. The water is now less acidic and unable to hold the same amount of limestone and the tiny crystals are deposited and slowly decorate the cave with speleothems.

Stalactite is a very common crystalline deposit that is found to be hanging from ceilings of caves. Its formation starts when water drops hang on the ceiling and have lost their carbon dioxide in the air before falling, and a very little quantity of calcium carbonate gets deposited in the ceiling as a crystal and it gradually takes the shape of a fascinating crystal known as stalactite, which can serve as an attraction to tourists.

When water drops on the floor a new crystal formation known as stalagmite forms, and in the passage of time, when it grows upward to join with a hanging stalactite, it forms a column-like crystal.

Caves are decorated by some other beautiful deposits known as flowstone, helicitites, etc.

9.13.2 Underground Inhabitants

Caves and sinkholes are developed in darkness and quiet except some rippling sounds of flowing water are heard. The author entered the second largest cave in Asia situated at Siju, Meghalaya, India. The cave entrance is a huge one, and the length of the cave is kilometers. The floor of the cave was filled with water of ankle depth, and many branches radiated on the both sides from the main tunnel and the sound of flowing water was heard.

Inhabitants of this cave were mainly bats and crickets; they are regular visitors and they make this cave their temporary nest to raise their babies. These animals are called trogloxenes.

About 500 sinkholes of various sizes were located in the reservoir area of the Kopili hydroelectric project in India. Some species of fish and cave beetles were noticed in sink holes. These animals which cannot live outside the caves are called troglobites, but regular visitors of these sinkholes were bears.

10

Concrete and Its Application in Concrete Gravity Dam

10.1 Introduction

Concrete is an artificial stone composed of crushed rock as its coarse aggregate and sand as its fine aggregate, which is then mixed with cement and water proportionally. All ingredients are mixed, batched, placed in the desired shape and size, and are compacted with vibrators to achieve the designed strength of the structure.

In other words, concrete is made up of the binder and the filler. Cement is the binder, and coarse and fine aggregates are the filler. Concrete is like water, which takes the shape of the container where it is kept. One of concrete's weaknesses is that it cannot retain its shape and size on its own. It needs formwork to be formed into the correct shape and size; with formwork, concrete will have character as well as beauty, while it loses its character without it.

$$\text{Concrete} = \text{Filler} + \text{Binder} + \text{Formwork}$$

10.2 Group of Concrete

As per its application, concrete is divided into 3 groups:

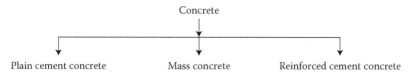

Concrete is weak in tension and strong in compression. Concrete can restrain 10–15% of tension being developed within the body of the structure. Steel bars, known as reinforcement, are provided to take care of tensile stress so as to enhance the performance; this type of concrete is known as Reinforced Cement Concrete.

Mass Concrete is defined in ACI116 R as any volume of concrete with dimensions large enough to require that measurements be taken to cope with generation of heat from hydration of the cement, as well as to counteract attendant volume change to minimize cracking.

10.3 Classification of Concrete with Type of Cement Used

Concrete is classified with the type of cement used in the concrete. Concrete is called Hydraulic Concrete when hydraulic cement is used, and concrete made with non-hydraulic cement is called Non-Hydraulic Concrete.

Additionally, concrete is classified in accordance with various parameters such as weight, strength, and additives; additional types of concrete are classified for their special characteristics.

10.4 Classification of Concrete as per Weight

Concrete is classified in accordance with the unit weight such as Light Weight Concrete, Normal Weight Concrete, and Heavy Weight Concrete.

Strength is the measurable index of concrete, with strength meanings compressive strength. Strength of concrete in different concrete grades is known as *characteristic strength*, which as defined by the IS is the strength of the material below which 5% of the test results are expected to fall under. Concrete is classified in accordance with strength and grouped in three categories based on its strength.

10.5 Grades of Concrete

There are many grades of concrete that designated as M15, M20, M25, and M30, where M refers to the mix and the number refers to the specified strength of the mix.

10.6 Some Special Concrete

Some concrete is manufactured by adding special materials such as fiber, silica fume, bitumen, and polymer; these materials are known as *additives*.

Fiber Reinforced Concrete (FRC) has materials such as steel, glass fiber, polypropylene, carbon, and cellulose mixed into it. Fiber is added to concrete to increase toughness, as well as improve the tensile strength and flexural strength of the concrete. It reduces shrinkage while also increase resistance to lateral cracking. Micro Defect Free (MDF) Concrete, Micro Silica (MS) Concrete, and Densified with Small Particles (DSP) Concrete are included in this category.

Some special categories of concrete used in industries for practical implications are Ultra High Strength Concrete (UHSC), which contains large quantities of silica fume and has a strength >150 MPa.

Polymer Concrete (PC) is made with resin and monomer, while Polymer Impregnated Concrete (PIC) is made by impregnating monomer and catalysts. Self Compacting Concrete (SCC) is a special type of concrete that is compacted by its own weight without using any sort of vibrator. It comes under the category of High Performance Concrete.

High Volume Fly Ash Concrete (HVFC) is made by using about 25–30% fly ash by weight of cement. HVFC is widely used in construction of gravity dams.

Concrete is also classified in accordance to operation such as Roller Compacted Concrete, which is considered a High Performance Concrete and Ready Mixed Concrete.

10.7 Classification of Concrete Mix

Types of concrete as indicated above are executed in two distinct methods.

Case 1 has ingredient portions that are predetermined and strength, type of cement, aggregates, Water-Cement Ratio, and content of all ingredients are specified and mixed by volumetric batching with a desire to achieve a specific strength of concrete.

Case 2 has ingredient proportions that are not predetermined, but concrete properties such as characteristic strength, exposure, workability, type of aggregate and its max size, type of structure, and its location and quantum concrete are indicated in addition to QA/QC methodology. Accordingly, a concrete mix is designed on the basis of all requirements and specifications.

Nominal mix: Nominal Mix Concrete is a concrete whose constituent proportions are predetermined, and the content of each to be used per cubic meter is fixed. Batching of this type of concrete is done by volumetric measure and is known as Uncontrolled Concrete. The following examples of Nominal Mix Concrete with grade are used in the industry.

1. M5   1:5:10
2. M7.5   1:4:8
3. M10   1:3:6
4. M15   1:2:4
5. M20   1:1.5:3
6. M25   1:1:2

Designed mix: Designed Mix is a concrete whose ingredient proportions are calculated and determined based on the results of preliminary tests confirming specifications in conjunction with exposure, type of structure, characteristic strength, workability, method of batching (weight basis), compaction and curing processes, and other factors, and is designed accordingly to achieve desired performance of concrete per the quality control system.

This is known as Controlled Concrete. The controlled and uncontrolled status depends on the level of quality control measures that are adopted during execution.

Construction of a gravity dam involves large quantities of concrete, especially in the core of dam. The core of the dam is constructed with Mass Concrete without any reinforcement. The design of a Mass Concrete structure like a dam is guided by the perception of economy, durability, water tightness, and thermal behavior of Mass Concrete. Heat of hydration occurs due to exothermic reaction, which raises temperatures in large levels within the concrete's mass if the energy is not dissipated. Significant tensile stresses are generated due to temperature may result in cracking of concrete, therefore requiring a high degree of quality control to have good performance of the dam concrete.

Concrete gravity dams have three characteristics:

- Weight, which acts against horizontal forces and tends to push the structure against it.
- Width, which covers almost two thirds of the height.
- A triangular shape that ensures safety and stability.

10.8 Parameters That Control the Character of Concrete

Safety and stability of a dam depends on the performance and character of the dam's concrete. Five parameters that control concrete characteristics are concrete strength, water tightness/permeability, unit mass, durability, and resistance to thermal behavior. Proper methods of mixing appropriate proportions and overseeing the placement of concrete is critical to achieve the designed strength, permeability, and durability of the dam's concrete.

The following text covers some important factors that have influence on properties and characteristics of concrete that must be discussed before taking up any design of concrete mix and adopting placement techniques of critical Mass Concrete in a dam's structure.

10.8.1 Water-Cement Ratio

Water-Cement Ratio is the most important factor that controls the concrete's properties. Water-Cement Ratio is the ratio of the weight of cement and weight of water. Water-Cement Ratio has tremendous influence on the strength of concrete. Compressive strength of concrete is inversely proportional to the Water-Cement Ratio. While the overall strength of concrete increases with a lower Water-Cement Ratio, low Water-Cement Ratios lead to high strength and low workability, while high Water-Cement Ratios lead to low strength but high workability.

$$\text{Water-Cement Ratio} = \frac{\text{Weight of water}}{\text{Weight of cement}}$$

For the selection of Maximum Water-Cement Ratio, exposure conditions are a vital consideration for the performance of concrete. The designer should assess the environmental conditions where the structure is to be constructed. Environmental exposure classification is determined by considering the mechanism leading to the deterioration of concrete. The code of practice of various countries around the world suggest make suggestions about Maximum Water-Cement Ratios and minimum cement content to be considered in accordance with the minimum specific compressive strength of the concrete under the particular class of exposure as a predefined performance criteria for concrete durability requirements.

Relevant code of practices suggest Maximum Water-Cement Ratios and minimum cement content in accordance with specific compressive strengths of concrete.

The Maximum Water-Cement Ratio is designated for each grade of concrete mix under specific exposure, and can be referred to for classified exposure such as mild, moderate, severe, very severe, and extreme conditions, with Maximum Water-Cement Ratios, minimum cement content, and minimum specific concrete compressive strengths.

10.8.2 Water Content

Water plays an important role in the formation of concrete. Concrete is a conglomerate of cement, sand, aggregates, and water. A reaction is initiated when water is mixed with cement. This chemical reaction is known as the hydration process. It is an exothermic reaction that liberates heat during the induction period, which includes the setting and hardening process. The heat generated during the process raises the temperature of concrete. Quantum water added shall be retained within the concrete during entire process of hydration; otherwise, concrete will not be able to achieve the designed compressive strength if any water escapes or evaporates.

Water added to cement makes a paste that coats the aggregates, fills the interstices between the aggregates, and balanced pastes generate a lubricating action in the concrete mix. But, it not possible to make concrete where all the interstices are filled with cement paste. Sometimes more water is added to concrete for the purpose of having workable concrete, which deteriorates the performance of concrete by creating voids or fissures within the concrete. These voids or fissures are sometimes interconnected, which ultimately become an easy passage for water within the body of concrete and affects the water tightness of the concrete. Water tightness of concrete is an important factor for the performance of dam concrete.

Potable water is generally used in the concrete, but water being collected from any source shall be tested before use in the concrete. Water should be clean and free of impurities like acid, alkali, salt, and other organic materials that are considered harmful to the health of concrete. The pH value of water to be used in the concrete shall not be less than 6. The quantum of water required for per cubic meter of concrete depends on the size, shape, and texture of aggregates, workability, additives, and chemical admixture.

10.8.3 Cement Content

Cement is the main constituent of concrete and acts as a binder. The contribution of cement is to provide designed strength to concrete, durability, and binds all constituents in a way to make hardened artificial stone. Types of cement shall be used depending on the type of structure to be constructed. In the case of the construction of a gravity dam and considering thermal behavior of concrete, the type and quantity of cement shall be selected. The higher the cement content, the greater amount of heat generation, which is vulnerable to the performance of dam concrete and will be subjected to cracking due to the development of temperature stress. The core of concrete gravity consists of Mass Concrete without reinforcement. Therefore, at least 20% of OPC shall be replaced with pozzolonic material, fly ash, or any other suitable additives to reduce temperature and control the possible cracking of concrete.

The quantity of cement is to be minimal for the control of temperature as well as overall economy. Some standard codes prescribe limited cement content shall be 236 kgs/m^3 due to heat controls.

The following cements are to be used in the construction of dam as per guidelines:

- Portland cement
- Blended cement

- Hydraulic cement
- Portland pozzolona cement
- Blast furnace slag cement

10.8.4 Aggregates

Concrete is a conglomerate of water, cement, aggregates, additives, and admixtures. The third constituent of concrete is known as aggregates. Aggregates are a granular, strong, hard, and durable material that are free from impurities and deleterious substances.

10.8.4.1 Classification of Aggregates

Aggregate is a broad term whose size varies from 150 mm to 4.75 mm and are used in construction industry. An informative classification of aggregate is mentioned below.

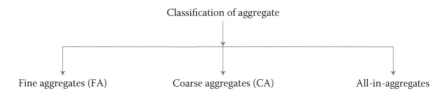

Fine aggregates are those particles that are passing through a 4.75 mm sieve, are retained in 75 µm, and are free from all impurities. River sand and quarried sand fall under this category.

Coarse aggregates are defined as those particles retained on a 4.75 mm sieve and free from any deleterious substances. Crushed stone and gravel fall under this category.

All-in aggregates are composed of fine and coarse aggregates, and there aggregates do not need to be separated as fine and coarse aggregates. During sieve analysis, if it is found to be non-uniform, single sized aggregates shall be added in the grading towards adjustment.

Aggregates should conform to certain specifications and standards for its use in construction. Aggregates should be angular and clean, hard and strong, durable and sound, free from deleterious substances such as pyrites, mica, alkali, and clay, which affect the durability of concrete. Aggregates should not contain a coating over it, and should be tested before it is used in the concrete because aggregates occupy almost 80% of concrete's volume and it is a vital constituent that provides strength and durability to the concrete.

Coarse aggregates are manufactured from igneous, metamorphic, or sedimentary rocks. Basalt is by far the best rock for making aggregate, which is fine grained volcanic rock, while granites are coarse grained plutonic rocks suitable for coarse aggregates. Quartz, schist, slate, and sandstone can be used if it is found strong, dense, and unlaminated.

Coarse aggregates are classified in three categories based on weight.

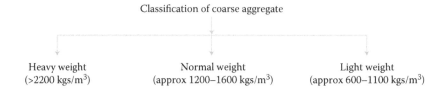

Generally, concrete is manufactured from normal weight aggregates, and unit weight of concrete varies from 2200–2500 kgs/m³.

10.8.4.2 Grading and Size of Aggregates

Coarse aggregates of various sizes from 10–150 mm are used in different types of concrete.

- 20 mm and smaller aggregates are used in all types of RCC construction where 20 and 10 mm are blended in appropriate proportions.
- 40 mm and smaller aggregates are generally used in Plain Cement Concrete and also preferred in the concrete of footing of large size and in Mass Concrete.
- 80 mm aggregates are used in Mass Concrete.
- 150 mm aggregates are used in Mass Concrete.

Economy of concrete depends on the max size of aggregates used in the concrete. Quantity of water and cement is required more in concrete where a smaller size of aggregates are used due to more surface area of smaller aggregates than that of bigger size aggregates.

The grading of aggregates are essential to produce compact concrete with good workability. Sieve analysis of coarse aggregates is conducted to check the gradation that each fraction of aggregate is within tolerance to avoid negative effects on strength, shrinkage, and durability of concrete as per the prevailing code of practice. It is decided by the series of sieves that have a standard opening and no sieves as per code of practice of different countries. Samples of sieve analysis performed are tabulated as indicative, which produced fairly workable grading of aggregates obtained from a particular stone crusher (Table 10.1).

Conception of grading of all-in-aggregates that was performed at site level for Mass Concrete grade A80 S185 (Table 10.2).

Grading of fine aggregates being conducted at site as per the limit of relevant code as tabulated in Table 10.3.

TABLE 10.1

Size of Coarse Aggregate, Sieve Designation, and Percentage of Passing

Sl No.	Size of Aggregates	IS Sieve Designation	Percentage of Passing
1	80 mm and above up to 150 mm	150 mm	80–100
		80 mm	0–20
2	40 mm and above up to 80 mm	80 mm	80–100
		40 mm	0–20
3	40 mm–20 mm	40 mm	80–100
		20 mm	0–10
4	20 mm and below	20 mm	80–100
		4.75 mm	0–10

TABLE 10.2

Grading of All in Aggregates of Mass Concrete

	IS Sieve	Percentage Passing of All-In-Aggregates		
Sl No.	Designation	40 mm Nominal Size	20 mm Nominal Size	10 mm Nominal Size
1	80	100		
2	40	90–100	100	
3	20	40–75	90–100	100
4	10	30–65	45–70	90–100
5	4.75	20–45	25–50	40–75
6	600	8–30	10–35	25–50
7	150	0–6	0–6	0–5

Note: Fineness modulus (FM) of coarse aggregates lie between the value of 5.5 and 9.

TABLE 10.3

Grading of Fine Aggregates

Sl No.	Size	Percentage of Passing
1	10 mm	100
2	4.75 mm	90–100
3	2.36 mm	85–100
4	1.18 mm	60–85
5	600 μm	30–60
6	300 μm	15–30
7	150 μm	0–15

Note: Fineness modulus of fine aggregates shall be between 2.3 and 3.

10.8.4.3 Mechanical Properties of Aggregates

The quality of concrete depends on the quality of aggregates utilized in the concrete. The following tests of coarse aggregates are to be conducted to ensure satisfactory performance of the concrete, with the approximate acceptance criteria of value tests of coarse aggregate mentioned in Table 10.4.

TABLE 10.4

Acceptance Criteria of Coarse Aggregates

Sl No.	Name of Test	Acceptance Criteria
1	Aggregate crushing value	Should not exceed 35% for concrete
2	Aggregate impact value	Should not exceed 40% for concrete
3	Aggregate abrasion value	50% for other concrete
4	Soundness of aggregates	For FA, 15% when tested with Na_2SO_4; 10% when tested with $MgSO_4$ For CA, 10% when tested with Na_2SO_4; 20% when tested with $MgSO_4$

10.8.4.4 Physical Property

Physical properties of aggregates are also very important for the good performance and serviceability of concrete. Physical properties are as follows:

- Specific gravity of aggregate is essential for calculation in mix design.
- *Porosity and absorption*: Porosity has influence on the strength of concrete. Porosity of cement paste increases with the increase of the Water-Cement Ratio, and reduces the strength of concrete.
- Durability of concrete depend on permeability. Low permeability improves resistance to percolation of water which carries deleterious substances like chlorides, sulphates, alkali, and other substances that may affect the health of concrete.
- Water absorption is also a vital parameter for the performance of concrete. An aggregate is said to be saturated when all pores are filled with water.
- An aggregate is said to be in Saturated Surface Dry Condition (SSD) when only the surface is dry but the pores are filled with water.
- Moisture content of aggregates is the water content when water is in excess of set SSD conditions and total water content can be calculated by adding absorption and moisture.

10.9 Mix Design and Its Requirement

Concrete is an essential construction material. Concrete is a conglomerate of cement; fine and coarse aggregates; water; additives like flyash, silica fumes, and Ground Granulated Blast Furnace; and various admixtures.

Projects of any kind and their nature comprise 30–35% of total project cost which should be implemented and monitored in a structured way to control costs and improve project economics.

Any hydel project involves a huge quantum of concrete and leads to the consumption of large quantities of cements, aggregates, and water. Optimal utilization of all constituents with a proportion conforming to the specifications to achieve the required concrete strengths. The mix design of concrete is one of the tools in civil engineering whose aims and objectives are to achieve the desired proportion of all concrete ingredients in an economic way.

10.9.1 Advantages of Mix Design

Mix design is done with a purpose to ensure the optimal usage of material like sand, aggregate, cement, etc., to have desired properties of concrete with following advantages.

- Optimal utilization of ingredients like cement, sand, aggregates, and water which will save up to 20–25% of cement consumption depending on the grades of concrete.
- Optimal utilization of material reduces the quantum of material, which solves logistics problems and reduces transportation costs of materials.
- Achieve designed strength of concrete, water tightness, durability, and resistance to thermal behavior of concrete.
- Economics.

10.9.2 Information Required for Mix Design

Concrete grades in the dam profile are not the same in every portion of dam. Multiple grades of concrete are used in various part of a concrete gravity dam (Table 10.5).

Mix Design shall be carried out by an engineering institution or design lab duly approved by the project authority.

A project management consultant (PMC) shall collect the sample of cement/aggregates/sand/water from approved sources, which will be sent

TABLE 10.5

Various Grade of Concrete Used in Concrete Gravity Dam

Sl No.	Location	Grade of Concrete	Remarks
1	Core of dam	A160 S185	A is size of aggregates
2	External part, training wall	A40 S210	S is cylindrical strength of concrete
3	Drainage gallery, inspection gallery, pier, sluice, bucket	A20 S210	
4	Blinding concrete	A80 S210	

to the approved design lab with a letter containing the following details and requirement for the Mix Design of concrete:

a. *General*
- Type of structure
- Location of structures
- Grade of concrete
- Type and grade of cement
- Type and maximum nominal size of aggregates
- Exposure
- Minimum and maximum cement content
- Maximum Water-Cement Ratio
- Workability such as slump, compacting factor, and degree of control
- Maximum temperature of concrete at the time of placement
- Batching process, transportation, and placement system of concrete, compaction, and curing methods
- Type of additive/admixture to be used and condition of use

b. *Material testing*
- Specific gravity of cement
- Specific gravity of fine aggregates and coarse aggregates
- Water absorption of fine and coarse aggregates
- Free surface moisture of ca and fa
- Sieve analysis of ca and fa

c. *Management of site equipment*
- Set up of batching plant and calibration
- Mix design full practical assessment
- Review of consistency or any adjustment

Many methods of concrete mix design are in practice which include ACI, USBR, Euro, and IS.

The mix design of concrete shall be done through various phases as indicated below along with a flow diagram of concrete mix.

- Selection of exposure and assume standard deviation
- Determination of target strength
- Selection of free Water-Cement Ratio
- Selection of water content
- Calculation of cementitious material content

- Calculation of coarse aggregates
- Calculation of fine aggregates
- Calculation of mix
- Trial mixes and checking of properties of concrete mix as per specifications
- Cubes are cast for each trial mix
- Testing of cubes to find out 28 days compressive strength
- Finalization of mix

10.9.3 Flow Diagram of Concrete Mix

Sequential steps of concrete mix design are represented in the Flow Diagram for easy understanding of the process.

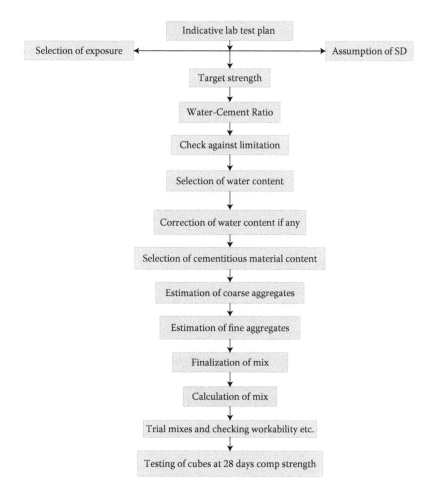

10.9.4 Procedure of Design Mix

Stage 1: Availability of test data as per the lab test plan. Test results of all ingredients as per the lab test plan shall be made available for conducting mix design.

Stage 2: Selection of environmental exposure. Exposure shall be selected carefully in accordance with the exposure condition suited to the location of the structure.

Proper selection of exposure ensures the durability of the structure so that concrete performs satisfactorily in anticipated exposures during its lifecycle. The right exposure will suggest minimum cement content and Maximum Water-Cement Ratio of the subject concrete with its characteristic strength. Mild, moderate, severe, and extreme exposures are classified in accordance with environmental conditions of the proposed site. A basic idea of exposure is required to determine under which conditions concrete shall be designed for durability (Table 10.6).

Stage 3: selection of standard deviation. In practice, each specimen of concrete gives different test results that hints at uncertainties of test results, and a judgment can be drawn from observing the range of differences between the specimens. So, a set of samples made of 50 specimens are tested and can be assessed for the differences of results between individual specimens and the average value of the result of the set of 50 specimens; this range is standard deviation (SD).

Standard deviation will be less if concrete testing is done by applying strict quality control measures, and the SD will be higher if quality control at the site is poor.

In the beginning of the project and before initial design of the concrete mix, the result of particular concrete grade will not be available at site; in these cases, the SD of a particular grade of concrete can be assumed from the code,

TABLE 10.6

Probable Environmental Exposure of Concrete

Sl No.	Environmental	Exposure Condition
1	Mild exposure	Concrete is to be cast in normal condition other than heavy rain areas, cold areas, or any area not exposed to salty climates.
2	Moderate exposure	Concrete is to be cast in an area having high rain fall and cold temperatures, where concrete is required to be protected from heavy downpours and freezing conditions. Concrete is poured under water with saturated salt, etc.
3	Severe exposure	Concrete is exposed to severe rain, severe cold, and highly polluted climate conditions like industrial flume, coastal environments, etc.
4	Extreme exposure	Concrete comes in contact with aggressive soil conditions and mixed with adverse chemical, land, and harmful liquids or solids.

and design can be implemented. Effort shall be made to get at least 30 samples as soon as the mix is in use, and the appropriate SD shall be compared and changed if needed. Standard deviation varies from 3 to 5 N/mm^2 for various grades of concrete.

Stage 4: Computation of target strength. Characteristic strength is the guiding parameter on which mix design shall be based upon. Considering various factors for safe performance of concrete, the mix is designed and is proportioned for higher strength than the characteristic strength of concrete, with some approved margin like SD and other numerical co-efficients. This enhanced strength of concrete to be considered for design is known as target strength.

$$\text{Target strength} = \text{Characteristic strength} + \text{Numerical co-efficient} \times \text{Standard deviation}$$

Stage 5: Selection of the Water-Cement Ratio. The Water-Cement Ratio is a key factor that has influence over the strength of concrete. But, the strength of concrete gets changed with different sizes, shapes, grading, and textures of aggregates. The strength of concrete also gets altered from the type and grade of cement, as well as different cementitious material under the same Water-Cement Ratio. So, the relationship between the strength of concrete and the Water-Cement Ratio can be established by testing all materials actually to be used in the work.

Preliminary Water-Cement Ratios can be obtained in various methods as explained in the following text:

- Free Water-Cement Ratios can be obtained from the drawn graph, with compressive strength of concrete and the Water-Cement Ratio as axis and abccisa, respectively. The strength of the approved cement shall be obtained at 28 days, and the Water-Cement Ratio shall be calculated from the graph plotted with target strength and Water-Cement Ratio with respect to strength of cement.

Stage 6: Selection of water content. The quantity of water required for mixing concrete depends on various factors. The size and shape of coarse aggregates play a vital role in the selection process of water content and other factors like workability, additives, and admixture dictate terms in selection of water content for any grade of concrete. Smaller aggregate sizes increase the demand of water, while use of larger aggregates and admixtures will decrease the requirement of water.

Water content can be selected from the specified table located in the relevant code and in accordance to the size of the aggregate. For example, maximum water content M3 under SSD is 175 kg for 80 mm aggregate.

The water content thus selected can be modified under different conditions, and revised water content can be established for the design of concrete mix.

- 15 kg of water can be reduced in case of usage of non-angular aggregates instead of angular ones.
- 25 kg of water can be reduced if gravel is used with crushed aggregates.
- 25 kg of water can be reduced if gravel is used.
- Water content specified in Table 10.6 holds for slum varying from 25–50 mm. Water content can be modified in case different workability (slump above 50 mm) by trial or increase water content by 3% for every additional slump of 25 mm.
- 5–10 and 20% of water can be reduced if water reducing and super plasticiser is used, respectively.

Stage 7: Computation of cement content. Cement content per cubic meter of concrete can be computed after the Water-Cement Ratio and water content are established.

$$\text{Cement content} = \frac{\text{Water content}}{\text{Water-Cement Ratio}} \qquad (10.1)$$

Calculated cement content shall be checked with minimum cement content specified in the relevant code, and the greater of the two shall be adopted for the mix.

Cement content calculated shall also be checked with the maximum limit of cement specified in the relevant code. High cement content usually harms the concrete instead of increasing the strength. High cement content develops shrinkage, cracking, and creep in the concrete. It may give rise to excessive cracking due to thermal behavior.

Stage 8: Calculation of coarse aggregates. Workability of concrete depends on maximum size and well as the gradation of aggregates. The grading of coarse and fine aggregates are vital factors for calculating the right volume of aggregates to be used in concrete mix. Table 10.7 will suggest the volume of coarse aggregates per unit volume of total aggregates for different zones of fine aggregates to achieve a Water-Cement Ratio of 0.50.

From Table 10.7, the volume of coarse aggregates shall be taken with respect of its maximum size and grading zone of fine aggregates already obtained. This volume of aggregates hold good for a 0.50 Water-Cement Ratio.

The volume of coarse aggregates shall be adjusted with respect to the Water-Cement Ratio selected for the mix in stage number five. The adjustment shall be done at ±0.01 for every ±0.05 change in the Water-Content Ratio to find the corrected volume of coarse aggregates. For pumpable concrete it should be reduced by 10%.

TABLE 10.7

Volume of Coarse Aggregates of Various Sizes for Different Zones of Fine Aggregates

Sl No.	Nominal Max Size of Aggregates	Volume of Coarse Aggregates per Unit Volume of Total Aggregates for Different Zones of Fine Aggregates			
		Zone I	Zone II	Zone III	Zone IV
	10 mm	0.55	0.50	0.48	0.46
	20 mm	0.66	0.63	0.62	0.60
	40 mm	0.78	0.75	0.73	0.70
	80 mm	0. 80	0.78	0.75	0.72

Stage 9: Estimation of fine aggregates. To estimate fine aggregates, absolute volume of cement, water, and admixture should be found. It can be obtained by dividing its mass by specific gravity, and the result of it shall be multiplied by 1/100. The result obtained shall then be subtracted from unity. This value shall be divided into coarse and fine aggregate in the proportion already obtained in Stage 8.

Stage 10: Calculation of mix. Mix calculation can be done by considering per unit volume of concrete and the volume of each ingredient to be calculated by dividing the mass of the ingredient by its specific gravity and multiplying it by 1/1000. In this process the volume of all ingredients are to be calculated to establish the correct mix proportion.

i.   Consider volume of concrete $= 1 \text{ m}^3$

ii. Volume of cement $= \dfrac{\text{Mass of cement}}{\text{Specific gravity of cement}} \times \dfrac{1}{1000} \text{ m}^3$ (10.2)

iii. Volume of additives

$$= \dfrac{\text{Mass of additives}}{\text{Specific gravity of additives}} \times \dfrac{1}{1000} \text{ m}^3 \quad (10.3)$$

iv. Volume of water $= \dfrac{\text{Mass of water}}{\text{Specific gravity of water}} \times \dfrac{1}{1000} \text{ m}^3$ (10.4)

v.   Volume of admixture

$$= \dfrac{\text{Mass of admixture}}{\text{Specific gravity of admixture}} \times \dfrac{1}{1000} \text{ m}^3 \quad (10.5)$$

vi. Volume of all in aggregates $= \{i - (ii + iii + iv + v)\} = X \text{ m}^3$ (10.6)

vii. Mass of coarse aggregate $= X \times$ Volume of coarse aggregate $\times$ Specific gravity of CA $\times$ 1000 kg   (10.7)

viii. Mass of fine aggregates $= X \times$ Volume of fine aggregates $\times$ Specific gravity of FA $\times$ 1000 kg   (10.8)

Stage 11: Mix proportion for trial mix. The volume of all ingredients are to be converted to kilograms, and proportions by weight shall be established and go ahead for trial mixes accordingly.

10.9.5 Design Procedure of Mass Concrete per ACI Method

Design and construction of Mass Concrete structures like gravity dams is very critical. This criticality relies on the phenomena of thermal behavior of Mass Concrete. ACI defines Mass Concrete as "any volume of concrete with dimensions large enough to require that measures be taken to cope with the generation of heat of hydration from the cement and attendant volume change to minimize cracking."

Mass Concrete shall be designed in such a manner that appropriate measures are to be taken to resist the consequence of heat generation due to exothermic reactions so as to minimize the development of cracks.

All ingredients required for Mass Concrete such as cementitious materials, fine aggregates, coarse aggregates, water, and admixtures are mixed in an ideal proportion, and concrete thus obtained will not exceed the permissible limit of temperature rise while simultaneously attaining the required strength, durability, and serviceability.

In the construction of a concrete gravity dam multiple mixes are used in various parts of the dam. One specific mix is recommended for interior concrete, which is known as Core Concrete, and the other is external concrete that is recommended for around the periphery of non-flow and overflow sections of dam that have a certain thickness, with some nominal reinforcement named as skin reinforcements for resistance to various exposures, and balance mixes are used as structural and blinding concrete.

Effects of temperature on the property of concrete shall be considered during design mix of mass concrete.

Step 1: Availability of test data as per indicative lab test plan

- Sieve analysis of fine and coarse aggregates.
- Specific gravity of aggregates.
- Particle shape and size of coarse aggregates.
- Water absorption of aggregates.
- Fineness modulus of fine aggregates.
- Specific gravity of cements like OPC, PPC, and any blended cement; additives.
- Physical and chemical properties of above including heat of hydration at seven days.

Step 2: Selection of slump. Slump is a phenomena in concrete by which workability, placeability, pumpability, and consistency of the concrete can be

TABLE 10.8

Maximum and Minimum Slump to be Used in Concrete

Concrete Construction	Slump, mm (Inch)	
	Maximum	Minimum
Reinforced foundation walls and footings	75(3)	25(1)
Plain footings, cassions and substructure walls	75(3)	25(1)
Beams and reinforced wall	100(4)	25(1)
Building column	100(4)	25(1)
Pavement and slabs	75(3)	25(1)
Mass Concrete	75(3)	25(1)

Source: Adopted from Table ACI A1.5.3.1 of ACI 211.1.91.
Note: Selected slump is 25–75 mm for Mass Concrete.

judged and measured. Slump is selected per size of aggregate, location, and type of structure.

The size of the coarse aggregate shall be 150 mm as per the ACI specification for Mass Concrete. The slump of concrete shall be selected with respect to size of the aggregate as indicated in Table 10.8.

Step 3: Selection of water content. Water content in a mix depends on various factors such as maximum size and shape of aggregates, grading of aggregates, concrete temperature, air entrained, quantity, and admixtures.

As the structure is not exposed to severe environmental condition the concrete can be considered as non-air entrained. So, water content Kg/m^3 shall be selected with regards to slump and size of aggregates as indicated in Table 10.9.

The quantity of water indicated in Table 10.9 for air entrained concrete is based on the total air content requirement for moderate exposure and applicable for well shaped angular aggregates. Graded aggregates conform to the specification and to make trial batches at 26°C. This quantity can be reduced by 18 and 15 kg for non-air entrained and air entrained concrete, respectively, in cases when rounded aggregates are used. In case of the use of water reducing admixture as per ASTM C 494, quantity y of water can be reduced by 5%.

Step 4: Determination of standard deviation. Standard deviation shall be established before selection of the Water-Cement Ratio, and can be established with available test records. These records should contain at least 30 consecutive tests or two groups of consecutive tests totaling 30 tests as per ACI 318. A modification factor is prescribed in Table 5.3.1.2 of ACI 318 as indicated in Table 10.10.

Required average compressive strength when data are not available to establish SD.

TABLE 10.9

Approximate Mixing Water and Air Entrained Requirement for Different Slump and Nominal Maximum Size of Aggregate

Water, Kg/m³ of concrete for indicated nominal maximum sizes of aggregate								
Slump in mm	9.5	12.5	19	25	37.5	50	75	150
Non Air Entrained Concrete								
25–50	207	199	190	179	166	154	130	113
75–100	228	216	205	193	181	169	145	124
150–175	243	228	216	202	190	178	160	—
Approximate amount of entrapped air in non-air entrained concrete, percent	3	2.5	2	1.5	1	0.5	0.3	0.2
Air Entrained Concrete								
25–50	181	175	168	160	150	142	122	107
75–100	202	193	184	175	165	157	133	119
150–175	216	205	197	184	174	166	154	—
Recommended average total air content percent for level exposure	4.5	4.0	3.5	3.0	2.5	2.0	1.5	1.0
Mild exposure	6.0	5.5	5.0	5.0	4.5	4.0	3.5	3.0
Moderate exposure	7.5	7/0	6.0	6.0	5.5	6.0	4.5	4.0
Extreme exposure								

Source: Adopted from Table A1.5.3.3 of ACI 211.91.

TABLE 10.10

Modification Factor

No. of Tests	Modification Factor for Standard Deviation
Less than 15	Use Table 10.9
15	1.16
20	1.08
25	1.03
30 or more	1.00

Source: Adopted from Table 5.3.1.2 of ACI 318.

From Table 10.10 average compressive strength shall be established, and shall be used to check the limit of the Water-Cement Ratio in the following step for the modification factor for SD when less than 30 results are available.

Step 5: Selection of water cement ratio. The Water-Cement Ratio shall be selected as per Table 10.11. This Water-Cement Ratio shall be converted into W/C+P and limits should be checked with permissible W/C indicated in Table A5.7 in conjunction with 28 days compressive strength.

The approximate compressive strength of air entrained concrete for various Water-Cement Ratios in accordance with the type of aggregates is reflected in Table 10.12.

TABLE 10.11

Maximum Permissible Water-Cement Ratios for Massive Section

	Water-Cement Ratio by Weight	
Location of Structure	Severe or Moderate Climate	Mild Climate with Little Snow or Frost
At the waterline in hydraulic or waterfront structures where intermittent saturation is possible	0.50	0.55
Unexposed portions of massive structures	No limit	No limit
Ordinarily exposed structures	0.50	0.55
Complete continuous submergence in water	0.58	0.58
Concrete deposited in water	0.45	0.45
Exposure to strong sulfate groundwater or other corrosive liquid, salt or sea water	0.45	0.45
Concrete subjected to high velocity flow of water (4 ofls) (>12 m/s)	0.45	0.45

Source: Adopted from Table A5.8 of ACI 211.1.91.

TABLE 10.12

Approximate Compressive Strength

Water-Cement Ratio by Weight	Approximate 28 Days Compressive Strength Psi (MPa)	
	Natural Aggregates	Crushed Aggregates
0.40	4500(31.0)	5000(34.5)
0.50	3400(23.4)	3800(26.2)
0.60	2700(18.6)	3100(21.4)
0.70	2100(14.5)	2500(17.2)
0.80	1600(11.0)	1900(13.1)

Source: Adopted from Table A5.7 of ACI 211.1.91.

This Water-Cement Ratio shall be converted to W/P + C when pozzolona will be used.

For severe exposure conditions the Water-Cement Ratio shall be kept low even though the strength of concrete is achieved with higher value. The limiting value is indicated in Table 10.13.

Step 6: Computation of cement content. Water content and the Water-Cement Ratio is obtained from the cement content that can be calculated:

$$\text{Cement content} = \frac{\text{Water content}}{\text{Water-Cement Ratio}} \qquad (10.9)$$

Step 7: Determination of quantity of coarse aggregates. The quantum of coarse aggregate for a particular mix of concrete can be estimated for a given size of aggregates and fineness modulus of fine aggregate selected for the mix.

TABLE 10.13

Water-Cement Ratio in Severe Exposure

Type of Structure	Structure Wet Continuously or Frequently and Exposed to Freezing and Thawing	Structure Exposed to Sea Water or Sulphates
Thin sections (railings, curbs, sills, ledges, ornamental work) and sections with less than 1 in. cover over steel	0.45	0.4
All other structures	0.5	0.45

Source: Adopted from Table A1.5.3.4b of ACI 211.1.91.

The volume of dry rodded coarse aggregates per volume unit of concrete can be obtained from Table A1.5.3.6 of ACI 211.1.91 with regards to the size of aggregate and FM of fine aggregates. The quantity obtained can be increased by 10% for less workable concrete such as pavement, and the quantity can be decreased by 10% if concrete to be placed by pumping (Table 10.14).

The value obtained from Table 10.14 shall be multiplied by dry rodded unit mass of coarse aggregate in Kg/m^3, which will be used in each cubic meter of concrete.

The approximate coarse aggregate content when using natural (N) or manufactured (M) fine aggregate (percentage of total aggregate by volume) is indicated in Table 10.15.

Step 8: Estimation of quantity of fine aggregate. The quantity of cement, water, and coarse aggregates have already been calculated, and the remaining materials in concrete are fine aggregates and air entrained concrete.

Calculation on a mass basis. In this calculation, an estimate of fresh Mass Concrete whose unit mass per cubic meter under the non-air entrained category can be adopted from Table A1.5.3.7.1 of the ACI as indicated in Table 10.16.

TABLE 10.14

Volume of Dry-Rodded Coarse Aggregate Per Unit Volume of Concrete

Nominal Maximum Size of Aggregate in mm	Volume of Dry-Rodded Coarse Aggregate Per Unit Volume of Concrete for Different Fineness Moduli of Fine Aggregate			
	2.4	2.6	2.8	3.0
9.5	0.50	0.48	0.46	0.44
12.5	0.59	0.57	0.55	0.53
19	0.66	0.64	0.62	0.60
25	0.71	0.69	0.67	0.65
37.5	0.75	0.73	0.71	0.69
50	0.78	0.76	0.74	0.72
75	0.82	0.80	0.78	0.76
150	0.87	0.85	0.83	0.81

Source: Adopted from Table A1.5.3.6 of ACI 211.1.91.

TABLE 10.15

Approximate Coarse Aggregate Content When Using Natural (N) or Manufacturer (M) Fine Aggregate (Percentage of Total Aggregate by Volume)

		Fineness Modulus							
		2.40		2.60		2.80		3.0	
Nominal Size of CA	Sand Type	N	M	N	M	N	M	N	M
150 Crushed		80	78	79	77	78	76	77	75
150 Rounded		82	80	81	79	80	78	79	77
75 Crushed		75	73	74	72	73	71	72	70
75 Rounded		77	75	76	74	75	73	74	72

Source: Adopted from Table A5.5 of ACI 211.1.91.

TABLE 10.16

Estimate of Concrete Unit Mass

Nominal Maximum Size of Aggregate in mm	First Estimate of Concrete Unit Mass, kg/m³	
	Non-Air Entrained Concrete	Air Entrained Concrete
9.5	2280	2200
12.5	2310	2230
19.0	2345	2275
25	2380	2290
37.5	2410	2350
50	2445	2345
75	2490	2405
150	2530	2435

Source: Adopted from Table A1.5.3.7.1 of ACI 211.1.91.

Select concrete unit mass of non-entrained concrete for 150 mm size aggregate in Kgs.

Quantity of fine aggregate = Unit mass of concrete − Unit mass of water (Net) − Unit mass of cement − Unit mass of coarse aggregate

Calulation on absolute volume basis

i. $\text{Volume of water} = \dfrac{\text{Mass of water}}{\text{Specific gravity of water}} \times \dfrac{1}{1000}$ (10.10)

ii. $\text{Volume of cement} = \dfrac{\text{Mass of cement}}{\text{Specific gravity of cement}} \times \dfrac{1}{1000}$ (10.11)

iii. $\text{Volume of additives} = \dfrac{\text{Mass additives}}{\text{Specific gravity of additives}} \times \dfrac{1}{1000}$ (10.12)

iv. Volume of coarse aggregates

$$= \frac{\text{Mass of coarse aggregate}}{\text{Specific gravity of coarse aggregates}} \times \frac{1}{1000} \qquad (10.13)$$

v. Volume of fine aggregates $= 1 - (\text{i} + \text{ii} + \text{iii} + \text{iv})$ \qquad (10.14)

Quantity of air is also to be calculated for air entrained concrete.

Adjustment for absorption on aggregates shall be done per moisture found by testing for both CA and FA, and the percentage of moisture shall be added to the weight of the aggregate.

10.10 Example of Mix Design of Mass Concrete

A concrete gravity dam is planned to be constructed in an area having mild environmental exposure. The project authority decided to use concrete with a grade of 20 MPa in the core of the dam, and decided to use nominal maximum coarse aggregates of 150 mm with the following mix.

- Grade designation of concrete: 20 mpa
- Size of coarse aggregate : 150 mm
- Exposure: mild
- Cement: portland cement type ii (moderate heat) as per astmc150
- Additives: fly ash class f
- Ca and fa as per astm125
- Specific gravity 150, 75 to 37.5, 37.5 To 19, 19 to 4.75 Is 2.75, 2.7, 2.7, 2.65
- Specific gravity of fa: 2.65
- Specific gravity of cement
- Specific gravity of fly ash
- Fm of fine aggregates: 2.6
- Water absorption
- Surface water

Step 1: Selection of slump. Slump is selected per Table ACIA1.5.3.1 of the ACI as 50 mm.

Step 2: Selection of water content. Exposure is mild, and this concrete shall be considered non-air entrained concrete. Water content is selected per Table A1.5.3.3 of the ACI for 50 mm slump as 113 Kgs.

Step 3: Standard deviation. Standard deviation is selected from Table A1.5.3.2.1 of the ACI where no record is available.

Average compressive strength $= 20 + 1000 = 20 + 6.9 = 26.9 = 27$ MPa.

Step 4: Selection of Water-Cement Ratio. The maximum permissible Water-Cement Ratio for massive sections is selected from Table A5.8 and is 0.55.

The calculated Water-Cement Ratio of 0.55 shall be checked per Table A5.7 with respect to average compressive strength. Per the table, the Water-Cement Ratio for crushed aggregates and average compressive strength of 27 MPa is 0.48, so the accepted Water-Cement Ratio is 0.48. Fly ash shall be added for Pozzolan Class F at 25% by volume.

Step 5: Air content. Air content for 150 mm size aggregate is 3–4% per Table A5.6.

Step 6: Computation of cement.

Cement content $=$ Water content/Water-cement $= 113/0.48 = 235$ Kgs.

Step 7: Volume of dry-rodded coarse aggregate per unit volume of concrete for FM of sand $(2.6) = 0.77$ as per Table A5.5.

Required dry mass $= 0.77 \times$ Unit weight of aggregate $= 0.77 \times 1600 = 1232$ Kgs.

Determination of absolute volume:

Per Table A5.6A/6A.6B/6A.6C of ACI 211.1.91

$$Vc + p = \frac{Cw}{Gc(1000)}, \tag{10.15}$$

$$Vc = Vc + p(1 - Fv), \tag{10.16}$$

$$Vp = Vc + p(Fv) \tag{10.17}$$

where
  Cw $=$ Weight of equivalent Portland Cement
  Gc $=$ Specific gravity of Portland Cement
  Vc $=$ Volume of cement
  Vp $=$ Volume of Pozzolona
  Vc $+$ p $=$ Volume of cement and Pozzolona
  Fv $=$ Percent Pozzolona by absolute volume of cement plus Pozzolona

$$Vc + p = \frac{Cw}{Gc \times \text{unit weight}} = \frac{235}{3.15 \times 1000} = 0.075 \text{ m}^3/\text{m}^3$$

$$Vc = Vc + p(1-Fw) = 0.075(1-0.25) = 0.0556 \text{ m}^3/\text{m}^3$$

$$Vp = Vc + p(Fw) = 0.075(0.25) = 0.019 \text{ m}^3/\text{m}^3$$

$$Vw = 113/1000 = 0.113 \text{ m}^3/\text{m}^3$$

$$VA = 0 - 035(1) = 0.035$$

Determination of absolute volume of CA and FA:

Volume of aggregate $= 1 - Vc + p - Vw - Va = 1 - 0.075 - 0.113 - 0.035 = 0.777$

Percent of total aggregate by absolute volume to be found out with regards to fineness modulus of fine aggregates per Table A5.5 of the ACI in Table 10.14.

Approximate Coarse Aggregate Content when using Natural (N) or Manufacturer (M) Fine Aggregate (Percentage of Total Aggregate by Volume) (Table 10.15).

Percentage is 77 with respect to FM 2.6 of crushed aggregate

Volume of coarse aggregate $= 0.777 \times 0.77 = 0.598 \text{ m}^3/\text{m}^3$

Volume of fine aggregates $= 0.777(1 - 0.77) = 0.178 \text{ m}^3/\text{m}^3$

Let us then convert the weight per unit volume of all ingredients:

1. Cement $= 0.056 \times 3.15 \times 1000 = 176.4 \text{ Kgs}/\text{m}^3$
2. Fly ash $= 0.019 \times 2.45 \times 1000 = 46.55 \text{ Kgs}/\text{m}^3$
3. Water $= 0.113 \times 1 \times 1000 = 113 \text{ Kgs}/\text{m}^3$
4. Air $= 35 \text{ Kg}/\text{m}^3$
5. Coarse aggregate $= 0.598 \times 2.7 \times 1000 = 1614 \text{ Kgs}/\text{m}^3$
6. Fine aggregate $= 0.178 \times 2.64 \times 1000 = 469 \text{ Kgs}/\text{m}^3$

Check Mortar content:

$$\text{Mortar} = Vc + Vp + Vw + Vs + VA$$
$$= 0.056 + 0.019 + 0.113 + 0.178 + 0.035 = 0.401$$

Mortar Content for 150 mm crushed aggregate per Table A5.6 is 0.39 m³/m³ with limit +0.01 and becoming $0.39 + 0.01 = 0.40$.

Absorption correction:

Correction for absorption shall be performed.

Suppose the moisture tested in CA is 1% and in FA is 2%.

1. Coarse aggregate (wet) $= 1614 \times 1.01 = 1630 \text{ Kgs}$
2. Fine aggregate (wet) $= 469 \times 1.02 = 478 \text{ Kgs}$

Absorbed water shall be excluded.

Surface water on CA $= 1 - 0.5 = 0.5\%$ and on FA $= 2 - 0.7 = 1.3\%$

Estimated quantity of water to be added $= 113 - 1614 \times 0.005 - 469 \times 0.013$
$$= 113 - 8.07 - 6.0 = 99 \text{ Kgs.}$$

Therefore:

1. CA wet $= 1630$ Kgs
2. FA wet $= 478$ Kgs
3. Water $= 99$ Kgs
4. Cement $= 176$ Kgs
5. Fly ash $= 47$ Kgs

Total Weight per m^3 $= 2430$ Kgs.

Batch of 0.2 m^3 for the trial mix shall be made.

1. Water $= 1.98$ Kgs
2. Cement $= 3.52$ Kgs
3. Fly ash $= 0.94$ Kg
4. FA $= 9.56$ Kgs
5. CA $= 32.6$ Kgs
   Total $= 48.60$ Kgs

The appropriate quantity of admixture shall be used to reduce water content without changing the Water-Cement Ratio for better workability. It may reduce water content 5–10%, and shall be used per the manufacturer's technical specifications for the particular product. After the trial mix, the yield of concrete shall be also checked.

10.11 Batching and Placement of Concrete

Concrete is an artificial stone composed of various ingredients having different properties and characteristics that provide strength and durability, workability, and serviceability. All high quality ingredients should be measured to ensure they conform to the proportion requirements when batched, mixed, transported, placed, compacted and cured to achieve the required strength and durability of the concrete.

10.11.1 Definition of Batching and Mixing

Batching is a process where all ingredients conforming to the specification are measured (volumetric or weight) individually as predetermined proportions and are placed in mixing unit that has specific revolutions per minute and batch timing so that a homogeneous concrete can be produced.

10.11.2 Concrete Production Flow Chart

The following flow chart represents sequential activities of concrete production.

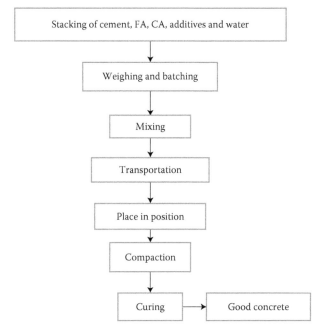

Two types of batching are generally adopted in production of concrete: volumetric and weight batching.

- *Volumetric batching*: Volumetric batching is done by mixer machines with various capacities where ingredients are measured individually by box and placed in the hopper. The hopper is in turn lifted, and the entire mix is put into rotating drum of the mixer. The size of the measurement boxes are made equivalent to the volume of one bag of cement, which is 35 liters or 0.035 cubic meters. Volumetric batching is only suitable for concrete of nominal mix.

- *Weight batching*: Concrete of mix designs shall be produced by the weight batching mechanism in the batching plant. All ingredients are weighed individually by the weighing mechanism and mixed in the mixing unit to produce a concrete of good workability, strength, and durability. In this batching, the quality control system is implemented for strict compliance and the end product is called Controlled Concrete. Weight batching can also be divided by category.

- *Dry mix*: As per requirement and demand, a dry mix is prepared and delivered in the transit mixer where the dry mix is converted into concrete in the transit mixer.
- *Wet mix*: All ingredients are mixed and concrete is produced and delivered in the transit mixer.

10.11.3 Batching Plant

The batching plant is a plant where all ingredients such as cement, sand, aggregate, additives, and water are weighed individually by a built-in weighing mechanism and mixed in a mixing unit Pan Mixer/Turbo Type/Twin Shaft Mixer, which guarantees the production of high quality, homogeneous, and durable concrete. The output of a plant is expressed in cubic meters per hour.

Different types of batching plants are available in the field as indicated below:

- Truck mounted plant
- Compact plant
- Mobile plant
- Horizontal plant

Flow chart of concrete mixing plant

10.11.4 Component of Batching Plant and Its Function

The batching plant consists of many components such as storage of sand and aggregate in bins located side by side. Materials are fed through a belt

TABLE 10.17

Base Plant CP-30

Sl No.	Particulars	Unit	Data
1	Pan mixer (filling capacity/compacted concrete)	M^3	0.75/0.5
2	Output volume (compacted concrete with 30 sec mixing time)	M^3/hr	30.00
3	Concrete discharge height	M	4.1
4	Mixer platform height	M	4.28
5	Aggregate weighing system (in line silo)	Kg	1250
6	Cement weighing system	Kg	250
7	Water weighing system	Kg	150
8	Cement type	Up to	3
9	Water supply	DN	50
10	Water pressure for operation	Bar	5–6
11	Aggregate type		4
12	Admixture		6 Kg
13	Dg requirement		75 KVA
14	Voltage		3 ph, 450 V, 50 Hz
15	Control system	PC Duel core processor with 1GB ram, 15.6 LCD monitor	

conveyor that hangs on the digital load cell-based weighing system. Materials are weighed individually and are transferred to the mixing unit through the system. Cement is transferred from the silo/hopper, weighed, and is then transferred to the mixing unit through screw conveyor. Water received from the pump is weighed as per proportions that have been set in the control panel before being transferred to mixing unit. Additives are transferred from the silo and weighed and transferred to the mixing unit. The mixing unit, either a Pan Mixer, Turbo Type Mixer, or Twin Shaft Mixer, mix all ingredients in the prescribed batch time to produce high quality homogeneous concrete, which is then delivered to the transit mixer.

The batching plant is featured with a quality assortment control system and weigh load cell with PLC modules and membrane key buttons. The key is built-in with indicator LEDs to give the status of current operation, as well as SMS for critical alarm and generation of production details slip through printing systems. The printout production slip indicates date, time, details of ingredients, and quantity. The technical specifications of CP-30 of the Schwing Stetter is listed in Table 10.17 for a better understanding of a compact plant.

10.11.5 Placement, Compaction, and Curing of Dam Concrete

In dam construction the placement of concrete involves many activities such as the preparation of horizontal surface, transportation and handling of concrete, the arrangement for required machinery and equipment, and the placement, compaction, and arrangement of monitoring functions for the concrete's temperature. A concrete pour shall be prepared as a part of placement, duly indicating the sequence of concreting in various layers to avoid the formation of possible cold joints, and subsequent construction joints should be accompanied by the required indicative test plan for required quality control measures.

Indicative test plans contain the Pour Card, which is a vital document for concreting activity. The first part of the Pour Card shall be filled in by site engineers and will indicate the date, details of the structure, quantity, concrete grades, details about the rebar, compaction methodology, and the curing arrangement. The second part shall be filled up after pour completion which includes the time of completion, the quantity of cement, recorded temperature of concrete, number of cubes taken, slump tests, and other vital records.

Aggregates ranging in size from 20–150 mm are used in dam concrete, with core concrete consisting of larger size aggregates of approximately 150 mm. The project authority should make the determination in the selection of transportation methods for concrete made with 150 mm size aggregates. There is a chance of segregation of 150 mm aggregates due to motion of inertia if the concrete is transported by transit mixer or any other usual mode of transport.

An effective transport system for this core concrete can be developed so that segregation of bigger aggregates can be avoided.

- A crane or cable way shall be installed, covering the best possible area of blocks of dam.
- One mast of the cable way should have a sliding facility to a certain angle so as to cover additional areas of blocks.
- The batching plant shall be installed in close proximity to the reach of the lifting system.
- The gap distance between the reach of the crane and the batching plant shall be connected by the rail track.
- A bucket with a 2–3 cubic meter capacity shall be fabricated with the ability to have a pneumatically opening top and bottom lid.
- A mini locomotive shall be placed on the rail with a mini career on which the bucket can be placed by crane or cable way and the locomotive will carry and place it below the delivery point of the batching plant.
- The locomotive will push the filled bucket back within the reach of boom radius of the crane or cable way.
- It will lift the bucket and transport it to the place of pouring, then lower it down.
- The bottom lid of the bucket shall be opened pneumatically to place the concrete in position.
- The crane or cable way will lift earlier empty bucket for refilling.

10.11.5.1 Methodology of Placement of Dam Concrete

The placement methodology varies from portion to portion of the dam. Dam concrete is classified into at least four categories depending on the position and utility of the structure. Each part of the dam plays distinctly different roles for the safety of the dam. The four categories are as follows:

- External Concrete
- Core Concrete
- Blinding Concrete
- Structural Concrete

Construction engineers working at the site should take the following precautions during the placement of concrete, with these instructions being applicable to any part of the dam:

- Concrete shall be placed in its final position.
- Avoid the rehandling of concrete. Rehandling causes loss of time, workability, and homogeneity of the concrete.
- The free fall of concrete should not be more than 1.5 m, as it causes segregation of aggregates.
- Concrete shall be laid in specified layer thicknesses to facilitate proper compaction.
- Freshly laid concrete shall be compacted within the initial setting time of cement, otherwise it will form cold joints and construction joints that will weaken the dam concrete.
- The vibrator needle shall be inserted vertically into the concrete.
- The vibration of concrete shall continue until the cessation of air bubbles coming from the concrete.
- Over vibration of concrete shall not be allowed in any case.
- Concrete should not be allowed to flow during compaction, and the flow of concrete shall be stopped by placing a stopper to arrest the flow of concrete.
- The vibrator needle shall be withdrawn slowly after compaction is over to avoid the formation of holes within the freshly compacted concrete.
- The dam is divided into many blocks. Concrete is done block wise. Each block consists of specific lengths and widths. The depth of the block is also limited in each pour to control heat generation and reduce the risk of possible cracks.
- Concrete shall be taken up block wise. Level differences between two adjacent block and level differences between the first and last block shall be maintained per specifications.

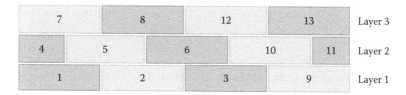

FIGURE 10.1
Sequence of pouring of concrete.

10.11.5.1.1 External Concrete

External Concrete is placed around the periphery of non-overflow and overflow sections as shown in Figure 10.1, with a specific thickness of 2–2.5 m per the designer to protect and provide stability to the dam. This concrete is not simple, plain cement concrete. Some reinforcements are provided in the form of skin reinforcement to take care of the stress being produced by temperatures. This concrete is richer than core concrete. The function of this concrete in the downstream portion of the overflow section of the dam is taken care of by the concrete done for the spillway. Separate External Concrete is not necessary to be done in the downstream of the dam's overflow section. This concrete is designated as A40 S210, with A standing for the size of the aggregates and S standing for the cylindrical strength of concrete at 28 days. External Concrete shall be taken up along with core concreting of dam.

10.11.5.1.2 Core Concrete

The core of dam is the central, innermost, and most critical part of a concrete gravity dam. The core part of the dam is the main contribution to the weight of the dam that is resistant to sliding. This part of the dam is built with Mass Concrete without any reinforcement. This concrete can be made with larger aggregates up to 150 mm. Aggregate size is preferred to be less than 80 mm. Aggregate greater than 80 mm may cause difficulties in compacting and in spreading, but adequate care shall be taken for the compaction of concrete using larger size aggregate. However, use of aggregate less than 80 mm in size will increase the cohesiveness of concrete and decrease the voids. This mix is designated as A150 S165.

Mass Concrete in the core of the dam is done block wise with specified lift thicknesses that varies from 1.5–3 m. This limit is fixed due to the generation of heat within the concrete from the hydration process. This designed thickness of concrete is done through multiple layers of thickness varying from 300–500 mm depending on the size of aggregates used in the concrete. The sequence of concrete layer placement shall be done per the pour scheme as demonstrated in Figure 10.1.

The surface of the block on which fresh concrete shall be laid is to be simultaneously cleaned with air and water jets to clean the area of dirt, dust, and other contaminants before pouring fresh concrete. The width of

External Concrete shall be marked on the surface from the inner surface of the formwork.

10.11.5.1.3 Core Concrete

Core Concrete shall start from the upstream side to the downstream side, providing a gentle upward slope. This slope will provide resistance to sliding as well as drainage. This slope facilitates cleanup activities of the completed layer for laying up for the next lift. The surface shall be cleaned by air and waterjets, removing some millimeters of film from the top of green concrete which can be termed as a green cut so as to have a good bond between the layers. Laying of External Concrete and Core Concrete shall be taken up together. Concretes of grade A40 S210 shall be placed in position within the marked area in the specified layer and compacted. Then, A150 S160 grade concrete shall be placed in position, and a stopper shall be placed to arrest the flow of concrete and be compacted while concrete is placed continuously to avoid the formation of cold joints and the subsequent formation of construction joints. This construction activity shall be supported by a temperature monitoring and recording system for the Core Concrete both at the site as well as in batching plants. Concrete temperature should be well below 18°C at the time of its positional pouring, while placement temperatures of concrete can be controlled by narrowing the difference between external and internal temperatures. The internal temperature of concrete increases in a 1:2 proportion with the external temperatures. In concrete, the main contribution for increase of temperature is CA and FA, which occupy about 75% and 35% of concrete, respectively, although their specific heat is low. The specific heat of water is high but occupies only 6%. Pre-cooling arrangements shall be made to control the temperature of concrete, and water temperature should be kept at 2°C below the concrete's temperature.

Measuring the temperature of all ingredients at the site, concrete temperature can be found out by the following formula (Section 11.5.1).

The maximum temperature difference within the Mass Concrete can also be monitored by placing sensors within the concrete.

10.11.5.1.4 Blinding Concrete

Concrete is constructed on rocks. The foundational level is reached by the excavation of hard rock by various methods such as open blasting/controlled blasting, breakers, and other means. Blasting operations are permitted up to minimum of 150 mm above the foundational level to avoid possible cracks being generated by blasting. A balance of 1,500 mm shall be excavated by breaker or chiseling. After this activity, foundation treatments shall be made as per professional advice from geologists. Foundation treatment includes dental treatment, cleaning of grout holes by flushing clay seams, and other preparatory actions before starting consolidation grouting in the area. Finally, water tests will be conducted to find out the water tightness of the foundation by judging the Lugeon value of the test.

The foundation area is to be cleaned to remove all dust, dirt, debris, and any foreign material that may be harmful to concrete by using air and water jets. Any scaling or coating of rocks or suspected areas shall be cleaned, and any part of the rock that is found to have a stiffer slope where anchor holes are to be drilled up to the required depth and reinforcement of required diameter shall be grouted. After cleaning, the area should be examined by geologists for geological clearance, after which the area is ready for concreting. As the area prepared for concreting is uneven, up to 300 mm in concrete shall be laid on the surface of rock to make it even with grade M20 with larger size aggregates of 40 mm, which is known as Blinding Concrete.

10.11.5.1.5 Structural Concrete

There are many other parts of the dam such as the drainage gallery, inspection gallery, sluice, spillway, bucket, training wall, and piers, all of which are constructed with reinforced cement concrete known as Structural Concrete.

The placement of Structural Concrete can be done by means already described in this work. However, the placement of the sluice and its bell mouth and flare curve at the discharge point is critical. Layout of the curve, formwork, and concreting shall be done effectively to pass the water smoothly downstream.

The most critical and vital part is the spillway. The spillway is a passage on the top of overflow section of the dam, one that has an ogee curve on the crest is known as ogee-crested spillway. Excess floodwater is allowed to pass over it, safely rolling downstream where the stilling basin initiates energy reducing actions and energy dissipation at the toe of spillway, creating hydraulic jump. Any deviation from the design or construction variances change the crest shape; this can change the flow properties, and will be detrimental to the performance of spillway. Therefore, the layout and setting of the ogee curve will have to be perfect, and concreting will be done perfectly for the smooth discharge of floodwaters.

10.11.5.2 Compaction of Dam Concrete

Freshly laid concrete contains many voids, and voids are filled with air. These air voids in concrete weaken the concrete until air is expelled from the voids. The expulsion of air from the concrete void is called compaction. Compaction of concrete is done by a vibrator. Vibrators of many types are available in the market, but immersion type needle vibrators are the most suitable and effective for dam concrete compaction.

10.11.5.2.1 Procedures of Dam Concrete Compaction

Mass Concrete in the core of the dam is done by blocks with specified lift thicknesses that vary from 1.5–3 m. This limit is fixed due to the generation of heat within the concrete during the hydration process. The designed thickness of concrete is done through multiple layers of thickness varying from 300–500 mm depending on the size of aggregates used in the concrete.

The compaction of Core Concrete in the dam is done through immersion type needle vibrators. The Core Concrete of a dam is composed of 150 mm size aggregates for which a 150 diameter needle vibrator will be most suitable for compaction.

The compaction of dam concrete shall be done in a systematic way. The covered area of compaction shall be calculated per the capacity of the bucket carrying the concrete. If bucket capacity is 3 cubic meters and the thickness of each layer is selected to be 400 mm, then approximately 7.5 square meters is to be covered. A strip 2 m wide and 3.75 m long can be taken up for compaction. The number of vibrator operators to be deployed is determined by considering the supply of concrete per hour and spacing of unloading of buckets should be fixed accordingly to avoid any overlap between the two heaps. One operator can compact about 30 cubic meters in one hour.

The needle of the vibrator shall be inserted into the concrete vertically in fixed spacing, and spacing shall be staggered to have an effect of vibration in each part of concrete being compacted. Otherwise, some parts of concrete will remain unaffected by vibration as shown in Figure 10.2. Concrete flow by vibrator is not allowed, and the flow of concrete shall be arrested by putting a stopper in the edge of the strip. Many air bubbles will initially appear on the top of the concrete. Care should be taken to compact the concrete up to its full depth, and the compaction process should continue until air bubbles cease to come from the void, which will indicate the completion of the compaction process. The tendency to over vibrate concrete shall be discouraged and the vibrator needle will be slowly withdrawn to avoid any formation of holes in

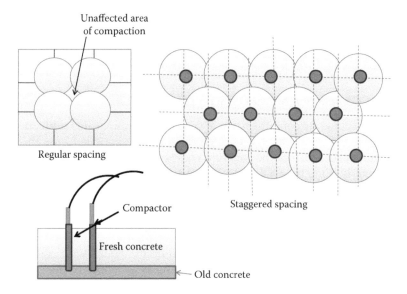

FIGURE 10.2
Spacing of needle vibrator for compaction.

concrete during withdrawal. The vibrator operator will level the concrete by removing any foot print, mark of needle, or other imperfection, and it will be done by placing wooden planks on the fresh concrete.

Vibrator operators should be provided with all necessary PPE during compaction process including gum boots, gloves, and safety goggles.

10.11.5.3 Curing of Dam Concrete

Curing is the process that prevents moisture from evaporating from the concrete's surface. Concrete achieves strength through the hydration process. This hydration process require fixed quantities of water that is added to concrete per the Water-Cement Ratio. If the quantity of water added is lost, the concrete will fail to achieve its desired strength. Excessive evaporation from the surface of concrete is the main reason for shrinkage cracks. Air, ambient temperature, wind speed, humidity, and concrete temperature will influence the evaporation rate. Curing is done to prevent this loss of water and enable the completion of the hydration process.

Mass Concrete is cured through the ponding of water. Concrete surface should be kept under water at all times, alternating wet and dry conditions during the curing period affects the strength of concrete.

In hot climates the curing process should start 4 hours after the concrete is cast, and in normal climate conditions it should start after 8 hours. The difference of temperature between curing water and concrete surface should not be more than 6–7°C. If the average temperature is below 5°C, the temperature of the concrete surface should be maintained at 10°C.

Mass Concrete should be cured for at least 14 days. If pozzolona and fly ash is used in concrete, curing shall be done for 28 days. Liquid membrane curing compounds are not the best way to cure concrete, but when moist curing is not possible, curing compounds can be used for curing.

Bibliography

Building code Requirement for structural concrete (ACI 318-05). American Concrete Institute.

Cement and Concrete Terminology (ACI116). American Concrete Institute.

Standard practice of selecting of proportion of normal, heavy and mass concrete (ACI211.1.91). American Concrete Institute.

11

Design Requirement for Temperature Control in Mass Concrete

11.1 Concept of Mass Concrete

A large volume of concrete is placed without reinforcement with all three dimensions of the deposit designed to optimally combat the generation of heat due to exothermic reaction, as well as decrease temperature due to differential cooling within the mass that causes volumetric changes in the volume of the concrete. Tensile stresses are being developed by volumetric changes, which becomes responsible for initializing cracking in the concrete.

Mass Concrete is defined in ACI116 R as "any volume of concrete with dimensions large enough to require that measures be taken to cope with generation of heat from hydration of the cement and attendant volume change to minimize cracking."

Any large volume of concrete can be called Mass Concrete if its tendency to develop cracks is being controlled and limited so as to attain desired durability, establish requisite water tightness, and to accomplish designed strength and economy. All of these parameters of Mass Concrete can be achieved by controlling the following characteristics:

- Maintaining placement temperature of concrete below 18°C
- Restricting quantity of pour of each lift to control the generation of heat
- Optimal design of the thickness of each shaft
- The pour scheme used
- Removal of heat through embedded cooling system
- Pre-cooling arrangement for all ingredients

11.2 Significance of Thermal Stresses in Mass Concrete

A large amount of concrete without having a main reinforcement but having skin reinforcement can be designated as Mass Concrete when its performance is judged based on its durability, water tightness, thermal behavior, economy, workability, ability to place where needed, serviceability, and strength.

Strength is not the prime criteria in Mass Concrete; it is secondary to other characteristics, and the success of Mass Concrete depends on all of the aforementioned factors.

Thermal behavior of Mass Concrete makes it vulnerable and susceptible, differentiating it from other forms of concrete. Exothermic reactions release energy in the form of heat to the surroundings; this reaction between water and cement increases temperature within Mass Concrete and the heat generated is lost to its surroundings. Ideal conditions make the amount of heat lost equal to the amount of heat gained. However, this situation becomes critical when heat lost is much less than heat gained within the concrete during the reaction process, and this high amount of unreleased heat get stored within Mass Concrete. Tensile stress and strain is developed within concrete due to changes of volume. Concrete is weak in tension, and is subjected to surface cracking. This thermal cracking affects the Mass Concrete, and Mass Concrete tends to lose its virtues such as durability, water tightness, and serviceability. Precautionary measures should be adopted to prevent Mass Concrete from cracking to improve performance and to retain structural integrity.

A dam without cracks in its concrete is unimaginable. This thermal cracking of concrete has thrown challenges to the technical world, as the length, width, and height of a dam is very large and would be uneconomical if not for its constructability when reinforcement is provided to take care of these temperature stresses. Therefore, the dam should be designed in such a way that tensile forces being generated should not exceed the permissible stresses.

11.3 Concept of Thermal Cracking

In Mass Concrete cement-like materials act as a binder and are mixed with water to start the hydration process. This reaction is exothermic and generates heat energy, increasing temperatures within the concrete mass. Dissipation of heat in the concrete plays a vital role in the initiation of cracking. Dissipation of generated heat within concrete is very slow due to low conductivity, and the dissipation of energy in the concrete's core is much slower than the external dissipation on the face of the concrete. This phenomena creates significant differences in temperature between the internal and external concrete faces, which builds a surface temperature gradient that causes volumetric changes while restraint creates tensile stresses that leads to surface cracking of concrete.

11.4 Parameters Initiates in Rise of Temperature

Factors that affect the hydration process include:

- Cement composition, fineness, and content
- Aggregate content and co-efficient of thermal expansion
- Section configuration
- Placement and ambient temperature

11.4.1 Configuration of Dam Section

The configuration of the concrete gravity dam is triangular, with its weight the main factor for stability of the structure and requiring a large volume to resist horizontal forces, resulting in the large volume-to-area ratio while experiencing the large rise in temperature caused by the hydration process in the concrete. The surface temperature gradient is created due to unequal dissipation of energy between the dam's surface and core, leading to thermal cracking on the surface of the dam.

11.4.2 Cement Composition

Composition of cement is one of the main reasons for the rise of temperature in concrete. Major compounds like Tricalcium Aluminates (C_3A), Tricalcium Silicate (C_3S), and Tetra Calcium Alumino Ferrite (C_4AF) present in the concrete cause the temperature to rise, and minimal use of these compounds in cement are prescribed by various codes so that the generation of heat can be controlled. Low and medium heat cement is used in Mass Concrete, so cement meant for Mass Concrete should contain less C_3A and C_3S to reduce and control the generation of heat.

11.4.3 Fineness of Cement

Fineness of cement indicates its grained size distribution of particles and its related specific surface area. Fineness of cement has an impact on the Rheology property, hydration, strength, and pore structure of concrete, as well as an effect on workability, ability for placement, and water content of the concrete.

The heat of hydration depends on the fineness of the cement, with heat increasing with fineness. The hydration process starts at the surface of the cement particle, with higher fineness providing more surface area and accelerating the reaction between water and cement and increasing the rate of heat dissipation.

11.4.4 Content of Cement

The hydration process depends on the content of cement, which has an effect on the generation of heat in the concrete. Mixes designed for dam concrete should have a minimum cement content. The control of cement quantity has a direct influence on thermal stresses, with approximately 120 cal/g worth of heat generated during full hydration; the heat of hydration increases internal temperatures by acting as an insulator and producing low thermal conductivity.

Supplementary cement-like materials are used in the concrete as a replacement of cement content to reduce the liberation of heat in the concrete because the pozzolonic reaction generates less heat than is generated in ordinary Portland cement. Fly ash is less of a contributor to heat in concrete, accounting for up to 30% of heat for the same weight of cement. Fly ash improves workability while reducing heat generation and permeability of the concrete. The content of cement can be controlled by using larger aggregate amounts in the concrete.

11.4.5 Aggregate Content

Coarse aggregates occupy approximately 75–80% of the volume of concrete, and it is occupied by 150 mm size aggregates. Larger size aggregates will provide a larger surface area and will reduce cement content, which in turn reduces heat generation within the concrete.

The coefficient of thermal expansion of aggregates influences the coefficient of thermal expansion of concrete. Aggregates with lower CTE will reduce thermal stresses, and lower CTE will provide more resistance to thermal stresses. Lower CTE tends to have higher thermal conductivity and heat will be released faster from the core of the concrete.

11.5 Computation of Temperature of Concrete

Mass Concrete specifications stipulates the max temperature of concrete at the time of placement. This limitation in Mass Concrete is specified to prevent thermal cracking in concrete which makes concrete undurable. Effort should be used to control the temperature of concrete, and temperature should be kept as low as possible because cracking occurs due to steep temperature gradients. It is mandatory to have a clear idea of concrete temperature before and after placement of concrete so that necessary equipment will be in place to conduct pre cooling and post cooling of the concrete.

11.5.1 Background Information

Before mixing the concrete in batching plant, the site engineer should have an idea about the temperature of each ingredient and its ambient temperature. The temperature of these ingredients are responsible for raising the temperature of freshly mixed concrete, which is the root cause of the formation of steep temperature gradients that cause cracks in the concrete. So, the engineer and designer should have the necessary background information before taking up concrete work.

a. *Maximum temperature* of concrete at the time of placement shall not be more than 18°C. The concrete may be affected when max temperature exceeds the designed temperature. This rise in temperature will affect the long term performance of concrete. Concrete cured in higher temperatures will have lower strength, larger pores, and higher permeability.

The temperature of concrete can be calculated per ACI 305R, which is described in the following text. The temperature of all ingredients shall be measured in an approved method and values are to be put in the formula.

$$T = \frac{\begin{aligned}0.22(T_{ca}W_{ca} + T_{fa} \times W_{fa} + T_pW_p + T_cW_c) \\ + T_wW_w + T_{ca}W_{cam} + T_{fa}W_{fam} - W_i(79.6 - 0.5T_i)\end{aligned}}{0.22(W_{ca} + W_{fa} + W_c + W_p) + W_w + W_i + W_{cam} + W_{fam}}$$

(11.1)

where

T	= Concrete placement temperature
T_{ca}	= Temperature of coarse aggregates
T_{fa}	= Temperature of fine aggregates
T_c	= Temperature of cement
T_p	= Temperature of Pozzolan
T_w	= Temperature of water
T_i	= Temperature of Ice
W_p	= Weight of Pozzolan
W_{ca}	= Dry mass of CA
W_{fa}	= Dry mass of fine aggregates
W_c	= Mass of cement
W_i	= Mass of Icc
W_w	= Mass of mixing water
W_{cam}	= Mass of free and absorbed moisture of CA
W_{fam}	= Mass of free and absorbed moisture of fine aggregates

b. *Larger temperature gradient or difference in temperature* occurs when the core is hot and ambient temperatures are cold. The maximum temperature difference causes a change in volume due to the thermal expansion or contraction, and cause the development of cracks if it is restrained by the foundation.

The maximum allowable temperature differential as per Gajda & VangGeem is:

$$\Delta T = \frac{\in tsc}{K \propto R} \tag{11.2}$$

where
ΔT = Maximum difference of temperature
$\in tsc$ = Tensile strain capacity
K = Modification factor for sustained loading and creep
$\propto$ = Coefficient of thermal expansion of maturing concrete (1/°C)
R = Restraint factor

The designer and construction manager should be interested in predictions of the temperature of fresh concrete during pouring. The reaction process developed within concrete is hydration, which is an exothermic reaction. This exothermic reaction generates heat that may lead to the development of thermal cracks even before the structure is actually loaded. Therefore, computation of temperature during pouring and curing are very beneficial.

The hydration process starts during the batching of concrete, and initiates an increase in the internal temperature of concrete. There are many parameters that are responsible for temperature increases in concrete, including the type of cement and its composition, fineness and content of the cement, aggregate content and the coefficient of thermal expansion (CTE), configuration of structure, methodology of placement, thickness of each lift, and ambient temperatures. After reaching its maximum temperature, the temperature of concrete decreases.

Pours with a large volume-to-surface area ratio are more susceptible to thermal cracking. Concrete shall be poured in blocks with designed area and thickness of lift. Cements used for Mass Concrete should have a low C_3S and C_3A content to reduce excessive heat during hydration. Pozzolona and fly ash shall be used to reduce heat generation. Cement with a lower fineness and slow hydration reduces increases in temperature. Cement content shall be as low as possible within the minimal permissible limit to have the designed strength of Mass Concrete, which will lower the heat of hydration and control the increase in temperature. A higher coarse aggregate content (85–90%) and up to 150 mm in size can be used to lower the cement content, thereby reducing temperature increases. The coefficient of thermal expansion of the coarse aggregate has influence on the coefficient of the thermal expansion of concrete. The coefficient of thermal expansion of the aggregate is at a rate at which the aggregate will increase in length with the rise of temperature, and thermal conductivity of concrete transfers heat from one point to another of point of concrete so that lower CTE aggregates will have higher thermal conductivity. Heat is released fast from the core, with lower ambient temperatures producing less temperature rises and lower volume-to-surface ratios producing less increases in temperature.

Contractors are supposed furnish temperature control plans for Mass Concrete before work starts.

Prediction of temperature shall be made per the following methods:

- Portland cement association method
- Schmidt method as per ACI 207.1R
- Concrete work software package

11.5.2 The Portland Association Method (PCA)

This method calculates the max temperature rise above concrete placement temperature as 12°C for every 100 kg of cement, and the method is applicable to concrete having a cement content between 300 and 600 kg per cubic meter. It provides no information about the time of max temperature, does not allow quantification of differences in temperature, and it deals with ASTM C150 Type I cement.

$$T_{max} = T_i + \left(12 \times \frac{W_c}{100}\right) + \left(6 \times \frac{W_{scm}}{100}\right) \tag{11.3}$$

where
T_{max} = Maximum concrete temperature in °C
T_i = Concrete placement temperature in °C
W_c = Weight of cement Kg/m^3
W_{scm} = Weight of supplement cement in Kg/m^3

11.5.2.1 Schmidt Method

Estimation of maximum temperature and differences in temperature can be computed by the Schmidt Method. A numerical solution was prepared based on the Fourier Law governing heat transfer.

$$\frac{d}{dx} \times \left(k\frac{dT}{dx}\right) + \frac{d}{dy}\left(k \times \frac{dT}{dy}\right) + Q_h = \rho \times C_p \times \frac{dT}{dt} \tag{11.4}$$

where
Q_h = Heat generation in W/m^3
ρ = Density in Kg/m^3
C_p = Specific heat in J/K/°C
T = Temperature in °C

Temperatures are calculated for time step

$$\text{Time step } \Delta = \frac{(\Delta x)2}{2 \propto} \qquad (11.5)$$

where $\propto$ is thermal conductivity in W/m/°C.

11.5.2.2 Concrete Work Software Package

Concrete works software packages measure continuous concrete temperature. In this method air content, slump, designed compressive strength, coefficient of thermal expansion of concrete, and its properties are required to be measured, which is time consuming and expensive.

But, there are some easy and inexpensive methods available for the prediction of concrete temperature during curing; these methods are Multivariate Regression (SPSS software) and Artificial Neutral Network (MATLAB®).

11.5.3 Experimental Procedure

In this procedure a cylindrical specimen shall be prepared and shall be manually compacted, with Thermister being located within the specimen along with various types of vibrating wire strain gauges that will be used to continuously measure temperature during curing. The measurement of temperature starts immediately after the concreting, and shall be recorded for 30 hours after concreting until the stoppage of temperature changes.

11.5.4 Practical Measure

The temperature of concrete during placement can be measured by specially made thermometers with copper tubing protection so that the thermometer does not get broken while being inserted into the concrete. The temperature of the concrete immediately after pouring can be measured, and placement temperature should not be more than 18°C. If the temperature at the time of placement becomes more than 18°C, then the lead of concrete transportation shall be reduced to keep concrete temperatures within limit.

Thermo couples are required to monitor concrete temperatures during curing. Thermo couples are tied in various locations of the concrete. It will give ideas about the hydration process being developed within the concrete, and also provide information about temperature gradients between the core and the outer part of the concrete and record the information in the recorder.

11.6 Methodology of Controlling Temperature in Concrete

Temperature of concrete rises due to internal and external influences. Controlling the temperature is essential because heat is generated due to the heat of the hydration process, and the temperature of concrete rises due to high ambient temperatures, which helps to raise the temperature of concrete ingredients such as aggregate, sand, cement, and water. Maximum temperature of concrete, maximum temperature of placement concrete, maximum temperature gradient, and maximum ambient temperature are prescribed to control the temperature of concrete.

The following controls shall be practiced to manage the temperature of concrete so as to avoid thermal cracking and loss of concrete strength:

   i. *Cement*: Cement to be used in Mass Concrete shall have low C_3A and C_3S content, which will reduce heat generation. Fineness of cement shall be low. Contents of cement shall be the minimum possible to achieve the designed strength. Low heat cement is one of the solutions to control temperature of concrete. Supplementary cementitious material such as fly ash, blast furnace slag, and pozzolan can be used to reduce the generation of heat. Approximately 30% of SCM can be used in ordinary cement to control the temperature. Ground Granulated Blast Furnace Slag Cement (GGBFS) can replace approximately 65–75% of normal cement to reduce generation of heat.

  ii. *Aggregates*: Larger aggregates shall be used in the mix, with maximum 150 mm size aggregates used; higher aggregate content shall also reduce the content of cement per cubic meter of the concrete. Almost 80% of aggregate content can be used to reduce cement content. Lower coefficient of thermal expansion of aggregate shall reduce the thermal stress. Lower coefficient of thermal expansion of aggregates induces higher thermal conductivity, resulting in the faster release of heat from the core of concrete and a decrease in the temperature gradient.

 iii. *Lower ambient temperature:* Concreting shall be taken up within the limit of allowable ambient temperature. Lower ambient temperature during concreting will reduce temperature rise.

  iv. *Placement temperature of concrete:* The temperature of concrete at placement shall not be more than 18°C. Lower temperature of concrete at placement will reduce development of thermal stress within the concrete. It is observed that with the reduction of concrete temperature at placement by 0.6°C decreases the max temperature by 0.6°C, making it clear that placement temperature of concrete shall be kept as low as possible to control the maximum temperature of concrete.

v. *Precautions to be adopted at the time of concreting*
- Aggregates shall be kept under shade to control concrete temperatures.
- Cold water shall be sprinkled over the stacked aggregates in order to reduce placement temperatures.
- Cooling of aggregates will have a significant effect because 80% of concrete is covered with aggregate.
- Water used for concrete shall be mixed with flaked ice to reduce the temperature of concrete in the summer.
- Sun shades shall be arranged to protect fresh concrete from external heat.
- Wind breaks shall be arranged to eliminate excessive evaporation.
- Liquid nitrogen can also be added to reduce temperature.

vi. *Post cooling arrangement:* Cold water is passed through pipes being embedded in the concrete to reduce the internal temperature of concrete. The arrangement of this post cooling system will have a higher initial cost, but is very effective if the size of pipes, spacing, and other parameters are designed properly.

vii. *Insulation of concrete surface after concrete:* External surface of the concrete shall be insulated to resist heat transfer. Insulated shuttering or insulation shall be used to stop the transfer of heat from the surface of concrete, which will reduce the temperature difference and minimize thermal cracking in the concrete.

Bibliography

Evaluation of Temperature Prediction Methods for Mass Concrete Members Article in ACI Materials Journal, September 2006. American Concrete Institute.
Hot weather concrete (ACI 305R). American Concrete Institute.

12

Joints in Concrete Structures

12.1 Introduction

Joints in concrete structures are inevitable. Joints occur in various types of structures due to many reasons. Some joints are formed due to construction methods, while some types are formed due to the change of volume and temperature. Some joints are formed due to delays in pouring activities. All of these joints pose challenges to the integrity of the structure, and the concrete becomes vulnerable because these are weak zones of the concrete structure. Large-mass concrete structures or concrete of considerable quantity that cannot be poured in a single operation will form a joint before the next pour is conducted.

12.2 Type of Joints

There are four type main of joints as indicated below.

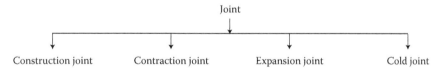

12.2.1 Construction Joint

In civil engineering, various type of structures are constructed with different configurations to suit and match utility as well as architectural aspects and technical requirements. Some multi-framed and multi-storied high rise structures are designed, or some structures with large masses are designed such as concrete dams. It becomes difficult and sometimes seems to be impractical to pour concrete in a single operation. The quantity that can be poured and cast depends on many aspects like capacity of the batching plant, manpower, time available, and if the concrete in the structure is cast in multiple operations. The discontinuity in pouring sequences and concreting is held up for a period

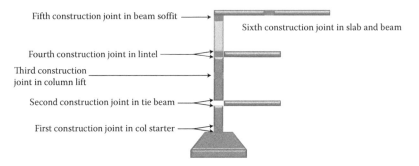

FIGURE 12.1
Various construction joints.

more than the initial setting time of cement, causing the formation of joints in the concrete structure that are known as construction joints.

Construction joints formed in concrete structures are treated as weak zones of the structure and should be tackled carefully by the site engineer in consultation with the design engineer. Many construction joints are created during the construction of high rise buildings due to the unavoidable circumstances indicated in Figure 12.1. The joints are both vertical and horizontal, but joints should be vertical to the main reinforcement as described in Figure 12.1:

- The first set of construction joints are provided on the top of footing with a key where the starter shall be placed to facilitate proper placement and alignment of columns.
- The second set of construction joints are provided for construction of tie beams in the plinth level, which also act as a plinth band to support the structure against earthquakes.
- The third set of construction joints are to provide for column lift from 1.5–2 m.
- The fourth set of construction joints are provided for continuous lintel and also act as a lintel band.
- The fifth set of construction joints are provided 50 mm below the beam soffit.
- The sixth set of construction joints are provided in the beams and slabs.

12.2.2 Type of Construction Joint

Construction joint occur due to the interruption of concrete pouring, and its type depends on the location of the joint in the structure, with care being taken accordingly.

- Construction joints provided from the top of footing to the beams soffit are horizontal and vertical to the main reinforcement.

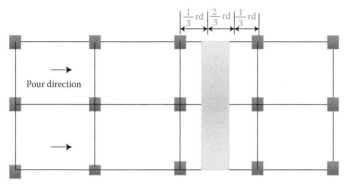

Construction joint shall be provided anywhere in the shaded area

FIGURE 12.2
Location of construction joint.

- Construction joints provided in beams and slabs are vertical and are vertical to reinforcements.

12.2.3 Location of Construction Joint in Beam and Slab

The location of construction joints in the slab shall be based on the concentration of shear force and bending movement. Construction joints are very weak in shear force, so the joint shall be placed where shear force is minimized and bending movement is maximized. The joint is preferred to be given in the middle of the span or within the middle third of the span. This joint is vertical and vertical to the main reinforcement. No horizontal joint is permitted in the beam because the beam is designed as the T beam. Refer to Figure 12.2 for locations of construction joints.

12.2.4 Sequence of Pouring

A pouring sequence shall be prepared as a relevant document for good construction joints. A good construction joint provides water tightness of the structure, while weak joints become a perpetual problem to the structure. Concrete shall be filled in the beam up to the slab soffit level, and the slab shall be cast accordingly. The surface of the construction joint shall be vertical and the edge of the slab in the joint shall be thoroughly compacted. Concrete mortar that sticks to reinforcement shall be cleaned.

12.2.5 Preparation and Construction of Joint

Site engineers should ensure proper bonds between old and new concrete, and should take up the following activities:

- The concrete face of the joint shall be cleaned by wire brush.
- All loose laitance shall be removed.

- Care should be taken that no aggregates are loosened during cleaning.
- The surface shall finally be cleaned by water jet which will remove all dirt and dust.
- Apply structural adhesive based epoxy resin, which is suitable for construction joints between old and fresh concrete. It provides good bonds and makes the joints monolithic.
- Fresh concrete shall be poured and compacted thoroughly to ensure proper bond between the old and new concrete.

12.2.6 Construction Joint on Floor Slab on Grade

Concrete floor slabs used in roads, airports, and industrial and commercial flooring are subjected to various type of forces due to the movement of heavy traffic that allows slab movement from one to another. This construction joint should allow horizontal transition, and should not allow vertical transition and rotation due to traffic movement.

12.2.7 Construction Method

The construction joint is a plane of weakness whether it is a Mass Concrete or Reinforced Concrete structure. Joints shall be treated properly to transmit tension, compression and share stress in the structure. Otherwise, this plane will remain a plane of weakness. A methodology of construction joint weakness treatment is described below.

- Material required
  - MS rebar 40 mm in diameter and 600 mm long to be used as dowel bars.
  - Steel/PVC sleeves of higher diameter than the rebar.
  - Filling material.

Dowel bars shall be placed at 300 mmc/c along the construction joint; 300 mm shall be inserted in the floor slab as shown in Figure 12.3, and the concrete shall be cast. The sleeve will be put in the next pour and aligned with the dowel bar in such a way that the 300 mm protruded dowel bar can be placed inside of

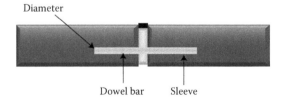

Diameter

Dowel bar   Sleeve

FIGURE 12.3
Construction joint with dowel bar.

the sleeve. Any movement of the slab due to contraction or expansion can be accommodated because the dowel bar will be able to move freely within the sleeve. Alignment should be correct, otherwise the formation of cracks is possible.

12.3 Construction Joint in Dam Construction

A dam is a massive structure of concrete, and the heat that is generated by exothermic reactions that lead to cracks is an acute technical problem. In order to avoid and control the phenomena, dam concrete placement is done block wise where the size of block is decided to suit to thermal condition. The depth of each pour is also decided to suit thermal conditions that generally vary from 1.5–2.5 m. A construction joint is created at the termination of each concrete pour, which is nearly horizontal and has a slight upward slope towards the downstream of the dam. The construction joint in the dam provides water passage through the dam, unless it is sealed and proper PVC water stops are used to stop seepage through the joints.

12.3.1 Preparation of Joints

When concrete pouring is completed the entire concrete area shall be uniformly finished. The surface of the concrete shall be cleansed by compressed air, which is known as a green cut, without loosening the aggregates of the concrete. It will remove all slurry with smaller particles from the top of the concrete surface. This phenomena will help make the proper bond between old concrete with fresh concrete being placed in the next pour. As necessary, epoxy latex adhesive may also be used for better bonding between old and new concrete.

12.4 Construction Joint in Tunnel Construction

Construction of a tunnel is a very difficult and critical construction methodology. Concreting in tunnels is done in three stages such as invert, crown and wall, which form horizontal and vertical joint in the section.

12.4.1 Longitudinal Joint

Tunnels in water conducting systems have various geometric designs such as circular, D-section, and horseshoe to support external loads from rock and water pressure. Dam sections are divided into invert, wall, and crown sections. In the process of construction, construction joints are formed in

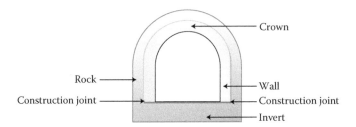

FIGURE 12.4
Typical tunnel section with longitudinal joint.

longitudinal and transverse directions. Concreting in the invert is the first operation, followed by wall and crown concreting that will generate longitudinal joints on both sides of tunnels and the length of the joint is generally maintained at up to 15 m. Water stops shall be used in longitudinal joints to tackle seepage problems whereever necessary (Figure 12.4).

12.4.2 Transverse Joints

In the construction of a tunnel section, transverse construction is provided at 15 meter per cubic centimeter to facilitate concrete availability. Rock is not homogeneous everywhere, with rock in some portions of tunnel alignment posing seepage problems due to geological conditions. In such conditions concreting can be cast without transverse joints to counteract seepage problems. Transverse construction joints are provided by approved quality water stops (Figure 12.4).

12.5 Construction Joint in Bridge

Like other structures, construction joints in bridge construction are also created by the disruption of concrete placement. In bridges, construction joint are either longitudinal or transverse. Construction joints are provided in decks and parapets of the bridge to facilitate construction activities. Locations of transverse construction joints shall be in the middle third of the span at the minimum shear force and maximum positive bending moment region, and the joint shall be filled up with approved sealant.

12.6 Contraction Joints in Concrete Structures

Contraction joints are deliberately provided in various concrete structures where the formation of cracks are expected due to volumetric changes of

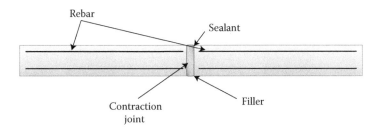

FIGURE 12.5
Details of contraction joint.

concrete under drying and cooling conditions of Mass Concrete. It is used to arrest the formation of cracks by releasing stresses. Location, spacing, and depth of contraction joints depend on the type of the structure under construction. Complete contraction joints are those that are present in the entire depth of the slab, and partial contraction joints are those that only interact with concrete.

12.6.1 Contraction Joint in Building

Tensile stresses are generated due to shrinkage and cooling effects in the walls of buildings, which is restrained by the foundation and resulting in the crack occurring on the wall. Contraction joints in the shape of grooves shall be provided at 9–10 m per cubic centimeter in the wall vertically to make the area weak so that cracks may occur and stresses can be released. There are two type of contraction joints: partial and complete. In construction, complete contraction joints are placed in the wall by providing grooves throughout the thickness of the wall through the discontinuity of reinforcement in contraction joint location (Figure 12.5).

12.6.2 Contraction Joint in Slab on Grade

Concreting of slabs on a grade is done where the conditions of the bottom and top surfaces of slab are quite different. This difference in exposure of both sides of the slab does not allow the entire slab to dry.

The top surface will be exposed to temperatures and wind that will help the top surface to dry faster than the bottom surface of the slab, resulting in cracks. These cracks can be controlled and minimized by proving contraction joints at specific intervals in the slab. The slab shall be cast in panels of less than 3 square meters. The concrete shall be done in alternative panels to reduce the formations of crack as per Figure 12.6.

12.6.3 Contraction Joints in Dams

The generation of heat in concrete gravity dams is an acute technical problem. This heat is generated by exothermic reaction. The temperature in the core of

FIGURE 12.6
Sequence of panel concreting in slab on grade.

Mass Concrete is higher than that of outer surface of concrete due to atmospheric conditions that create a temperature gradient and volumetric change, leading to the development of cracks in restrained conditions. Contraction joints are deliberately provided to control these cracks and relieve the stresses. The entire length of the dam is divided into several blocks, and the width of each block is designed to suit thermal conditions; these joints are vertical and parallel to the axis of the dam, and are known as transverse contraction joints. These joints do not provide any bond between the blocks. Generally, bituminous emulsion paint is provided on the surface on the old block before placing a new block. Any joint in the dam provides passage for water from the upstream side to the downstream side. PVC strips shall be provided in upstream side between the block to stop any possible seepage as shown in Figure 12.7.

Longitudinal contraction joints are provided along the length of the dam by dividing blocks and spacing, and longitudinal contraction joints are

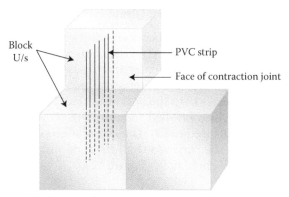

Block joint with arrangement PVC water bar

FIGURE 12.7
Contraction joints in dam.

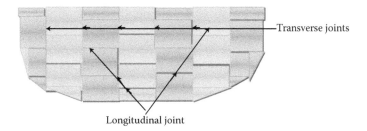

FIGURE 12.8
Blocks indicating transverse and longitudinal contraction joints.

decided by the designer to suit the structure's thermal conditions and are subsequently grouted to make it monolithic (Figure 12.8).

The thermal behavior of Mass Concrete is a technical problem of concrete gravity dams, and produces plenty of undesirable cracks that affect the structural integrity of the dam. However, it is not always advisable to divide the dam into several blocks to control thermal complexities by providing transverse and longitudinal contraction joints. Dams are a retaining structure, and construction and contraction can create free passage for the passage of water and make the dam vulnerable and susceptible. The water seepage within the body of the dam exerts additional upward pressure that alters the stability of dam and may make the dam unstable; close monitoring of dam stability shall be conducted to ensure the safety of the dam. Emphasis should be placed on the activity of pre cooling all ingredients in order to produce concretes of low temperature. Additionally, post cooling methodologies shall be improved instead of providing many joints in the structure (Figure 12.9).

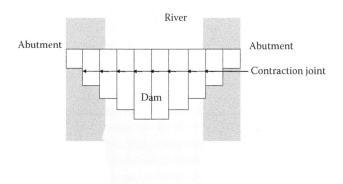

FIGURE 12.9
Transverse joint in dam.

12.6.4 Contraction Joint in Water Retaining Structure

Liquid retaining structures also experience the same phenomena of volumetric change and are subjected to the formation of cracks. The formation of these cracks weakens the structure and also creates a public health hazard, as water stored for human consumption becomes contaminated due to the seepage through these cracks. Care shall be taken to control the formation of cracks by providing contraction joints at suitable locations in regular intervals in the structure with a purpose of accommodating the contraction of the concrete.

Contraction joints are nothing but a zone deliberately weakened by reducing the thickness of the wall and permitting cracks to occur in this zone. The thickness of the wall is reduced by providing V groove(s) of designated depths at regular intervals as specified by the designer and do not exceed 10 meters. Contraction joints may be full or partial depending on the requirements.

Water stops, elastomeric sealants, and other materials shall be used to prevent water seepage.

Contraction joints with both concrete and interrupted steel and contraction joints with interrupted concrete are indicated in Figure 12.10.

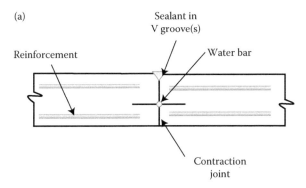

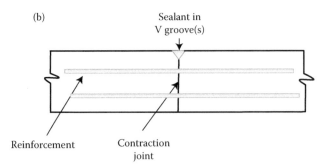

FIGURE 12.10
(a) Contraction joint with both concrete and steel interrupted. (b) Contraction joint with only concrete interrupting.

12.7 Expansion Joints

Expansion joints are a movement joint that are provided in the structure with a clear gap of approved size and are between both concrete and reinforcements. It isolates the structure with complete discontinuity of concrete and steel to accommodate for the expansion or contraction of concrete, while isolation joints isolate movement between the structures with no reinforcement existing between the structure.

12.7.1 Expansion Joint in Building

Construction of multi-storied structures contain huge length and width ratios and pose problems due to the change of temperature because the structure is restrained. The changes in volume of the structure occur because of temperature variations, and it induces stresses between the points where the structure is restrained. The expansion joint is provided to relieve these stresses by accommodating the expansion or contraction of the separate unit of the building without causing any damage to the building. Expansion joints are vertical as well as horizontal in the building, and spacing is decided based on structural needs. The length of the building without providing expansion joints can be found with regards to temperature, but in practice it is provided at least 45 m center-to-center.

Joint width (50 mm) is generally provided, and larger widths may be used to take care of any horizontal displacement of the building that occurred due to seismic effects.

During construction twin columns are built on a single footing to maintain the designed gap that accommodates for expansion as indicated in Figure 12.11. Fibrous, compressible, and flexible boards are used in the joint, and joint sealant shall be used on it to make the joint water tight.

12.7.2 Expansion Joints in Bridges

Bridges are subjected to expansion and contraction with the change of seasons. The changes in configuration generates stresses that deform the bridge structures because it is restrained. Gaps that are provided in suitable locations to accommodate for these expansions or contractions are known as expansion joints. These expansion joints provide bridge safety measures by taking care of stresses being produced by thermal causes, external loads, and other factors. Expansion joints are a movement joint, and movement of the bridge depends on the size of the bridge and type of bearing used in the bridge. Figure 12.12 indicates expansion joints for small movement.

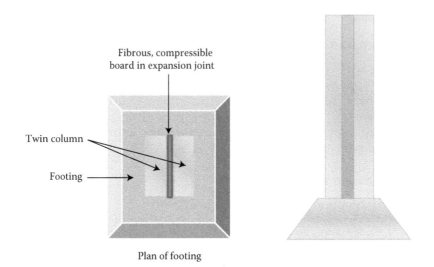

Fibrous, compressible
board in expansion joint

Twin column

Footing

Plan of footing

FIGURE 12.11
Expansion joint.

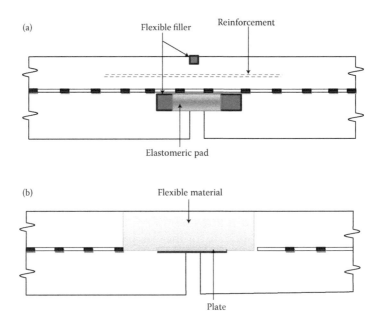

(a)

Flexible filler   Reinforcement

Elastomeric pad

(b)   Flexible material

Plate

FIGURE 12.12
(a) Buried joint. (b) Asphaltic plug joint.

12.8 Cold Joints in Concrete

Joints in concrete like expansion. Contraction and expansion is inevitable, but these joints are deliberately made in suitable and predetermined locations in the concrete to accommodate and relieve stresses that are caused by thermal effects, environmental effects, or any other external force in order to avoid possible cracks. The correctly designed, properly located, and technically constructed joints do not affect the structures and protect the concrete from experiencing cracks.

A joint in concrete which is not deliberate but is formed in a batch of concrete due to the interruption or delay in compaction or delivery of the next load of concrete is known as a cold joint.

Cold joints in concrete are considered a poor bond that affects the strength of concrete, and precautionary measures should be taken against the formation of cold joints in the concrete.

In big pours such as Mass Concrete in concrete gravity dams that consist of multiple layers placed in a single pour operation, a concrete pour scheme shall be formulated with sequential placement of each and every layer.

The sequence of pouring shall be arranged in such a way that each layer of concrete is placed and compacted while the previously compacted layer is in its plastic form and is to be compacted by inserting the vibrator needle into the old concrete surface to have the requisite bond between the new and original concrete.

12.8.1 Causes of Cold Joints

Cold joints are a critical phenomena in concrete. Causes of formation of cold joints are mentioned below.

- Absence of proper pour schemes of concrete with the sequence of concrete placement.
- Delays in preparation of concrete mix.
- Delays in supply of concrete due to improper concrete transportation arrangements.
- Inadequate arrangements of spreading concrete in position.
- Delays in compaction of concrete.
- Lack of effective supervision.

12.8.2 Treatment of Cold Joints

Cold joints in concrete are almost inevitable, but they should be avoided as far as possible. The following methodology is adopted for the treatment of cold joints.

- Due to some unavoidable circumstances, concrete that has been laid has hardened beyond the plastic stage before the next load of concrete is placed in position. The top surface of the concrete shall be raked properly to remove the loose laitance, and shall be cleaned thoroughly without affecting the homogeneity of the already compacted concrete. A sand cement mortar of the same proportion of concrete shall be spread on the surface of the concrete, and fresh concrete shall be placed on it and compacted. Care should be taken during compaction so that the needle of the vibrator does not penetrate the old concrete.

- When concrete has set and hardened, the surface of the concrete shall be scarred and all loose laitance shall be removed. The surface of the concrete shall be cleaned by water jet in order to remove all dirt and dust. An approved adhesive shall be placed over the concrete surface. A fresh load of concrete shall be placed on the old concrete, and the concrete shall be compacted thoroughly to have the necessary bond between the layers.

- If concrete with cold joints is very close to the upstream of the dam where the possibility of percolation is obvious, the following precautionary actions should be taken:
  - Conduct percolation tests to assess the porosity of the concrete.
  - Drill some grout holes of required depth in the affected areas.
  - Take up low pressure grouting to seal all pores.
  - After grouting is over, drill some test holes.
  - Conduct additional percolation tests.

12.8.3 Effect of Cold Joints in Concrete

Cold joints have several effects on concrete. Cold joints reduce shear resistance and increase water permeation. Any grade of concrete, whether it is RCC or Mass Concrete in the dam, will have a huge effect on water permeability in cold joint concrete. The pores in the concrete are the main route for the ingress of harmful ions (chlorine and carbon dioxide) and water permeation. This will cause corrosion in the RCC of different types of structures, including concrete gravity dams where reinforcement is used in external concrete, buckets, piers, sluices, spillways, training walls, and around the gallery. Corrosion staining is sometimes visible from the outer face of concrete. This corrosion phenomena reduces the area of steel in the concrete and also reduces the bond strength, resulting in the collapse of the structure.

In Mass Concrete cold joints affect shear strength more severely and reduce the durability of the concrete. Researchers found that the ingress of harmful ions through cold joint areas are faster than at areas of solid concrete; these findings initiated additional studies on carbonation, chloride attacks, and permeability in cold joints. Permeability affects the Mass Concrete when water enters the body of the dam through pores created by cold joints. It creates additional uplift within the body of dam that destroys the stability analysis done during design, making it mandatory to recheck the design of the dam for stability and safety of the structure.

12.9 Sealant Material

Joints are designed to accommodate movement of the structure due to various reasons. These joints are to be protected from adverse environmental exposures and external loading conditions.

Sealant is a material that is applied to fill and protect the joint that exists between the two surfaces. It is used to seal the gap so that water cannot permeate and corrode the concrete and embedded steel. It also increases the bond between the surfaces.

12.9.1 Type of Sealant

There are different types of sealants used in the construction industry, some of which are explained below.

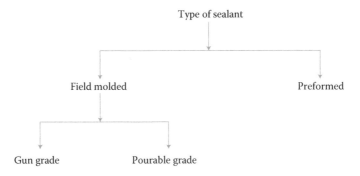

Field molded sealants are in liquid or semi-liquid form and take the shape of the joint where it is applied. It requires both tension and compression, and changes their shape without changing the volume. Field molded sealant is available in two parts: gun grade or pourable grade sealants.

Preformed sealants are in relatively solid form, and require compression to change the shape with the change of width.

Sealants that are used in construction work:

- Silicone
- Polyurethane
- Sulphides
- Acrylic
- Synthetic rubber
- Thermoplastic
- Polysulphide
- Mastic

12.9.2 Selection of Sealant Material

Selection of the right type of sealant depends on the type of joints, their location, and expected movement of the joints. Location of movement will suggest its environmental exposure such as chemical, bacterial attack, and type of external force that will be encountered. Additionally, care should be given on application procedure of sealant and the following factors:

- It should be easily molded in the field and easily applicable
- It should be free from creep, slump, and bleeding
- It should not shrink after application
- It should have good cohesion
- It should have the ability to accommodate movement of any type
- It should not react with concrete
- It should be resistant to chemical aggression, abrasion, etc.
- It should be repairable

12.9.3 Various Type of Sealants

Various types of sealant can be used in specific locations to strengthen the structures.

- *PVC water stop*: PVC water stop is a preformed type of sealant that is extruded from flexible PVC material and ribbed with the center bulb. It has two identical halves extended on both sides of the bulb, with the center bulb accommodating movement. This profile can be used in movement and non-movement joints. It is a long lasting, customized, polymer based material that has high tensile strength and high elasticity. It is resistant to weathering, chemicals, and to high hydraulic pressure. During construction half of it is embedded in the concrete and the other half is extended and will be buried in concrete in the next pour. Splicing can be done by heating (Figure 12.13).
- *Hydrophilic water stops*: Hydrophilic water stops are also a preformed type of sealant that is made of modified chloroprene rubber. It expands three times of its original volume when it comes in contact with water, and this expansion is three dimensional. Swelling pressure is developed on the side that comes in contact with water. It exerts pressure on the surrounding surfaces and seals the joint tightly. It can be placed conveniently in position and does not get displaced when concrete is cast. The profile of this water stop is treated with special delay coating to prevent this water stop from reacting with fresh concrete and expanding before the concrete is placed. A groove shall be made on the concrete surface, and it will be placed in the groove already formed and the next layer of fresh

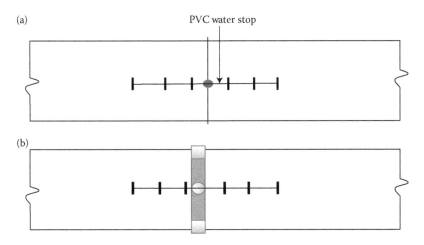

FIGURE 12.13
(a) Construction joint. (b) Expansion joint.

concrete shall be laid. It will be compressed by the concrete, and will rebound when the concrete shrinks during curing.

The profile of water stop can be spliced by cyanoacrylate adhesive, which will be applied on the two cut ends and shall be held together for 40–50 seconds to allow the adhesive to set. Some profiles are available with pick and stick adhesive backing. During installation the paper shall be removed and the profile shall be firmly pressed on the surface with fresh concrete placed on the profile.

This hydrophilic water stop can be used in the construction joints of concrete dams, tunnels, reservoirs, swimming pools, and other applications (Figure 12.14).

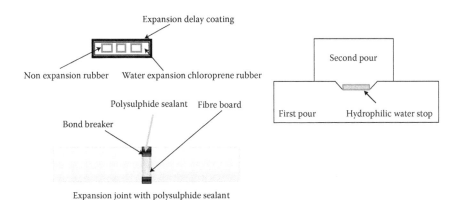

FIGURE 12.14
Hydrophilic water stop in construction joint.

- *Polysulphide sealant*: Polysulphide sealant is a polysulphide based elastic sealant which is a field molded, gun grade sealant. It can be used in horizontal and vertical joints that have a joint width configuration that varies from 8–40 mm and a width-to-depth ratio of 2:1 for butt joints and 1:1 for lap joints. It can also be used in overhead joints because it is non-sag.

    Polysulphide sealant is resistant to UV and to chemicals such as alkali, acid, vegetable oil, and petroleum products.

    Application of this gun grade sealant is easy. The surface shall be cleaned and should be free of dust and dirt. The edge of the joint, if broken, shall be repaired. The joint shall be filled by fiber board, and the filled board shall be raked to the correct depth. The sealant should adhere to both surfaces of the joint, but not to the base. Bond breaker shall be used, and the surface shall be protected by masking tape. The joint should be dried where sealant shall be used, and dryness of the surface shall be thoroughly checked. Wet surfaces will generate vapor if hot application is used, and adhesion will be reduced if cold applications of the sealant are used.

    A compatible primer shall be laid on both surfaces of the joint, and a gun shall be used to apply the sealant. The gun should be used at a 45° angle, should be pressed against the surface to avoid trapping air bubbles, and sealant can be applied in several passes. A wooden tool shall be used to compact the sealant and to remove any trapped air bubbles, and will create a slight concave surface (Figure 12.14).

12.10 Repair of Cracks and Grouting Procedures

Cracks occur in concrete due to various reasons. It may be due to thermal effects, poor quality of materials used, or may be bad quality of work done on site. Cracks are vulnerable or weak zones of concrete structures, and are signs of early distress that may lead to disaster if neglected. Any crack that develops on the surface of concrete shall be examined thoroughly and diagnosed properly whether it is structural or non-structural. Focus shall be given to structural cracks, and remedial measures are to be selected depending on location and type of cracks to strengthen the structure. Grouting is the one of the most popular and successful processes adopted all over the world to strengthen concrete.

12.10.1 Causes of Cracks

Concrete has various stages in its lifecycle which starts from mixing, transporting, placement, compaction, and curing. These stages force concrete to move from the plastic to solid stage. During this journey, it is exposed to different environmental conditions, experiences critical internal changes due to chemical reaction, suffers from differential changes of volume between core and outer surface temperatures, and in many occasions is affected by external forces. Under these circumstances, concrete inherits various type of cracks during hardening and also in posthardening stages of its life cycle.

Cracks are classified as dormant or active. Active cracks change their shape, size, and direction with the passage of time, while dormant cracks remain unchanged.

12.10.2 Cracks in Concrete Gravity Dams

A concrete gravity dam is made of a huge mass concrete that is subjected to various type of cracks due to many reasons like thermal phenomena, wind effects, etc. Various types of cracks are narrated below.

12.10.2.1 Plastic Shrinkage Cracks

Construction of a concrete gravity dam is generally taken up in remote areas that have wide open areas that provide open space for wind to blow fiercely in all seasons. It makes an unfavorable combination of wind, temperature, and humidity, which proves to be a deadly combination for Mass Concrete. Dam concreting is done in blocks with specified lifts that cover fairly large surfaces. The surface of fresh concrete will be exposed to wind, temperature, and humidity. Moisture from the top surface of concrete will evaporate faster and dry more rapidly in relation to concrete below the surface. The surface of concrete shrinks inducing tensile stress in the concrete because it is restrained at the bottom and resulting in cracks of varying depth from shallow to full depth and in many forms and patterns with longer lengths.

This type of plastic shrinkage crack is a nightmare to dam construction engineers, who should study and diagnose the reasons that this type of crack occurs; these cracks can be due to differential volume changes in fresh concrete in plastic conditions that can be controlled by the reduction of volumetric change between concrete layers. The reduction of volumetric change is possible if rapid evaporation is minimized from the top surface of concrete.

As a precautionary measure, wind shields with sprinklers shall be provided in order to stop the blowing of hot and dry wind over the surface of fresh concrete. Hessian cloth shall immediately be spread over the finished concrete to minimize evaporation from the top surface of plastic concrete because plastic shrinkage occurs before the concrete sets.

12.10.2.2 Drying Shrinkage Cracks

Another type of crack occurred in concrete is drying shrinkage cracks. Volumetric changes in concrete due to loss of moisture from cement paste constituents is a characteristic of concrete. The contraction of hardened concrete due to loss of capillary water is a reason for drying shrinkage. This change in the volume of concrete takes place under restrained conditions that will generate tensile stress in the concrete. This restraint is provided by the established concrete below the surface of the newly poured concrete. A crack of varying depths will occur when tensile stress is developed due to the restraint exceeding the tensile strength of concrete. If concrete is allowed to have volumetric change without restraint, no crack will develop in the concrete.

Massive structures like concrete gravity dams have wider surfaces where tensile stress is caused due to differential volume changes between outer and inner surfaces of concrete, resulting in cracks that may go deeper over time if they are not treated and will provide passage to water within the body of the dam that may create safety problems for the dam's structure.

The degree of drying shrinkage depends on the quantity of aggregate and water content used in the concrete. Shrinkage reduces with increased use of aggregate and decreased use of water content. Drying shrinkage can be intercepted by providing contraction joints in the concrete, and also by helping construction joints that can be designed in such a way that joints are provided as a weak zone where developing stresses can be neutralized.

12.10.2.3 Crazing

Crazing is a type of surface crack that has closely spaced cracks with a maximum depth of 5 mm. It is a type of drying shrinkage crack that forms due to water content on the surface being higher than the inside portion of concrete. It is frequently found on the surface of block concrete in dams. It is observed that overvibration of concrete brings more slurry to the top of concrete surface and subsequently leads to crazing, requiring overvibration of concrete to be controlled.

12.10.2.4 Thermal Cracks

Thermal phenomena of Mass Concrete in dams are a significant factor in the concrete technology and a matter of great concern. Temperature rises in Mass Concrete due exothermic reactions between water and cement, and heat is not dissipated quickly. Core concrete temperatures rise, and heat is generated where the dissipation of energy is less and the outer surface of concrete that is exposed to environment cools down and dissipates more heat than the concrete core. This condition creates a temperature gradient that initiates differential volume changes, and restraint is being provided by the adjacent cooler concrete while developing tensile stress in the concrete. Thermal cracks occur in the external concrete surface when developed tensile strength exceeds tensile stress of concrete. This tensile stress is proportional to the temperature gradient, coefficient of thermal expansion, modulus of elasticity, and restraint.

Cracks that occur due to thermal causes can be reduced by controlling the temperature of core concrete. This control will reduce the temperature gradient, and subsequently reduces differential volume changes in order to minimize cracks.

12.10.3 Remedial Measures

Temperature controlled can be exercised by selecting cement friendly materials and additives with smaller cement contents, both of which can lessen heat generation. Low heat cement and additives like pozzolan or fly ash can be used, while larger size aggregate up to 150 mm shall be used.

Pre-cooling the ingredients of concrete will produce low temperature concrete at placement. Post-cooling will limit the rise of temperature of already laid concrete. Care must be taken to reduce concrete transportation times and placement procedures to ensure changes in concrete temperature at the time of placement is at a minimum and should not be more than 18°C.

Cracks can be controlled by providing well designed contraction joints to release the energy.

12.10.4 Crack due to Reactive Ingredients

Aggregates are chemically inert, but some aggregates react with alkali and other deleterious materials and deteriorate concrete.

12.10.4.1 Alkali-Silica Reaction

Silica, chlorides, sulphates, alkali, and carbonate are in the list of deleterious materials that harm and deteriorate concrete. Concrete ingredients shall be carefully selected to create durable concrete, tested for deleterious materials, and will be kept within the specified limit.

Silica present in aggregates may be attacked by alkali originating from cement and other sources such as curing water or ground water. The chemical reaction between silica and alkali is an expansive one that forms a swelling gel. It draws water from the adjacent part of concrete and causes concrete expansion that results in the formation of stresses, and these tensile stresses develop cracks in the concrete.

12.10.4.2 Precautions

Use non-reactive aggregate and cement that has a low alkali content. Use fly ash and Granulated Blast Furnace Slag cement as a replacement for OPC.

12.10.5 Cracks and Deterioration of Concrete due to Corrosion

Corrosion in concrete is an electro-chemical reaction between embedded reinforcement and the surrounding environment. In this process, physical and chemical reinforcement properties are lost due to the constant interaction of environment with rebar. Corrosion causes rust to form on the rebar, which has a greater volume than the original rebar but with no positive mechanical properties. This process also produces pits and holes on the surface of rebar, reducing overall strength and cross-sections of rebar.

Reinforcement in concrete is usually in non-corroding and passive conditions. Passive layers protecting the rebar are destroyed by chlorides that cause rebar to rust and initiate cracks that radiate the from rebar through the cover of concrete, causing easy passage for many other deleterious materials and ultimately leading to structural collapse.

12.10.6 Precautionary and Remedial Measures

Some precautionary and remedial measures are described below.

- Galvanization of rebar
- Cathodic protection of rebar
- Electro-chemical chloride migration
- Re-alkalization

12.10.7 Grouting of Concrete Structures and Repair

Grouting in concrete is an injection process to fill up the cracks, voids, and fissures with select grouting materials under specified controlled pressure to strengthen the structure.

Material that is injected to fill up the gaps, cracks, and voids in the concrete is known as grout. This grouting material will by plyable, plastic, and non-shrink so that voids can be completely filled. It should be totally solvent free and completely solid, with the capacity to expand when it contacts water.

12.10.8 Classification of Grouting

Concrete consisting of voids, fissures, and cracks can be grouted to become strengthen in the following methods.

- Grout voids within the concrete of the water retaining structure to increase water tightness up to the permissible limit.
- Grout and fill the cracks of concrete to strengthen the structure.
- Grout anchor bolts of static and dynamic machine foundations.

12.10.9 Repairs of Cracks by Epoxy Grouting

Epoxy is a polymer-based synthetic resin with balanced mechanical, chemical, and adhesive strength properties. Epoxy resin can be used in the grouting of cracks to repair concrete structures such as dams, bridges, and aqueducts.

Epoxy is a unique grouting material that can be applied under any condition to any structure. It is quick setting, experiences low shrinkage, enjoys high strength and high adhesion, has a low viscosity, and has good resistance to all chemicals. This versatile material can be effectively used on cracks developing in concrete dams due to either thermal effects, plastic shrinkage, or drying shrinkage. Epoxy resins can penetrate hair cracks easily due to its low viscosity, and it can be used as injection grouting to strengthen the structure. The material composition is epoxy resin and hardener. A filler material shall be used if cracks are wide because epoxy is an expansive material.

- *Equipment required*
  - Electric or pneumatic drilling equipment
  - Grouting equipment with necessary gauges and control valves
  - Air compressor
  - PVC or metal pipe
  - Generator
- *Subject*: Longitudinal crack on the surface of concrete on blocks of concrete gravity dams.
- *Procedures*: Procedures includes surface preparation, making V groove(s) in the crack, filling cracks with epoxy material, the drilling of holes, application of epoxy grouting, and finishing.
  - Measure the length, width, and depth of crack.
  - The area of grouting shall be cleaned and should be free of dust and dirt, oil, grease or any loose material.
  - Cracks shall be cut in a V groove(s) by any suitable method and up to 10 mm deep.
  - Cracks shall be cleaned thoroughly by compressed air.
  - Grooves shall be filled with epoxy and shall be allowed to set for one day.
  - Mark the spacing of holes as required by the specific cracks.
  - Drill 10 mm diameter holes along the length of crack.
  - Fix all injection ports in the holes and fill all holes except one adjacent to first grouting hole.
  - Mix epoxy resin with hardener to prepare the grout. Material and quantity shall be fixed with regards to overall life of epoxy to avoid waste.
  - Injection of grout is done through a nozzle connected to the compressor or by a suitable gun, and minimum pressure of 3 kg/cm^2 shall be maintained. Grouting shall continue until epoxy comes out

from the adjacent hole and shall be plugged. Grouting procedures shall continue with a desired increased pressure for 2–3 minutes. The injection nozzle is then withdrawn and the hole is filled with epoxy mortar.

- The same procedures will be completed until all holes are filled.

12.10.10 Repairs of Defective Concrete by Epoxy Concrete

High strength epoxy concrete is a solvent free, epoxy-based repair concrete that has a compressive strength of 90 N/mm^2 at 7 days, a flexural strength of 20 N/mm^2, and a bond strength of 3 N/mm^2.

This concrete is supplied in three component systems, and coarse aggregates of 10 mm and smaller shall be added on site as prescribed by the manufacturer. Epoxy concrete is made up of the base, hardener, fine filler, and coarse aggregate, all of which shall be mixed by weight as required.

Base:Hardener:Fine Filler:Coarse Aggregates = 4.54:2.2:20.8:30

- *Place of application*: Epoxy concrete is suitable for the repair of defective concrete surfaces of dams, bridges, foundations, runways, and other structures.
- *Mixing procedure*: The base and the hardener shall be stirred thoroughly, poured into a container, and mixed properly for 3 minutes. Fine filler and coarse aggregate shall gradually be added to it in a small mixer to create a homogeneous mix.
- *Surface preparation*: Identified defective concrete shall be removed, and the surface shall be cleaned and should be free of dirt, dust, grease, and any loose laitance. The surface shall be roughened by grit blasting.
- *Application*: The primer is supplied in two components as the base and the hardener. The primer shall be applied to the surface in two layers. Mixed epoxy concrete shall be laid on the surface while the primer is still sticky, and shall be leveled and finished with a steel trowel.

12.11 Case Study

12.11.1 Repair of Bauxite Secondary Crusher Foundation

The bauxite secondary crusher foundation was cast in a single pour including a raft and pedestal that stand 8.5 m high. It was cast in a single pour to avoid construction joints for dynamic loading conditions. The 8.5 m pedestal consisted of many pockets for providing anchor bolts. However, it was a critical but difficult foundation to be cast in a single pour.

After removal of form work, the vertical surface of the pedestal was found to be full of honeycombs. The project authority doubted the structural integrity of the foundation, and instructed the dismantling of the foundation.

The PMC and contractor inspected and studied the defective foundation in detail visually and through testing, and submitted a proposal for repairing the structure instead of dismantling the structure in the following steps:

- Conduct non-destructive tests by performing ultrasonic pulse velocity tests to ensure structural integrity of the core concrete of the pedestal.
- Remove up to 200 mm of defective concrete from the surface of the pedestal or until a good and firm concrete surface is encountered without damaging the reinforcement of pedestal.
- Reinforcement exposed shall be cleaned, straightened, and re-lapped if required.
- The newly exposed concrete surface shall be cleaned by grit blasting and all loose lattice shall be removed.
- Place a wire mesh on the reinforcement.
- Apply a two-component epoxy resin bonding agent on old concrete by spraying it on.
- Shotcreting onto an old surface after the epoxy is applied.
- Finish the surface, ensuring it is properly levelled and vertical.
- Conduct final ultrasonic pulse test for acceptance.
- *Observation*: Ultrasonic pulse velocity tests were conducted, and the pulse velocity obtained through testing was more than 4.5 km/sec. Quality grading of existing concrete was interpreted as excellent, so the project authority decided not to dismantle the foundation.
- *Action initiated as per submitted proposal*
  - Entire defective concrete from the surface of pedestal was removed up to 200 mm. Sound and firm concrete surface was encountered.
  - Reinforcement of pedestal was rearranged and straightened. Surface of reinforcement was cleaned and made free of any concrete.
  - Surface of old concrete was cleaned by grit blasting.
  - A solvent-free epoxy resin and a bonding agent (two component) that had a high compressive strength, tensile strength, shear strength, and adhesion strength was applied on the grit blasted surface of old concrete via spray.

- Wire mesh was fixed immediately after the epoxy resin was applied.
- Dry mix shotcrete was applied within 90 minutes of bonding epoxy resin application while under tacky conditions.
- Dry mix shotcrete was adopted because this process provides high bonding, and operation can be controlled at the nozzle by the operator as required. Cement and aggregates were mixed and directly delivered to a pneumatically operated continuous feed gun. The gun ensured continuous flow of mix through the delivery hose fitted with nozzle at the end. The nozzle is fitted with an internal water injecting arrangement that uniformly injected water into the mix.
- Finished and curing.
- Final ultrasonic pulse velocity test was conducted to ensure the stability of the dynamic loading foundation.

Bibliography

Se-Jin Choi, Suk-Pyo Kang, Sang-Chel Kim, and Seung-Jun Kwon. Analysis technique on water permeability in concrete with cold joint considering micro pore structure and mineral admixture. 2015. *Advances in Materials Science and Engineering*, 2015, Article ID 610428, 10 pages.

13

Quality Assurance and Quality Control (QA/QC)

13.1 What Is Quality?

Quality is a standard of work which is not negotiable. Quality of work or services is measured by its condition which is free from defects and deficiencies, flaws and faults, variations and violations with respect to specifications and standards. A work or product or service can be defined as a quality work/product/service when it will be able to satisfy the customer's needs and requirements. Quality is nothing but understanding the customer's needs and fulfilling the commitments or promises within the timeline by a group, and it is achieved through a collective approach and getting everyone to do what they have agreed to do in the right time. It is perceptual and perpetual, conditional and habitual.

13.2 Quality Assurance

A set of procedures is established to ensure that the products/services as promised will be done conforming to predetermined specifications and requirements.

13.3 Quality Control

A set of procedures formulated to ensure that products/services are accomplished in accordance with specifications, standards and requirements. It is an operational technique and set of activities rendered to fulfill the specified requirements.

13.4 Quality Policy

Quality Policy is a company's controlled document that communicates the directives of top management in connection with the quality of its products, services and processes with a commitment of continuous improvement of quality management service.

Quality Policy decodes the commitment to comply with the relevant ISO 9000 requirements, comply with the customer's requirements and continuous improvement of the quality management system.

As per the International Organization of Standardization, the objectives of the quality policy shall be aligned with the customer's need and to satisfy the customer's requirements.

13.5 Quality Manual

The Quality Manual is an official document created by a company and circulated amongst personnel with various deliverables to ensure that products/services/ procedures/processes are reliable, credible, authentic and fit, and is continually addressed to meet the customer's needs and requirements.

The Quality Manual includes the following:

- Quality policy
- Objectives and goals
- Organogram with roles and responsibilities
- Procedures
- System
- Document control
- Record control
- Internal review
- Non-conformance
- Corrective action
- Preventive action

13.6 Job Procedure

A job procedure is a step-by-step description of how to accomplish a work from start to finish in accordance with specifications, standards, and drawings, duly adhering to quality norms as per the quality management system along with safety norms in conjunction with related item descriptions of SOR of the contract.

It is an elaborate work plan with all sequential activities involved to complete a particular work of importance. It is a tool that facilitates site personnel to execute work effectively.

Quality Document No 13.1: Job Procedure

NAME OF COMPANY		DOC NO	
		DATE	
PROJECT	TITLE: Job Procedure of Mass Concrete of Gravity Dam		
Job Procedure of Mass Concrete In Concrete Gravity Dam			

1. *Scope:* The scope of this procedure is applicable to all types of concrete related to construction of a concrete gravity dam.
   - Core concrete
   - External concrete
   - Structural concrete

2. *Purpose:* The purpose of this job procedure is to ensure that all categories of dam concrete shall be cast in accordance with relevant technical specifications, engineering standards, and IFC drawings interpretations in conjunction with the description of items available in the SOR of the contract.

3. *References:*
   - Specification
   - Standard
   - IFC drawings
   - Relevant codes
   - Quality manual
   - HSE manual

4. *Surveying, surface preparation and foundation treatment:*
   - Surveying, laying out dam axis and levels with respect to coordinates and benchmark provided by PMC appointed by the project authority.
   - Removal of over burden to expose rock surface in the dam area.
   - Hard rock excavation by blasting operation and breakers as and when required to reach hard foundation as approved by the geologist.

- Chiselling of the top surface of rock shall be carried out to remove any fractured surface of rock that occurred during blasting operations. The foundation area shall be inspected thoroughly and any slope of rock foundation found to be severe shall be protected by providing a dowel rod for stability or a key shall be provided as required.
- Geological logging of the foundation area shall be carried out by the resident geologist.
- A drawing will be issued by the geologist duly incorporating location, diameter, depth and direction of grout holes in grid fashion for taking up consolidation grouting.
- Drilling of grout holes in block as per drawing by the drilling equipment.
- Flushing and washing of grout holes under controlled water and air pressure and connectivity between holes shall be established to take out clay seams, if any, from the rock.
- Consolidation grout of block shall be carried out under prescribed grout pressure with grouting equipment to make foundation impervious and strengthened.
- Water tests shall be carried out to ensure the water tightness of the foundation after consolidation grouting and to be approved by the geologist.
- Foundation preparation work shall be carried out to dispose grouting spill over and any scale of rock or any hair crack, etc., from the surface of rock and the entire block area shall be cleaned by application of air and water jet to have a surface area free from any dirt and dust, etc.
- Foundation area shall be inspected, and is to be released for blinding concrete, and duly sign off the concrete pour card as per quality norm.
- Blinding concrete shall be carried out by batching concrete of a specified grade with requisite proportion in the batching plant and laying of blinding concrete to level the area of the foundation and compaction of the same by needle vibrator, and curing is done by sprinkling water as and when required.

5. *Core and external concrete:*
   - *Marking:* Marking of the axis of the dam and edge of four sides including upstream and downstream of the dam.
   - *Form work:* Special form work for dam concrete is designed with a cantilever bracket including a suitable anchoring system in old concrete, etc., is submitted and approved by PMC and appointed by the project authority.

The dam form work shall be fitted on the on the edges with a cantilever bracket with an anchor to transfer the pressure of fresh concrete to old concrete. Form work is easy to adopt to sloping surface, etc., and having a working platform complete. The PMC appointed by the project authority shall check the physical measurement and stability and give clearance.

- *Reinforcement:* Reinforcement shall be tested as per quality requirements and test results shall be submitted to the PMC.

    A Bar Bending Schedule (BBS) for external concrete shall be prepared as per relevant code and submitted to the PMC for approval. Reinforcement shall be straightened, cut, and bent as per approved BBS.

    Reinforcement shall be fixed in the correct position as per drawings. A cover block of thickness equal to the cover of concrete shall be provided and properly supported to be in position during concreting.

    Horizontal and vertical reinforcement shall be tied with 20SWG black annealed binding wire in each crossing.

    All chairs, spacers and laps provided shall be measured.

    Reinforcement shall be within the tolerances as specified in the relevant code.

- *Fixing of joint sealant in construction and contraction joint*: Approved joint sealant shall be applied in respective joints.

- *Concreting*

    - *Mix design:* Mix design of specific grade of concrete for external, core and structural concrete shall be done in a laboratory or engineering college duly approved by the PMC. A letter from the PMC addressed to the lab indicating the details of different grades of concrete with all parameters required for designing the mixes along with samples of all ingredients of concrete shall be submitted to the lab to design the mixes. Mix design shall be carried out in accordance with the relevant code of practice of the country.

        Mix design of external, core and structural concrete shall be submitted to the PMC along with all test results for approval.

    - *Batching & mixing:* Batching of various mix designs shall be done in the batching plant. Capacity and number of BP shall be fixed with respect to the quantum of concrete that shall be done. The batching plant should have cement and additive storage silos which should be connected with a conveyor system.

        The batching plant shall be calibrated by the standard weight and measure agency approved by the project authority. The calibration certificate shall be displayed in a suitable spot of BP so that it can be renewed in due course of time.

All ingredients such as cement, additives, fine and coarse aggregates, etc., shall be mixed on a weighted basis. Admixture can be mixed separately and shall be mixed.

A suitable arrangement of controlling the temperature of the concrete shall be made at BP for such cooling of aggregates and water in summer and heating arrangements in winter because Mass Concrete is affected by thermal phenomena.

A scheme of concrete pour with sequence shall be prepared and submitted to the PMC for necessary approval. External and core concrete shall be done sequentially with a different grade of concrete during concreting, so clear communication of the sequence of pouring shall be made available with supervisory staff both at site and BP to avoid any confusion.

Samples of cubes of each grade of concrete shall be taken as per norm and slump to be checked in the concrete lab at BP.

- *Transportation:* Transportation of concrete shall be carried from BP to site by a transit mixer and a print out of each batch shall be handed over to the TM to pass on to site engineer for confirmation of grade, quantity, time etc., for necessary documentation. Concrete being transported will be delivered at site either in a boom placer, concrete pump or directly in the place of pouring. The first load shall be of slurry if concreting is to be done by the boom placer or pump for necessary lubrication to avoid choking of equipment.

  Necessary cubes shall be taken at the site as per norm and slump tests shall be conducted.

- *Placement of concrete:* Placement of concrete can be done in various methods depending on location of each block.
  - Direct unloading in location
  - Placement by boom placer in location
  - Placement by concrete pump
  - Placement in position by pneumatically operated bottom opening concrete bucket lifted by crane
  - Same bucket being lifted by cable way and place in position

  Concrete shall be placed in position as soon as possible so as to keep the temperature of the concrete below 18°C. Arrangements shall be made to measure the temperature of the concrete at the time of pouring.

  A vertical drop of concrete shall be limited to 1.5 meters when concrete is unloaded from a bucket. Concrete shall be laid in layers of specified thickness. Concrete shall be done continuously up to the approved construction joint. Care shall be taken to avoid formation of a cold joint in the Mass Concrete.

- *Compaction of concrete:* Concrete shall be thoroughly compacted by vibrator and by any other approved means. A vibrator needle shall be inserted into the concrete vertically. Concrete should not be allowed to flow. A suitable stopper can be used in core concrete to stop the flow of concrete. The flow of concrete generates segregation of aggregates. Compaction shall continue until cessation of air bubbles. Overvibration of concrete shall be avoided. The needle of the vibrator shall be withdrawn gradually to avoid the formation of holes within concrete. Spacing of the vibrator shall be done in such a manner as to overlap each insertion. The vibrator needle should penetrate the entire thickness of the layer and also penetrate the previous layer in the plastic stage to have a better bond and homogeneity between the layers in order to avoid the formation of cold joints in Mass Concrete. The vibrator should not touch the reinforcement.

- *Green cut:* After completion of concreting, green cut activity should start by using an air jet which will provide a better bond when the next lift of concrete shall be laid.

- *Curing:* Curing of concrete is a critical activity to gain the strength of concrete. Curing of dam concrete shall be done by ponding for 7–10 days. Curing should start 8 hours after the pouring of concrete in normal atmospheric conditions and 4 hours after in very hot conditions. Arrangements for continuous sprinkling of water shall be done by using hose pipe and other arrangements.

  Hessian cloths shall be spread over the surface to keep the concrete surface moist.

- *Precaution:* Mass Concrete is prone to shrinkage cracks. Care should be taken during concreting in case wind blows at that time by providing a wind shield to stop the loss of moisture from the surface of concrete. Hessian cloth shall be used to cover the surface immediately after the concrete is poured to stop moisture loss from the surface of the concrete.

  The temperature of the concrete shall be controlled during concreting.

  - Aggregates shall be kept cool by the sprinkling of water in hot climate

  - Water to be used in the concrete shall be kept cool by using ice blocks

> 6. *Removal of shuttering:* Shuttering of blocks can be removed within 24 to 48 hours after completion of pouring the concrete.
>
> 7. *Inspection:* Inspection of concrete shall be conducted immediately after removal of shuttering. Defective work or small defects, if any, shall be rectified.
>
>   If honey comb, surface bulging, etc., is observed, it shall be brought to the attention of the engineer in charge and necessary corrective measures shall be taken up as per discretion of the EIC.

13.7 Indicative Inspection and Test Plan

The Inspection and Test Plan (ITP) is a quality assurance document that ensures quality of a particular work is achieved. It is a tool that can help and ensure a quality of work. ITP indicates and identifies all activities.

13.7.1 ITP for Consolidation Grouting in Foundation Area of Gravity Dam

Quality Document No 13.2: Indicative Test Plan

Sl No.	Activity	Contractor	Client/PMC
1	Fixing the axis of dam and taking level of dam site with respect to BM provided by client	WC	HP
2	Removal of over burden to expose rock surface in dam site	WC	WP
3	Taking rock level before excavation of hard rock by blasting	WC	HP
4	Hard rock excavation by blasting	WC	WP
5	Taking level of finished rock surface	WC	HP
6	Drilling of grout holes	WC	WP
7	Flushing and cleaning grout holes by air and water pressure	WC	WP
8	Consolidation grouting	WC	HP
9	Maintaining grout register and cement consumption record, etc.	WC	R

Hold point (HP): An critical activity which is to be inspected and verified 100% by Client/PMC and to be accepted by them, and the contractor should not process this activity further without written approval of Client/PMC.

Witness point (WP): An activity which requires witnessing by Client/PMC and will be accepted as per relevant specification/drawing/standard/SOR, etc.

Review (R): To examine the inspection records to confirm compliance with requirements.

Quality Document No 13.3: Concrete Pour Card

Document No			Pour Card No		
Pour Information			Date of Offer		
Pour Location: Blinding/Block/Gallery/Sluice/Spillway/Training wall/Bucket/Pier etc.					
Section-1			**Concrete Information**		
	Comment		Technical Requirement	**Comment**	
Activity	Contractor	PMC		Contractor	PMC
A. Block Concreting				1. Details of Mix design	
1. Block No.				2. Grade of Concrete	
2. Category of Block				3. Grade of Cement	
3. Block Size				4. SCM if any	
4. Block Depth				5. Estimated Quantity	
5. Level of Block				of Concrete	
6. Status of Green cutting				6. Designed Slump	
of previous level				7. Admixture	
7. Foundation Treatment				8. Predicted temperature	
8. Reinforcement				of concrete	
9. Location of joint				9. Placement Method	
10. Installation of				i. Pump	
water bar				ii. Boom Placer	
11. Shuttering				iii. Pneumatic Bucket	
12. Cover				with crane/cable	
13. Others				way	
B. Gallery				iv. Any other method	
1. Alignment				10. Compaction	
2. Level				equipment	
3. Size				11. Method of Curing	
4. Installation					
Water Stop at Joint					
5. Rebar				Do	
6. Shuttering					
C. Sluice					
1. Layout					
2. Level					
3. Size					
4. Curve for Bell mouth				Do	
5. Reinforcement					
6. Shuttering					
7. Water bar					
D. Spillway					
1. Layout					
2. Level					
3. Layout of curve					
for Ogee				Do	
4. Reinforcement					
5. Shuttering if any					
E. Training wall/Pier etc.					
1. Layout					
2. Level					
3. Dimension					
4. Shuttering					
5. Reinforcement				Do	
6. Others					

Work has been made ready as per drawing and specification and is offered for inspection.
Signature of contractor with date

Section-2	**Pre-Pour Report**
1. Weather 2. Ambient Temperature 3. Slump 4. Cube testing 5. Rectification if any Certified that work has been checked as per IFC drg no .... and release or concreting after duly liquidating the check list. Signature with date	

Section-3	**Post-Pour Report**
1. Date 2. Location 3. Actual Quantity 4. Nix ID 5. Start of concreting 6. Finish of concrete. 7. Slump 8. Ambient Temperature 9. Temperature of concrete 10. No of cubes taken	

13.7.2 Pour Card

A pour card is a history of the concrete pour. It is a certificate for release of the concrete pour. The pour card is initiated by the contractor, and shall be submitted to the PMC who will ensure that the work is done in accordance to IFC drawings and specifications. It will have at least three sections, of which the first section is for the contractor, the second section is for the inspection, and the third part is for the post pour report. All parts shall be signed by the concerned authority as per the Quality Manual.

13.8 Records Are to Be Maintained at Site

As per the QA/QC manual, records of each activity shall be maintained at the construction project. These records are evidence of work done as per drawing and specification, and all work documents shall be subjected to annual quality reviews being conducted as specified in QA/QC manual.

The following records are to be maintained on site:

- QA/QC manual
- Technical specification of work

- All IFC drawings with latest revision
- All superseded drawings shall be marked as "superseded" and shall be kept in an isolated place
- All records for deviation and extra work
- Job procedure
- All indicative test plans
- All pour cards of concrete
- Clearance of foundation of each block of dam by geologist
- Record for foundation treatment like consolidation grouting, dental treatment, etc., duly signed by geologist/engineers as per QA/QC manual
- Water percolation test of the foundation
- Result of all tests being conducted per the QA/QC manual
  - Aggregates
  - Sand
  - Water
  - Cement
  - Plasticizer
  - SCM
  - Slump test
  - Cube test
- Cement register with brand, quantity, date, etc.
- Daily consumption record of cement for making reconciliation statements at the end of the project
- Record of steel receipt, issue and use
- Record of chairs, laps, etc.
- Test result of steel, etc.

14

Health, Safety, and Environment

14.1 Introduction

The modern world is highly competitive, with firms and individuals competing against one another. It creates growth and progress that drives positive changes in the environment. Many opportunities are created by creating employment in various sectors. The construction world is one of the sectors where millions of people participate and come from different cultural, economic, and social backgrounds to earn their livelihood. They work as ordinary, semi-skilled, skilled, or expert employees in different categories in the construction industry. They sometimes work in very vulnerable situations that are hazardous, full of risks, and require their protection by their employers. Employers should be committed to employee safety at all times. Employers should communicate their intentions and protective mindsets to their employees. It must be communicated and implemented by a solid system known as Health, Safety, and Environment, or HSE. Safe working practices are not the action of an individual; it is a collective approach that is continuous and habitual.

14.2 Purpose

Accidents are an unfortunate occurrence during construction that lead to physical disabilities and mental disorders. Occupational accidents cause serious social and economic problems including injuries, loss of life, and loss of property. Statistics show that fatal incident rates are very high all over the world due to inadequacies in safety measures at many construction sites. Occupational accidents cause large financial and human losses in companies and countries, which act as a barrier to progress and overall growth of a nation.

Safety management systems are designed to improve the employee and employer's abilities to identify hazards and reduce health and safety risks at workplace.

14.3 Safety Policy

A safety policy is a documented statement of commitment of an organization for extending protective and safe practices to support the health and safety of their workforce in their work place, with the ultimate goal of having an accident-free working environment while providing safeguards to employees on-site. It is a commitment to protect against the loss of life and property, as well as manage other aspects of safety performance.

- Commitment to the prevention of injury and ill health and continued improvement in occupational health and safety performance.
- Create environment of learning and practicing the organization's HSE system amongst all employees and business associates.
- Commitment to continual improvement in the prevention of pollution.
- Utilize resources in an effective manner so that generated waste is reduced.
- Commitment to comply with applicable legal aspects in regard to environmental hazards and risks.
- Provide a framework for establishing and reviewing environmental, health, and safety objectives.
- Understand and circulate the organization's intention within the organization to increase awareness of corporate health and safety obligations.
- Communicate the intention of the company to all who are working with or on behalf of the organization.
- Training and validating employees and business associates on health and safety practices.
- Encouraging the use of less harmful substances to save the environment.
- Provide adequate training to personnel on injury prevention, adverse health, and emergency response for tackling any harmful incident.
- Build strong organizational support through the internal safety department.

14.4 Safety Oath

All employees of an organization are responsible and accountable for maintaining a safe working atmosphere by adhering to company's safety policy. All employees should take an oath to promote and display safe working attitudes on site.

SAFETY OATH

We take an oath to protect all workers working under us from injury and health hazards, to make the workplace accident free, and to protect against the loss of life and property by adhering to company's safety policy.

Modes of communication to all company's personnel, business associates, and other stakeholders shall be through the following activities:

- HSE meetings
- Induction training programs
- Displaying safety slogans at different locations
- Safety bulletins
- Safety workshops
- Safety slogan and essay competitions

14.5 Action Plan

A detailed action plan states various actions to be undertaken including monitoring and reviewing for strict implementation of safe working practices at the worksite.

14.5.1 HSE Meeting

Safety is everyone's business. Employees should understand what is expected of them, and employers should communicate to employees their intentions with the company's safety policies. Safe working practices can prevail on site if these two stakeholders collaborate. Effective communication methods shall be in place to communicate situations that are possible through various level of meetings. Safety meetings will be able to provide information, awareness, training and help for everybody to adopt safe working practices.

14.5.2 General Safety Meeting

The General Meeting (Safety) shall be held between the company president, all construction managers, and safety leadership. This meeting will be held once during the mobilization of manpower and machinery by contractors where management will communicate their views about safety compliance on site.

14.5.3 Agenda of the Meeting

Safety measure to be taken for safe working practice at site:

- Standard work procedures
- Safety rules
- Company safety policy
- Action plan of each CM about strict implementation of safety policies

14.5.4 Monthly Safety Review Meetings

Monthly Safety Review Meetings shall be held between the president, CMs, and safety leadership every month to discuss:

- Monthly safety report of all sites shall be distributed among attendees.
- Safety leadership will brief the safety status of each site to the president and highlight flaws, shortcomings, and improvement ideas for all sites.
- Discussions about recent incidents and accidents, including why it happened and future actions to prevent additional accidents.
- Training program.
- Any other safety issue raised by CMs.
- A report shall be made and circulated to all concerned.

14.5.5 Weekly Safety Meeting at Site Office

This meeting will be held between the CM and Resident Construction Manager of the contractor and safety leadership every week to discuss:

- Safety performance including accidents, injuries, and near misses (if any).
- Root cause analysis and remedial measures against reoccurrences.
- Any violations of safety norms.
- Creation of further safety awareness amongst workforce.
- Display of sufficient poster and safety slogan in key places of the worksite.
- Safety competitions.
- Imposition of penalties for safety violations.
- A report shall be made and circulated to all concerned.

14.5.6 Tool Box Talk

The Tool Box Talk is a meeting between safety engineers and workers to be held every day in a particular spot at the worksite. A favorable spot will be chosen by the safety engineer which is known as the Assembly Point. Every day, the work force will assemble in the selected spot before moving on-site.

The safety engineer will briefly talk about safety and compliance with safety norms for the betterment of the employee, as well as personal protection equipment (PPE) use and brief about incidents that happened at other sites due to workers not using their PPE.

Employees are reminded again and again that their families wait for their safe return home, and to practice safe work habits on-site. Records of Tool Box Talks shall be maintained in formats required by the organization.

14.5.7 Induction Training Program

It is mandatory for all workers who take up construction work to attend construction induction training, which is training where a new worker is provided with information concerning hazards, risks, and their control measures. This training provides new workers joining the worksite with basic knowledge of health and safety norms and company safety requirements.

Induction training is conducted in various ways depending on the company and magnitude of project.

For major projects workers are trained in the training center by different methods such as Tool Box Talks, demonstrations, and videos. Sometimes on-the-job instruction is also given, including:

- Various safety requirements.
- PPE use and their benefits.
- Housekeeping.
- Cautions against un-barricaded, open excavation pits.
- Use of safety belts and harnesses while working in an elevated position.
- Who to report safety concerns and incidents to.
- Layout of working place including entrances and exits.

An induction checklist shall be maintained during induction training to ensure that all relevant points are covered.

14.5.8 Displaying Safety Slogan in Key Locations

Workplace safety is a critical requirement for any project authority. The display of safety slogans in key locations of the site is a successful method of raising the safety awareness of the workforce. A good, catchy, and humorous slogan reminds workers about safe working practices, with slogans encouraging workers to follow safety rules on-site and to create a happy working site.

Periodic safety slogan competitions shall be held to encourage the workforce to promote and create safety awareness at working sites.

14.5.9 Safety Bulletins

A safety bulletin is a short safety report periodically published or as required with the goal of acting as a safety notice regarding any alert notice such as potential hazards including inclimate weather, timeframes for hazards, and the appropriate safety precautions that should be taken on site to avoid any accidents. For example, in the case of high winds:

- All cranes with high booms shall be lowered
- No worker should work at elevated positions
- Workforce should evacuate the worksite

The safety department may also use safety bulletins to pass on non-urgent safety messages on-site for implementation.

14.5.10 Safety Workshops

A safety workshop is a brief educational program focusing on adopting safe on-site working practices. Workshops can be organized to educate the workforce and create safety awareness by displaying different safety flaws that caused accidents but could have been avoided had proper safety precautions been taken. Safety workshops can also be organized with different groups.

14.5.11 Safety Slogan and Essay Competitions

Safety slogan and essay competitions also create awareness and interest within the workforce. This competition extracts creativity of workers while pertaining to safety, making them think and write about safety, thereby causing safety to become part of their everyday life during the competition.

14.5.12 Competence and Awareness Training

The company should identify the competency of personnel in regard of safety knowledge and activity to make record of personnel for future references.

- Identification of competency of personnel affecting HSE activities.
- Feedback shall be given to the employee in regards to their attitudes toward the company and how performance is affected.
- Induction training programs shall be arranged to meet the company's requirements.
- Actions taken shall be evaluated and records shall be maintained.

14.6 Risk and Hazard Assessment

A hazard is a source of potential damage, harm, and adverse health, whereas risk is the probability or threat of damage, injury, liability, loss, or any negative occurrence that is caused by external or internal vulnerabilities.

Risk assessments help to identify specific work conditions activities, and habits that may cause physical harm to the workforce or damage to assets.

14.6.1 Purpose of Risk Assessment

A construction site is full of hazards and risks. A systematic process is being adopted to assess potential risks that may jeopardize critical activities at the work site, and the purpose of this risk assessment is focused as under.

- Create awareness about hazards and risks.
- Identify who may be at risk.
- Determine those hazards that pose significant risks to employee health and safety.
- Determine if existing control measures are adequate or need strengthening.
- Prioritize hazards and control measures.

14.6.2 Procedure of Hazard Identification/Risk Assessment

Risk assessment procedures consist of four stages that include the identification of hazards, assessment of the risks, implementation of risk controls, and the review of control measure.

14.6.2.1 Identification of Hazard

The responsible site engineer should visit and inspect the site as well as consult with the workforce to know about any past incidents, accidents, or near misses that should be examined. He should examine upcoming jobs for the following:

- Position of work whether on the ground, elevated, or accessible.
- Nature of activity and modus operandi.
- Equipment and material to be used.
- Planned safety measures.
- Workforce and management.

14.6.2.2 Assessment of Risk

Examine the event and critically consider the hazardous conditions and likelihood of occurrence.

- Judge the severity
- Adequacy of existing control measures
- Type of actions and duration of activities

14.7 Controlling Measures

Risk assessment becomes successful if all identified hazards are eliminated. Groups should learn the means and methodology for eliminating hazards by incorporating engineering changes to eliminate the hazard or arrange for induction training to enhance workforce skills, planning for safe work practices or safe work permits, and the usage of PPE.

After detailed assessments, the EIC may raise a Site Change Request (SCR) to the engineering group with detailed descriptions of the request with reasons and possible alternatives to remove the identified hazard. Engineering will look into the matter and will study the impact of the change on schedules, production, and overall cost so that an engineering solution can be made to eliminate hazards.

If SCR does not work, the EIC may adopt the process of conducting workforce training, Toolbox Talks, and plan for safe work practices in vulnerable situations. Some expert workforces are trained to safely tackle all sorts of hazardous work, and a safe work permit shall be issued with critical safety obligations.

Use of PPE is the last measure. Usage of PPE cannot eliminate hazards, but it can reduce the effects of hazards. It can eliminate injury and also reduce the severity of injury.

Proper PPE quality shall be issued to the workforce to safeguard the worker from identified hazards. It should be comfortable to wear and facilitate comfortable work through easy movement of the employee's limbs under adverse conditions.

14.7.1 Details of PPE

Details of PPE to be used are as follows:

- *Safety helmet*: Safety helmet shall used and tied properly for head protection.
- *Safety shoes*: Safety shoes shall be used as foot protection.
- *Safety belt*: It is mandatory for people working in elevated sites to wear a safety belt and secure it to a safe anchorage point or with a lifeline as a protection against falls.
- *Gloves*: Hand protection.
- *Safety goggles*: Eye protection.
- *Ear plugs*: Ear protection.

- *Mask*: Respiratory protection against dust and other contaminants.
- *Life line*: It is a system adopted on site when workers work at high elevations. This system is used to arrest the worker in the event that they fall. The system consists of a long rope that is stretched over head at a suitable height and both ends are anchored firmly. The worker working at height with the safety belt will have a long cord, which is to be mounted on the rope being stretched over head. The worker will hang in the air in case of an accidental fall because the cord of the safety belt is mounted on the life line.

14.8 Hazard Identification and Risk Assessment of Dam Work

Hazard identification and risk assessments were conducted for the erection of formwork for block concrete in concrete gravity dams, and formwork was supported by cantilever brackets duly anchored in previous concreted blocks at height. Erection was done with the help of a crane. A risk assessment was done for the crane and working at an elevated position separately.

The Concern Group visited and inspected the work spot on the top of the completed block and made notes on the following items for the identification of hazards and risks:

- Work spot was at an elevation of 18 m above ground level.
- Form work to be erected on the upstream face of the dam where cantilever bracket was to be anchored to support the entire form work, because this 2.5 m height shutter cannot be supported by the opposite formwork.
- River water of knee depth was exactly below the block due to the presence of a coffer dam.
- The foreman of the carpenter group and the crane operator were consulted on the possibility of any risk they had identified and information on the same type of previous reviews were collected in regard to any slippage of safety norms as well as any incidents, accidents, or any near miss incidents.
- Inspected formwork materials, the maximum weight of the single piece erection, and the anchoring system.
- Inquired about average health of the workforce.
- Inspected the health of the crane by checking the crane's fitness certificate, load test, and load chart.
- Inspected and assessed the maximum weight of element and checked with the load chart for boom radius with permissible boom angles.
- Checked the guy rope of crane.

FORMAT 14.1

Erection of Formwork in U/S of Dams

Health, Safety and Environmental Risk Analysis

Title			Risk Assessment			
Company						
Project			Site			
Ref:			Revision		01	Date

Hazard and Risk Identification					Risk Evaluation and Control Measure			
Sl No.	Activity	HSE Aspects	HSE Impact	Initial Risk Rating L / S / R	Control Measure	Law Compliance	Residual Risk Rating L / S / R	Engineer in Charge
	Analyzed by			Reviewed by			Approved by	
1	Erection, placement and fixing of formwork in position of upstream face of dam with supportive cantilever bracket. Bracket to be anchored to already completed block to transfer load of fresh concrete to it.		i. Fatal injury ii. Serious injury iii. Fall of worker from height iv. Fall of material on other worker		i. PPE shall be replaced by new gear ii. Toolbox Talk iii. Life line shall be made normal procedure iv. Toolbox Talk v. Safety net to be replaced vi. Proper housekeeping to be done vii. Toolbox Talk viii. Worker should put tools in bag ix. Handle of hammer to tied with thread and other end of thread to be put around wrist x. Platform and barricading to be done properly	Yes		

FORMAT 14.2

Erection of Formwork in Block by Crane

Health, Safety and Environmental Risk Analysis

Risk Assessment

Title			
Company			
Project		Site	
Ref:		Revision	01 Date

Hazard and Risk Identification | | | **Risk Evaluation and Control Measure**

Sl No.	Activity	HSE Aspects	HSE Impact	Initial Risk Rating			Control Measure	Law Compliance	Residual Risk Rating			Engineer in Charge
				L	S	R			L	S	R	
1	Erection of formwork in block of dam by crane	i. Collapse of crane boom due to improper boom length and angle of boom ii. Faulty operation iii. Fall of load due to faulty slinging iv. Collapse of guy v. Severe wind	i. Serious Injury ii. Loss of life iii. Toppling of crane				i. Erection scheme shall be made considering Load Chart of the crane ii. Licensed and trained operator shall be engaged iii. Guy and sling shall be checked and tested by load periodically iv. Erection shall not be permitted during severe wind v. Qualified Rigger shall be engaged	Yes				

Analyzed by | Reviewed by | Approved by

FORMAT 14.3

Hard Rock Excavation by Blasting

Health, Safety and Environmental Risk Analysis

Title				Risk Assessment			
Company						Site	
Project						Revision	
Ref: ACI/PRL/						01	Date

Hazard and Risk Identification / **Risk Evaluation and Control Measure**

Sl No.	Activity	HSE Aspects	HSE Impact	Initial Risk Rating			Control Measure	Law Compliance	Residual Risk Rating			Engineer in Charge
				L	S	R			L	S	R	
	Analyzed by				Reviewed by					Approved by		
1	Excavation of Hard Rock by Blasting Operation	i. Improper storage of explosives ii. Improper transportation of explosives iii. Improper priming and stemming iv. No area cordoning v. Movement of vehicle over loaded holes vi. No warning siren before blasting operation	i. Cause injury ii. Loss of Life iii. Injury caused to public				i. Gelatin, detonators, etc., should not be stored in one place ii. Explosive shall be stored in explosive magazine under Explosive Act iii. Explosive cannot be stored more than 24 h in day box iv. Explosives shall not be carried in clothes v. Smocking should be prohibited vi. Inflammable item shall not be allowed	As per Explosive Act of the country				

(Continued)

FORMAT 14.3 (*Continued*)

Hard Rock Excavation by Blasting

	Health, Safety and Environmental Risk Analysis						
Title			Risk Assessment				
Company							
Project					Site		
Ref: ACI/PRL/					Revision	01	Date
Hazard and Risk Identification				Risk Evaluation and Control Measure			
Sl No.	Activity	HSE Aspects	HSE Impact	Initial Risk Rating	Control Measure	Law Compliance	Residual Risk Rating	Engineer in Charge
				L | S | R			L | S | R	
		vii. No all clear siren after blasting viii. Excessive flying of blasted materials ix. Misfire x. Smocking during loading of blast holes xi. Inadequate spot inspection after blasting operation xii. Re drilling near unblasted holes			vii. Proper warning sign shall be given as explosive act viii. Area shall be inspected after blasting and any misfired hole shall be marked with color red ix. Controlled blasting shall be conducted or kentiledge shall be used over the rock area to stop flying ANFO shall be used less. Blasting operation shall be carried out by Licensed Blaster as per Explosive act.			
	Analyzed by			Reviewed by		Approved by		

15

Fabrication and Erection of Steel Structure and Penstock

15.1 Fabrication

Fabrication is a process which is comprised of cleaning, grinding, cutting, bending and making the fit up, and finally, by welding the fit up structure in accordance with the shop drawing to form a complete structure.

It needs an expert fabricator who will be able to do the accurate layout on the ground and be able to make a correct template of the structure in accordance with the shop drawing so as to go for mass fabrication.

Efficiency and expertise drive the success of the workshop fabrication, which ultimately leads to successful completion of work within the timeframe and budget.

Many modern fabrication shops are equipped with computer aided design connected to a CNC machine with CNC programming. Requisite data is fed to the program which generates actual details of components and directs the machine tool, the CAM machine, to manufacture the goods.

What is Erection: After completion of fabrication of each unit and it is then lifted and placed in position with respect to the IFC drawing, maintaining alignment, level, etc., and duly grouted.

15.2 Material

Material used in this work is called structural steel. Structural steel is available in the market in various sizes and shapes per American, Euro, and Indian codes of practice. These materials are medium and high tensile structural steel manufactured in a rolling mill with a hot rolling process under controlled or normalizing rolling. References code of practice are IS 2062, ASTM A36, A992, A588, A529-50, A572-50 EN10025, BS4360, DN17100, etc.

These structural steels are suitable and compatible to welding, bolting and riveting works and for other engineering work.

General technical delivery requirements shall be covered under ISO 6929 and the inspection document shall be as per ISO 10474.

15.2.1 Section

The standard open section of structural steel in different countries is in different nomenclatures like ISMB (Indian Standard Medium Beam), ISMC (Indian Standard Medium Channel), ISA (Indian Standard Angle), etc. The Universal Beam (UB), Universal Channel (UC) and Universal Angle are defined in respective relevant codes.

Standard American beams are known as J Beam, S Beam and I Beam as per relevant standards.

15.2.2 Physical Property

Physical properties of structural steel irrespective of its grade may be taken as:

a. Unit mass of steel, $p = 7850 \text{ kg/m}$
b. Modulus of elasticity, $E = 2.0 \times 10 \text{ s N/mm}^2$ (MPa)
c. Poisson ratio, $p = 0.3$
d. Modulus of rigidity, $G = 0.769 \times 10 \text{ s N/mm}^2$ (MPa)
e. Co-efficient of thermal expansion $cx. = 12 \times 10'/oc$

The principal mechanical properties of the structural steel important in design are the yield stress, fy; the tensile or ultimate stress, fu; the maximum percent of elongation on a standard gauge length; and notch toughness. Except for notch toughness, the other properties are determined by conducting tensile tests on samples cut from the plates, sections, etc., in accordance with various codes.

15.2.3 Classification

Structural members are classified in various fashion in different countries as narrated below.

- American Code Practice
  - Standard American beam designated as J Beam, S Beam and I Beam have a tapered flange as per relevant standard
  - Wide flange steel beam non tapered W and H beam as per relevant standard

15.2.4 Chemical, Mechanical, and Physical Property

Materials brought to the site shall be tested as specified in relevant code before use. Random samples to be obtained from the lot shall be sent to an approved testing lab as per specification and contract document. Any material not conforming to specification shall be rejected. Materials are to be tested for the percentage of main constituents like carbon, phosphorous, sulphur, manganese, copper, tensile strength, etc.

- Indian Code of Practice
- *Classification of beam, channel,* etc.
  - Indian Standard Junior Beams (ISJB)
  - Indian Standard Light Weight Beams (ISLB)
  - Indian Standard Medium Weight Beams (ISMB)
  - Indian Standard Wide Flange Beams (ISWB)
- *Columns/Heavy weight beams*
  - Indian Standard Column Sections (ISSC)
  - Indian Standard Heavy Weight Beam (ISHB)
- *Channels*
  - Indian Standard Junior Channels (ISJC)
  - Indian Standard Light Weight Channels (ISLC)
  - Indian Standard Medium Weight Channels (ISMC)
  - Indian Standard Medium Weight Parallel Flange Channels (ISMCP)
- *Angles*
  - Indian Standard equal leg Angles (ISA)
  - Indian Standard unequal leg Angles (ISA)

15.2.5 Designation

Beam, columns and channel sections shall be designated by the respective abbreviated reference symbols followed by the depth of the section, for example:

a. MB 200—for a medium weight beam of depth 200 mm

b. SC 200—for a column section of depth 200 mm

c. MC 200—for medium weight channel of depth 200 mm etc.

Some samples of beams, channels and angles were collected from the warehouse and were tested in the approved lab. Indicative test results of chemical, mechanical, etc., as per specification is furnished (Table 15.1).

TABLE 15.1

Chemical and Mechanical Property

Description	Beam (300)	Channel (250 × 250)	Angle (150 × 75 × 10)
Carbon	0.25%	0.23%	0.25%
Manganese	1.45%	1.5%	1.45%
Sulfur	0.04%	0.035%	0.045%
Phosphorous	0.04%	0.045%	0.04%
Silica	0.35	0.3%	0.35%
Yield stress MPa	255		
Tensile MPa	425.00	260	270

15.3 Plans and Drawings

Plans, drawings and cut sheet shall be prepared according to relevant standard.

15.3.1 Plans

The plans (design drawings) shall show the sizes, sections, and the relative locations of the various members. Floor levels, column centers, and offsets shall be dimensioned. Plans shall be drawn to a scale large enough to convey the information adequately. Plans shall indicate the type of construction to be employed, and shall be supplemented by such data on the assumed loads, shears, moments and axial forces to be resisted by all members and their connections, as these may be required for the proper preparation of shop drawings. Any special precautions to be taken in the erection of the structure for design consideration shall also be indicated in the drawing.

15.3.2 Shop Drawings/Fabrication Drawings/Cut Sheet

Fabrication/shop/cut sheets are the back bone where a huge quantum of structural work are involved. Fabrication drawing is very much essential for fabrication of each component of the structures. Shop drawing shall be in reference of the design drawing of the steel structure. Designs do not provide detailing of structural elements.

Shop drawings, giving complete information necessary for the fabrication of the component parts of the structure including the location, type, size, length and detail of all welds and fasteners, shall be prepared in advance of the actual fabrication. Shop drawings will indicate connection details of structural components with gusset plates along with details of welding, duly indicating length and size of weld. They shall clearly distinguish between shop and field rivets, bolts and welds. Shop drawings shall be made to conform with IS 962. A marking diagram allotting distinct identification marks to each separate part of steel work shall be prepared. The diagram shall be sufficient to ensure convenient assembly and erection at site.

Symbols used for welding on plans and shop drawings shall be in accordance with relevant codes. This cut sheet is to provide the bill of material in the prescribed format.

15.3.3 Tolerance

Engineers should consider tolerance seriously. Tolerance prescribed in various codes seem to be very small or negligible but create huge problems when a fabricated structure is interfaced with other structures if tolerance was not considered or neglected by both groups. This negligence will cause delays and subsequently may lead to huge expenditure.

There are three different tolerances involved in total activity of structural work as per various codes of practice.

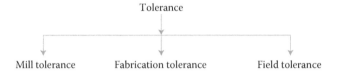

Mill tolerance of hot rolled sections shall be per relevant code.

Fabrication tolerance shall be considered per practice and tolerance of welded fit up with respect to relevant code of practice.

Field tolerance consists of site preparation and erection tolerance. Tolerance for the anchor and foundation bolt shall be per prevailing practice.

Tolerances for fabrication of steel structures shall conform to relevant code. Tolerances for erection of steel structures shall conform to code. For general guidance on fabrication by welding, reference may be made to specification.

15.4 Fabrication Procedure

15.4.1 Preparation

Some preparation work is required before starting actual fabrication activities. Layout of the structure shall be done on the shop floor to make a template of the structure, and it shall be inspected and be checked by the client for necessary approval.

During transportation and handling of steel sections, structural sections like beams, channels, angles and plates bend.

Fabrication of the structure shall be taken up in sequence as narrated below.

15.4.2 Surface Cleaning

Structural members shall be cleaned and should be made free from dust, dirt, scales and rust.

15.4.3 Straightening

Structural materials during transportation bend and buckle to some degree, which shall be straightened or formed to the specified configuration. This straightening of members shall be done by an approved method that should not affect the properties of the material.

Straightening and flattening of the member can be done by application of mechanical pressure or by application of heat in some area as per ISO 10721 Part 2. The temperature should not exceed 590°C or 20°C below the tempering temperature of steel, whichever is less.

15.4.4 Cutting

In the fabrication shop, cutting of structural elements such as beams, channels, plates, etc., means severing or removal of material from the surface of members. Members may be cut by mechanical means or by chemical process. Cutting shall be done by sawing, shearing, thermal cutting or laser cutting process.

Sections shall be cut to required length and plates shall be cut in the required size. Length of the member can be cut by circular saw to the required length.

The edge of cutting shall be smoothed and free from cracks. The edge of cutting shall be carefully inspected so that up to a minimum of 25 mm of parent material shall be free from unevenness and cracks to have quality welding during jointing. The surface of the cut shall be grinded and machined to have a smooth surface. Any irregularities or small holes shall be repaired by welding and shall be smoothened to flush with the existing good surface.

- *Shearing*: Plates having a thickness more than 20 mm shall be stress relieved to release internal stresses being generated prior to manufacturing the material. Flame cutting and shearing can be done where the fatigue quality is indicated by the designer. The use of sheared edges in the tension area shall be avoided in locations subject to plastic hinge rotation at factored loading.
- *Gas or flame cutting*: Gas and flame cutting is used to cut the plates. Flame cutting is an oxyacetylene or oxyfuel gas cutting process which uses pure oxygen and fuel gas to preheat a metal to 700–900°C and an

exothermic reaction is developed between metal and oxygen, resulting in the formation of oxides and slags on the surface. An oxygen jet blows away the slags and helps the jet to pierce and cut the plate.

- *Holing*: There are holes of various configurations like round, slotted and counter sinking. Round holes for bolts can be made by drilling or punching. Slotted holes can be made by punching. They can also be made by drilling two holes which can be completed by hand flame cutting to the requirement. But hand flame cutting of the bolt hole cannot be permitted.

Formation of holes through compound sections like a plate girder flange can be taken up after the compound structures are assembled and tightly clamped. It can be made by drilling, punching or machine flaming.

15.4.5 Punching

A punched hole shall be permitted only in material whose yield stress does not exceed 360 MPa and where thickness does not exceed (5 600/~Y) mm. In cyclically loaded details, punching shall be avoided in plates with thickness greater than 12 mm.

For greater thickness and cyclically loaded details, holes shall be either drilled from the solid or sub-punched or sub-drilled and reamed.

The die for all sub-punched holes or the drill for all sub-drilled holes shall be at least 3 mm smaller than the required diameter of the finished hole.

15.4.6 Fit Up and Clearance of Fit-Up

All components being fabricated as required by the cut sheet are placed on the layout made on the shop floor and fitted up temporarily by bolt or tack weld. Necessary splice plates and gusset plates shall be placed in position along with other fittings. This fit up shall be offered for inspection, and after inspection if any minor error or defect is noticed it shall be corrected and released for final welding.

15.5 Welding Procedure

Welding is a process through which two metals can be joined at their surface of contact by fusing two metals upon heating or pressure on some metallurgical condition.

The efficiency of a weld joint depends mainly on the weld joint preparation. The joint shall be prepared wide enough to give easy access to the welding folder which will ensure requisite weld thickness to achieve the designed strength and proper examination of the weld.

Weldability is the ability of the metal to be welded with the parent material with quality. Weldability signifies strength, effort and cost. It depends on the procedure of welding, preparation and fabrication condition.

There are two major groups of welding processes:

- Pressure welding
- Fusion welding or non-pressure welding

There are many classifications of welding procedures based on the source of energy as per various codes:

- Gas welding
- Arc welding
- Resistance welding
- Solid state welding
- Thermochemical welding
- Radiant energy welding

15.5.1 Welding Procedure Specification (WPS)

Before taking up any welding work at shop or field, a welding procedure specification shall be made which will provide all information in connection with welding activity and suggest the welder weld the joint in such a way so that the welded joint meets the design parameters such as properties, soundness, and strength in accordance with code requirement. Recording of WPS may be done as per relevant codes of practice.

15.5.2 Welding Procedure Qualification (PQR)

The contractor involved in welding work shall furnish a detailed welding procedure with a test sample of the concerned work. The test sample shall be welded and tested as per WPS to make sure that the test results conform to standard and specification, and it shall be recorded in the standard format of the approved authority.

15.5.3 Welding Qualification Test Procedure (WQPT)

The contractor's welding engineer and PMC shall inspect the welding shop before conducting the WQPT. They will check the test sample and its dimensions along with preparation of the sample. They will check the type of electrode to be used and its property. The welding shop should have the proper facility of lighting for visual check and the welding machine shall be checked and its status of calibration. They will check the availability of approved WPS for the welder, etc.

After inspection, the fit up shall be taken up and it shall be checked in conformity to approved WPS. The position of the weld and number will be marked on the sample and the filler material shall be checked. Preheat the test sample up to the temperature indicated in WPS and check the preheat temperature prior to welding, during welding and after welding. All parameters indicated in WPS shall be adhered to such as cleaning up, preheat temperature, interpass temperature, width of weld pass, thickness of weld, sequence and number of weld passes per layer, etc.

Upon completion of welding, a visual test shall be conducted and afterward NDT shall be conducted as per WPS.

15.5.4 Welder Qualification Test

The welder qualification test is conducted to find out the ability of the welder, whether the welder shall perform and be able to make the weld in accordance to WPS. The welder qualification test record shall be maintained and accordingly the successful welder shall be issued a license for conducting welding work.

WQTR 15.1

Welder Qualification Test Report

	COMPANY'S NAME		
	NAME OF PROJECT		
	NAME OF PMC		
	NAME OF CONTRACTOR		
	DOCUMENT NO		
	Particulars		Details
1	Welding Procedure Specification (WPS) No		
2	Welder's Name		
3	Address Locality, Street No, City, State, ZIP		
4	Welder's Id & Mob. No.		
	Sample Information & Welding Information		
5	Welding Process		
6	Welding Position		
7	Material Specification		
	Plate Thickness		
	Base Metal		
	Type of Joint		
	Type of Weld		
8	Filler Matal Specification		
9	Test Material Thickness		
10	Filler Metal Classification		
11	Fillet Size		
12	Electrode Diameter		
13	Machine Name		
14	Electrical Characteristics, Current, Amperage, Vol.		

(Continued)

WQTR 15.1 (*Continued*)

Welder Qualification Test Report

15	Direction of Welding	
16	Atmospheric Temperature	
17	Preheat Temperature	
18	Backing Shielding Gas	
19	Flux Type	
20	Cleaning Method	
21	No. of Passes	
22	Visual Inspection Result	
23	Radiography Result	
24	Other Test	
25	Bend Test	
26	Special Test	
	Test Conducted by	
	Test Witnessed by	
	Lab Test No. and Date	
	Test Approved by	
	Remarks	
Date		**Signature**

15.6 Shop Fabrication

15.6.1 Setting Out

The positioning and levelling of all steelwork, the plumbing of stanchions and the placing of every part of the structure with accuracy shall be in accordance with the approved drawings and to the satisfaction of the engineer while in accordance with the deviation permitted below.

A level of hard platform by concrete shall be made. The fabrication drawing will indicate a working point (WP). WP shall be marked on the surface and mark all elements of the structure as per the drawing. Marking shall be done by paint and all cut members shall be placed on the marks along with gusset plates. All the placed elements shall be tack welded to have the exact configuration of the structure as per drawing and measurements of all members shall be checked as per drawing.

After that, welding shall be completed and a template shall be made so as to take up work in full swing at the shop.

15.6.2 Surface Cleaning and Painting

The surface of the structure is to be cleaned before application of painting. Surface cleaning shall be done as per requirement depending on the type of structures and its applicability, it is done with sand/grit blasting.

All surfaces, which are to be painted, oiled or otherwise treated, shall be dried and thoroughly cleaned to remove all loose scale and rust.

Shop contract surfaces need not be painted unless specified. If so specified, they shall be brought together while the paint is still wet.

Before the surface is painted, a primer shall be applied on the surface with the requisite rate of application as per manufacturer's instruction and afterwards, two coats of paint shall be applied over the primer with the paint per square meter of area and dry film thickness (DFT) measured by elcometer.

Structures ready in shop shall be transported to site by convenient transport depending on size, type and weight of the structures. A trailer of required capacity shall be selected (20 ton/40 ton/60 ton etc.), and materials shall be carried to the site and stacked in a suitable location in close proximity to the foundation where the structure shall be erected.

15.6.3 Erection

Before erection, foundation on which the structure is to be erected shall be checked and an as-built drawing shall be made per the physical condition of the foundation whether it is made as per IFC (Issued For Construction) or not with in permissible tolerance. After confirmation of the same, an erection scheme shall be prepared duly indicating the type of crane, boom length, angle of boom, etc., along with the location of placement of the crane with all safety norms. The boom length and angle of boom, also known as lift range and lift angle, shall be ascertained from the load chart of the crane, and erection shall be taken up as shown in Figure 15.1.

Trusses up to 30 feet in length should be lifted using two pick up points positioned so that the distance between them is approximately one half the length of the truss. The angle between the two cables should be 60 degrees or less to reduce the tendency for the truss to buckle laterally during the lift. A tag line should be fastened to one end to prevent the truss from swinging and causing damage to other parts of the work or to the truss itself.

After placing the truss on the foundation, having anchor bolts will hold down the truss. Levelling of the truss shall be done by providing shim plates both flat and wedge shapesd between the top surface of the foundation and the base plate of the structure and grouted as required after attending alignment. The bolt shall be tightened by a nut and lock nut.

15.7 Glimpse on Penstock Fabrication

A penstock is a high pressure tunnel made of steel which connects the turbine scroll case with the surge shaft. The penstock shall be hydraucally efficient and structurally prepared to withstand the high pressure of the water head

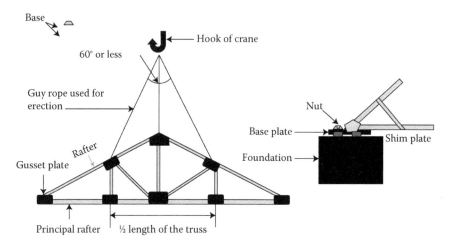

FIGURE 15.1 Erection procedure of structural component.

to prevent accidents. The penstock can be fabricated from many types of materials but steel is found to be the most friendly and suitable material for its strength and flexibility.

Location of the penstock depends on the type of dam, location of intake and outlet. The penstock is designed to carry water to the turbine with the least head loss.

The diameter of the penstock is decided by keeping an eye on both technical as well as monetary considerations.

Head loss: Hydraulic loss in the penstock reduces the effective head in proportion with its length. There are different types of head losses between the reservoir and turbine.

- Trash rack loss
- Entrance loss
- Friction loss in pipe
- Loss in bend
- Loss in valves and fittings

15.7.1 Magnitude of Loss

Magnitude of loss depends on several factors. Magnitude of entrance losses depend on the shape of entrance. Bell mouth entry reduces the loss due to its proportionate shape. Head loss occurs due to friction in a pipe. It depends on the velocity of flow, viscosity of fluid and condition of the inside surface of the pipe.

Bend loss depends on the shape of the bend and the condition of the inside surface of the pipe. Mitered bend causes more losses than smooth bend. Fittings should be fabricated smooth and streamlined inside to reduce the losses.

15.7.2 Water Hammer and Thickness of Penstock

A pressure wave is produced in the penstock if the turbine gate is opened and closed rapidly. This pressure wave is called a water hammer. In this phenomena, the velocity of flow is disturbed. A water hammer condition shall be determined for a unit operating at a rated head and under static maximum head. The highest total head, consisting of static head and water hammer head, shall be used to decide the thickness of the wall of the penstock.

15.7.3 Pipe Shell

The penstock is to be designed considering total head, which includes static head and water hammer head. The thickness of the shell shall be computed considering working stress under normal operating conditions, and it is preferred to consider yield stress for the determination of shell thickness under emergency conditions. The effect of temperature shall be considered for a penstock being supported on a pier in an open channel or on the ground. Accordingly, hoop stress due to internal pressure shall be calculated to find out thickness of the plate. However, irrespective of this stress, minimum pipe thickness is recommended for a large pipe for better handling for its rigidity, and a thinner plate can be used if it is strengthened by providing stiffeners.

15.7.3.1 Support

Penstocks in an open channel or on the ground are supported. A long pipe with multiple supports due to longer span acts as a continuous beam but at the expansion joint it loses continuity.

15.7.3.2 Joints

Joints are generally found as mainly horizontal and circumferential. 100% radiography is recommended for joint efficiency. The expansion joint is provided to accommodate expansion of the pipeline due to temperature change.

15.7.3.3 Material

Low carbon steel is considered to be the most suitable material for a penstock in hydropower. Carbon steel is very friendly for fabrication, welding and has a high ductility property. Steel conforms to ASTM specification A285 Grade B or C or A 201 Grade B, and shall be decided with respect to wall thickness of the penstock. Flanges, fittings, valves etc., are generally forged or rolled.

15.7.3.4 Penstock Accessories

- Temporary support
- Piezometer
- Air inlet
- Drains
- Manhole
- Walkway

15.7.3.5 Equipments and Machinery Required

- Rolling machine
- Presses
- Flame cutting machine
- Welding machine
- Radiography arrangement
- Testing lab
- Hydra
- Cranes
- Trailers etc.

15.7.3.6 Fabrication

Fabrication of the penstock can be done either in shop or in a convenient place near the dam site. The penstock is fabricated in a length which is comfortable for easy transportation and erection in position. It is generally fabricated in lengths varying from 6 to 12 m and the number of probable joints are also taken into consideration.

15.7.3.7 Straightening and Levelling the Plates

Plates brought to the fabrication site need to be straightened and levelled before taking up any action, and some 10% of plates shall be tested for an integrity check by conducting NDT to ensure that flaws, fissures or any cleavages that exist are within the permissible limit. At least 10% of the quantity shall be subjected to the integrity check.

15.7.3.8 Cutting the Plates to Required Dimensions

Plates are to be cleaned from dart, rust, etc., and to be cut for exact dimensions.

15.7.3.9 Preparation of Edge

The surface and edge of the plate shall be cleaned of dirt, dust, paint or any grease present on the surface, and it should be cleaned by mechanical means or by prescribed chemical means. The type of edge preparation of base metal depends on the type of welding procedure. Edges should be smooth, uneven and free from cracks to have quality welding of the joint.

15.7.3.10 Rolling and Bending

Rolling and bending is done on plate to have the required diameter.

- *Tack welding*: When plates are bent and rolled into the desired shape, size and dimensions, joints are tacked as required to facilitate flawless final welding.

15.7.3.11 Welding Procedure

Before taking up welding activities welding procedure qualifications (WPS), WQPT and welder's qualification tests shall be conducted. The welding test plate and testing of specimens shall be conducted in the presence of the welding inspector and welding of the product shall be done as per plan.

After welding of joints are completed, testing of joints are to be taken up with radiographic inspection if specification prescribes it. Plates of thickness more than 25 mm used in the work shall be subjected to preheating before welding which reduces cooling rates as required.

15.7.3.12 Inspection of Welding Job

Welding conducted in joints, both horizontally and circumferentially, as per approved WPS shall be checked and tested by the inspection engineer. This inspection is beyond the purview of the construction engineer. The inspection engineer will check the welding process regularly.

After completion of the butt weld, joints that are completed to be tested by means of radiographic process, either by X-Ray or Gamma Ray in order to detect defects and any flaws present in the completed weld joints. The defects are generally observed in the nature of cracks, under cut, porous, incomplete fusion, entrapped bubbles, slag, etc., which make the weld weak. Joints are radiographed, and may be 100% conforming to specification of work. The inspection engineer is to interpret and judge the acceptability of radiography. All films shall be marked and preserved as per specification.

There will be some joints where radiographic inspection is ruled out due to some constraints, and dye penetration tests, magnetic particle or ultrasonic tests can be conducted as required.

15.7.3.13 Hydro Test

It is preferable to do a hydro test after installation. It can be done in parts. It should be done with a pressure as specified in the specification and ensure that the plates and weld joints are able to withstand pressure for safety. Test pressure is designed considering the thickness of the plate, internal diameter and allowable hop stress.

15.7.3.14 Quality Assurance Plan

Quality of work, also known as fabrication, shall be carried out in accordance with the agreed and approved Quality Assurance Plan including inspection of material, fabrication, welding, erection, etc.

Bibliography

A Water Resources Technical Publication, Engineering Monograph No-03, Steel Welded Penstocks, Bureau of Reclamation, United States Department of the Interior.

16

Environmental Assessment of Hydropower Projects

16.1 Introduction

The world's population is growing in leaps and bounds. This rapid population growth presents challenges to meet the energy needs of the population. One of the safest energy sources that can control the emission of greenhouse gases is hydropower. Hydropower is a renewable energy and an alternative to fossil oil, and it is now widely used to protect the environment. Hydropower contributes to almost 20% of total power generation worldwide. Huge efforts are under way to leverage this energy in many developing countries where these resources remained unexplored.

Hydropower is technically and economically feasible. Its initial cost is high, but maintenance costs are low. Hydropower is a technological blessing, but has some adverse effects on environment. Hydropower projects cover large areas for the formation of reservoirs, requiring water conducting systems including dams, low pressure tunnels, surge shafts, high pressure tunnels or penstocks, power houses, and tail race tunnels. Additionally, hydropower coverage of large areas affect environmental resources and has a tremendous impact on the natural environment. Fresh water ecology is already in crisis, with many rivers having already been fragmented due dam construction. Vast amounts of water are being drawn for irrigation purposes while rivers are diverted. All of these features that are part of hydropower projects damage the ecosystem and subsequently affect social systems and harm both environment and the socioeconomic structure of the area. Hydropower projects require environmental impact assessment studies before they get environmental clearance. It is mandatory for any hydropower project to get environmental clearance if it has an adverse effect on the environment.

16.2 Environmental Impact Assessment (EIA)

EIA is an essential and key factor for the planning, development, and execution of any hydropower project, and shall be taken up before implementation of the project.

The probable effects of the activities pertaining to the development of any hydropower project on natural environment due to the utilization of natural resources are known as environmental impacts.

The assessment and evolution of environmental effects of these activities being undertaken by human beings for development is called the environmental impact assessment (EIA).

This environmental assessment process is a challenge to reduce and prevent negative impacts on the environment due to development. This assessment is not designed to slow development but to guide the project authority to ensure that the construction of the project will have the least amount of effect on the natural environment.

There is worldwide awareness about environmental assessment. It was started in the United States, and the United Nation Environmental Program (UNEP) believes that EIA is a tool that can identify environmental, social, and economic impacts of a project before it is implemented.

UNEP is a principal UN body in this field, and has also implemented a project about environmental and human rights with OHCHR concerning good practices in the field of human rights and the environment.

The idea is to strengthen the environmental policy by using human rights obligations as a link between human right and environmental law.

16.2.1 Environmental Impact Assessment and Strategic Environmental Assessment

Environmental impact assessments and strategic environmental assessments (SEA) are the only tools required by law in many countries.

EIA and SEA is a well formulated method to collect, assess, and evaluate information regarding the environment before making any decision in the development process of the project. This information is nothing but predictions of how the environment will get changed due to the project's impact and how best to manage the environmental change if alternatives are implemented. Environmental impact assessments focus on proposed physical development such as highways, power stations, and water resource projects or hydropower projects, while SEA focuses on proposed actions at a higher level such as new and amended laws, policies, programs, and plans.

Strategic environmental assessments are formal, systematic processes that analyze and address the environmental effect of policies, plans, programs and other strategic initiatives. It extends the original goals to a higher level of decision making where major options are still open, and allows the problems of environmental deterioration to be addressed in the policy and plan making process rather than mitigating their downstream symptoms.

16.2.2 Aims and Objectives of EIA

The aims and objectives of EIA vary from country to country and lead to sound decision making on effective environmental protections. It takes place before major decisions are taken while feasible alternatives and options to a proposed action are still open.

- To facilitate a systematic consideration of environmental issues as a part of development.
- To analyze potential environmental effects on specific project proposals.
- To indicate the process of how the potential effect can be prevented or reduced before making a final decision.
- EIA systems suggest that the approval process may be linked to regulatory authorization including the issuance of permits and licenses required for the project's forward progression.

16.2.3 Scope of EIA

Crude oil refineries, major thermal power plants, chemical plants, waste disposal incineration, major roads, nuclear fuel, smelters for iron and steel, water disposal installations, dams and similar installations, and pipelines transporting oil are examples of projects where ETA must be carried out.

Quarries, waste water treatment, coke ovens, agriculture, aquaculture, energy industries, mining industries, metal, and textile projects are also examples where ETA shall be carried out if the project has an impact on the environment.

16.2.4 General and Specific Principles of EIA Application

Effectiveness of objectives can be improved by using general principles and guidelines.

- EIA shall be applied as a tool in helping to achieve sustainable development.
- EIA should be applied as a tool to implement environmental management rather than as a report to gain project approvals.

- EIA should not be a one time process. It should be linked with the project lifecycle so that correct information can be provided to authorities at the right time and to ensure that changes can be incorporated to avoid adverse impacts as far as practicable in the project.
- EIA should be applied to all proposed actions that will cause adverse effects on the environment and human health.
- EIA should include the analysis of feasible alternatives to the proposed actions.
- EIA should involve public participation.
- EIA should provide information on social, economic, and biophysical impacts.

Some specific principles and guidelines:

- EIA should identify which project should be subject to EIA as per scope of EIA.
- Terms of reference (TOR) for EIA and EMP should be prepared.
  - Project details like coverage, master plans, topographical maps, scope, and estimated costs.
  - Required resources and manpower should be identified.
  - A description of the environment, land and soil environment, water and noise environment, the socioeconomic environment, and the biological environment should be evaluated.
  - Anticipated environmental impacts and mitigation measures should be made.
  - Environmental monitoring plan.
  - Public consultations.
  - Risk assessment and disaster management plans.
  - Natural resources conservation.
- R&R action plan.
- EMP.
- Environmental cost benefit analysis.

16.2.5 Some Key Factors of EIA Systems

The following are key factors for improving environmental assessment for hydropower projects.

- Political support and commitment.
- Legal basis for regulation and guidelines.
- Public involvement.

- Activities that have significant environmental effect.
- Processes, procedures, and mechanisms for quality of EIA reporting.
- Technical and professional ability to carry out EIA.

16.3 EIA Process

The EIA process is a quality control and assurance system which starts from conceptualization and pre-feasibility and progresses to feasibility, design and engineering, construction to monitoring, and final evaluation of the completion of a particular project. It is a step-by-step process that covers the background and details of the project with the goal of supporting the development process and collecting information on the environment. All of these studies assist in determining ETA scope per its classification, with the applicability of legal and administrative framework acting as guidelines. The identification of key impacts on the environment, its predictions duly defining the measures of avoiding or reducing effects, and mitigation measures and public involvement is essential in the process. The EIA report shall be reviewed for quality in accordance with TOR and with necessary follow up and inspection procedures.

- Examining the project details such as location of the project, type of proposed development, and the magnitude of investment to determine whether the project requires environmental clearance per statutory requirements of the EIA. This process is known as screening.
- Scope is defined in accordance to the category of project available in the category list. Quantifiable impacts shall be assessed on the basis of magnitude, frequency, and duration, while non-identifiable impacts shall be assessed through socioeconomic criteria, with alternative considerations also being completed. This process is known as scoping.
- Study legal and administrative frameworks and their applicability.
- Study the collection of existing environmental information and describe environmental status of identified study areas like air, noise, water, land, biological, and socioeconomic conditions.
- Preparation of TOR with clarity of requirements.
- Identifying key impacts.
- Predictions and mapping of environmental consequences of the project's significant aspects.

- Assessing impact significance and adverse effects in sensible areas and their magnitude through best impact analysis procedure.
- Defining mitigation measures with options of avoiding and reducing the effects.
- Public consultations and participation for suggestions.
- Preparation of environmental statements.
- Submission of environmental statements and applications for approval.
- Environmental statements shall be reviewed. The quality of report shall be assessed in accordance with the requirements indicated in the TOR, with the information reviewed as to whether it is adequate for decision making.
- Modifications are negotiated and incorporated.
- Implementation decisions are made.
- Oversee implementation and compliance.
- Regular inspections to be conducted to ensure that all stipulations are implemented.
- Monitoring and audits manage the status of predicted impacts and any unanticipated impact that are encountered.
- Review and reassessment with remedial action.
- Final reporting.

16.4 Roles in EIA Process

Many people and groups are involved in the EIA process with various responsibilities.

The bodies are as follows:

- Project proponents
- Project management consultant (PMC)
- Environmental consultants
- Environmental scientists
- Political scientists
- Politicians
- Pollution control boards or committees
- Public

- Impact assessment agencies
- Economists

16.4.1 Project Proponent

A project proponent is an individual, group, government, or private institution that carries out the project. The project proponent decides the type of project to be undertaken. It is the responsibility of project proponent to prepare detailed project reports, with project details incorporated in the EIA.

The proponent has to approach the concerned authorities for NOC and to hold the public hearings. After the public hearings the proponent submits the application for environmental clearance.

16.4.2 Project Management Consultant

A project management consultant is a company that is responsible for the execution of the project from start to completion.

16.4.3 Environmental Consultant

The environmental consultant should have adequate knowledge of existing and procedural requirements for EIA clearance. The EC is responsible for obtaining environmental clearance for the proposed project. The consultant is contractually obligated to guide the proponent through the initial screening and scoping of the project, determining whether EIA studies are required for the particular project and finalizing the scope of such studies. The EC will conduct EIA studies and collect all information. The environment related information required for the EIA will help the project proponent justify all findings in the EIA and EMP during the meeting with the expert authority groups for clearance.

16.4.4 Environmental Scientist

The environmental scientist makes predictions about the environment, environmental changes, and consequential effects, reductions, and alternatives to properly mitigation adverse issues.

Political science: A hydropower projects covers vast areas and responds to concerns about environmental change.

Public: These are stakeholders and they should be consulted on the EIA prepared by the project authority. They may be directly or indirectly affected by the project. This may lead to certain complications such as translating the documents and presenting the information to the public. Any upgrade or addition incorporated in the report shall be notified and brought to the public in a transparent way and in the language they understand.

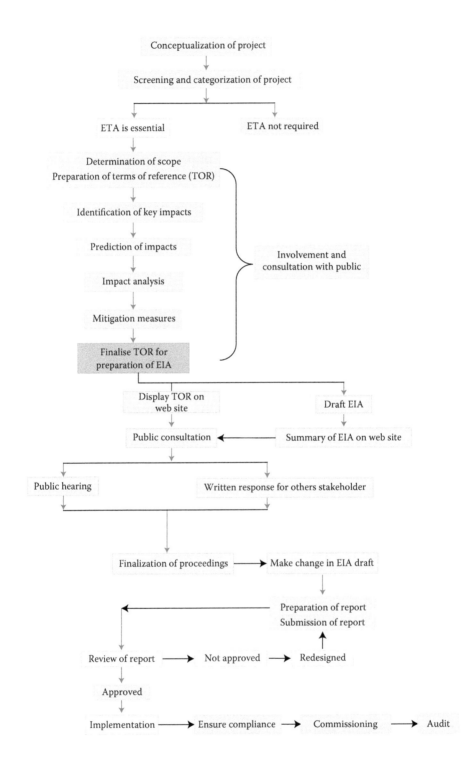

16.5 Effectiveness of EIA

The EIA should not be treated as a medium to obtain project clearance but to ensure the all terms and conditions are implemented in order to cause minimum impact on the environment and to help acquire the maximum amount of benefits from the project.

The EIA's effectiveness depends on the following:

- Type of legal assistance rendered by the country.
- Manner in which the ETA is conducted.
- Quality of the ETA report.
- Knowledge, skill, and experience of the professional who prepared the report.
- Competency and skill of the review committee.

16.6 Benefits of ETA

Major government and private clients accept the benefits of environmental assessment to improve project design. EIA has major benefits, including:

- Ease of approval procedure.
- Make project economically effective by taking all adverse effects into consideration.
- Enhance productivity of workforce.
- Financial institutions more readily accept proposals.

16.7 Cost of ETA

Expenditures will be incurred for both carrying out the ETA and the subsequent preparation of the report. Approximate expenditures should be considered to be up to 10% of capital costs depending on the location and size of the project.

16.8 Environmental Impacts due to Hydropower Project Construction

Hydropower construction projects cover a large area. This vast area is required to accommodate the project's reservoir, water conducting system, power house, tail race tunnel, diversion of the main stream, and large amounts of deforestation. These activities change the entire environmental structure of the project area.

The impact of hydropower projects on the environment depend on the size and type of project. Hydropower does not emit any greenhouse gases, but the large reservoirs and diversion of natural water flows result in the fragmentation of many rivers.

Here are some environmental and socioeconomic impacts due to the construction of hydropower projects.

- *Land use and land cover*: Hydropower projects cover vast areas. The vastness depends on the head of the river and the topography of the area. The area of the reservoir is directly proportional to the head of the river, with reservoir area increasing with larger project heads. The amount of area required for a reservoir in flat terrain is larger than the area required in hilly terrain, and results in additional loss of wild land, wetlands, wildlife habitats, and human habitat.

  Blasting operations involved in hard rock excavation and quarrying generate noise and dust, frightening wildlife and compelling it to leave its territory. Blasting initiates future earthquakes in the area due to the release of strained energy.

- *Before impounding of reservoir*: The entire reservoir area shall be cleaned, and deforestation in the reservoir area will occur before impounding, destroying wildlife habitats and agricultural land.

- *Impounding of reservoir area*: Vegetation in the reservoir area decomposes and forms carbon dioxide and methane.

  Reservoir impounding imposes additional loads on the surface and it makes substructures vulnerable. Impounding has an impact on wildlife within the reservoir area and the area in the down stream of the facility.

  Reservoir water becomes stagnant, and this stagnant water will have high sediment and nutrients resulting in the formation of algae and other aquatic weeds on the lower surface of the reservoir. Sometimes, silting occurs and decreases the depth of water resulting in the reduction of storage capacity.

  Water becomes less in the portion of river downstream of the dam. The drop in water levels will harm plant and animal life. Reservoir water contains low dissolved oxygen and the water is colder than flowing river, which will have negative impacts on plants and animals. Water also becomes toxic and harmful for fish. Fish migration gets blocked. A earthquake may be triggered if any dam is constructed near the vicinity of any active fault and also due to frequent filling and emptying of reservoir. The sudden release of water from the dam causes flooding downstream of the facility and also causes landslides in the bank of reservoir. Socioeconomic effects include the displacement of people to other locations which creates additional burden on existing infrastructures such as hospital,school etc. It also has an impact on local people and on their cultures and traditions.

*   *Public health*: Water born diseases and spread of disease through migration of labor force from different parts of the country.

16.9 Terms of Reference of Hydropower Projects

The terms of reference is a systematic study of the details of proposed hydropower projects to fix the environmental priorities for carrying out EIA and TOR, and should be able to guide the preparation of the EIA report and subsequent submission to the appropriate authority with the following details.

*   *General information*: Proposal with background on the project along with details of the developer, needs and purpose of environmental assessment, etc.
*   *Detailed description of proposed project*
    *   Location and coordinates
    *   Installed capacity in MW
    *   Name of river and tributary
    *   Average annual rainfall
    *   Maximum flood level
    *   Periphery of reservoir
    *   Minimum water level in reservoir
    *   Full water level in reservoir
    *   Dead storage of reservoir
    *   Type of dam, length, height, and slope of upward and downward slope
    *   Crest level
    *   FRL
    *   HFL
    *   Type of energy dissipation
    *   Maximum discharge
    *   Water intake structure, including its type and size
    *   Low pressure tunnel, including its length, shape, diameter, wall thickness, and discharge

- Type of surge shaft, including height and size
- High pressure tunnel or penstock, including length, type, diameter, and wall thickness
- Power house, including size and height
- Installed capacity
- Type of turbine, capacity, and number of turbines
- Tailrace tunnel, including length and size
- Switch gear

- *Legal and administrative framework*: Different countries have adopted environmental protection acts for wildlife and plants to preserve the biological diversity and pollution control in air, water, and the environment.

  In the United States
  - Clean Air Act
  - Clean Water Act
  - Endangered Species Act
  - Comprehensive Environmental Response Compensation and Liability Act
  - Resource Conservation and Recovery Act
  - Safe Drinking Water Act

  In India
  - Prevention and Control of Pollution of Air, Water
  - Wildlife Protection Act
  - Forest Conservation Act
  - National Environmental Appellate Authority Act
  - Hazardous Waste Management Rule
  - National Environmental Tribunal Act
  - Convention on National Trade in Endangered Species of Wild Fauna and Flora
  - Convention of Biological Diversity

- *Baseline environmental data*
  - *Land use*: Land covering the entire project area shall be mapped in detail via standard methods. This mapping includes details of forest, cultivable land, lakes, rivers, glaciers, etc., with special attention in the area around the dam and power house.
  - *Physical feature of land/physiographic*: Location of the proposed dam site on the river that includes details of the river and its

tributary and catchment area, low pressure and high pressure tunnels, etc.

- *Geology and seismology of terrain*: Detailed geological features of the region shall be described along with active faults, folds, clay seams, classification of rock strata, and categories of seismic zones with seismic activity.

- *Soil characteristics*: Description of soil characteristics in the entire project area that includes catchment areas, dike areas, and borrow areas.

- *Air quality and noise pollution*: Pollution in project areas is due to construction activities and the movement of construction equipment. Noise pollution is due to frequent blasting operations for quarry and excavation in hard rock. Change in ambient levels and ground level concentration and their effects on soil, water, vegetation, and human health shall be assessed.

- *Water quality*: Study of flowing water shall be carried out in summer, winter, and rainy seasons, with changes in quality and salinity assessed.

- *Biological*: Study of the effect of deforestation on animal habitats, effects on flora and fauna, aquatic flora and fauna such as algae, aquatic vegetation, invertebrates, fish, and endangered species shall be assessed.

- *Socioeconomic environment*: Study of effect of the hydropower project's effect on the surrounding area's demography, health, education, culture, heritage, and people.

- *Report preparation*
  - The timeframe shall be fixed for carrying out EIA, which should be indicated as per the established line chart or PERT network so the study can be completed within the stipulated time.
  - *Budget*: Total cost of the work shall be estimated which includes resettlement plans, mitigation plans, audits, and environmental management plans.

- *Likely environmental consequence*: Already explained

- *Mitigation measures, environment management plan, and audit plan*: Mitigation measures shall be summarized and the required budget shall be estimated. Roles and responsibilities of concerned group shall be indicated.

- *Monitoring plan*: This part of TOR includes the monitoring details of all construction and operation management of the project. Roles and responsibilities of each concerned agency shall be indicated. A detailed monitoring schedule shall be made, and expenditures to be incurred shall be indicated.

16.10 Anticipated Impact, Mitigation Measures and Environmental Management Plan

Hydropower project construction covers large areas and consists of impacts to air, water, noise pollution, displacement, loss of livelihood, education, culture, health, heritage, and forest biodiversity. There are many methods to accomplish assessments through questionnaires, checklists, and network methods. A simple, logical, and systematic approach shall be applied to identify impacts.

Mitigation is a method or process which prevents or reduces negative impacts to a certain level of acceptance. Mitigation measures adopted for various adverse impacts are documented in what is known as the environmental management plan (EMP).

The EMP should reflect the mitigation measures to be adopted for each impact within scheduled timeframes and with details of personnel responsible for implementation, budget, and monitoring.

Mitigation measures are adopted for projects in various ways depending on site conditions as indicated below.

- Change of site location
- Change in design such as type of scheme, lowering height of dam, change in reservoir area, etc.
- Providing compensation or incentive schemes
- Rehabilitation scheme
- Providing health programs and better infrastructure
- Education and training facilities
- Employment

16.10.1 Various Environmental Management Plan

The environmental management plan is made for create, implement, and monitor environmental protection measures during the execution and post commissioning of the project.

16.10.1.1 Catchment Area Treatment (CAT) Plan

The catchment area treatment plan is adopted for the betterment of the catchment area. The improvement of this area can be done by stopping the erosion of soil through actions taken during construction and operations.

Detailed surveys shall be conducted to identify the areas vulnerable to erosion and various functions like slope, type of soil, and other factors are classified to draw a treatment action. Some biological impact shall be adopted with due consultation with the forest department.

The total free area and total affected area shall be estimated along with estimated costs for erosion control.

Necessary erosion control measures to be adopted:

- Building of check dams on slope
- Building of bund on the contour
- Plantation
- Landscape development
- Nursery development

16.10.1.2 Management of Air and Noise Levels

The main purpose of this plan is to control pollution. Various practices should be adopted to keep pollution levels under control and costs shall be estimated.

- Transportation of materials in trucks shall be done with proper cover.
- Sprinkling of water on road to reduce dust.
- Speed control of moving equipment shall be adhered to.
- Proper arrangements shall be taken in stone crushers to reduce dust.
- Musk shall be given to all workers.
- Ear plugs shall be issued to all workers.
- Crushers, rock drilling machines, and other heavy equipment shall be appropriately covered to reduce noise levels.
- Machine maintenance shall be done.
- Development of green belts.

16.10.1.3 Water Management

During construction, appropriate measures will be taken for controlling water pollution.

- Flow of water from dumping area and area from concreting shall be controlled by providing check dams in steep slopes.
- Proper sewage treatment plants shall be arranged for labor and staff colony to control water pollution.

16.10.1.4 Public Health Management Plan

The purpose of this plan is to provide medical help to the construction workforce, local population, and affected people.

- First aid box shall be provided at site.
- Setting up health care centers in various locations.

- Qualified doctors and nurses are to be recruited to provide treatment.
- Dedicated vehicles shall be stationed on site for emergency medical help to transport affected worker to the hospital.
- Surrounding and vulnerable areas shall be kept clean to stop the breeding of mosquitoes.
- Adequate immunizations and vaccinations shall be provided.
- If necessary, a mobile clinic facility shall be provided.
- Costs shall be estimated.

16.10.1.5 Waste Management Plan

Proper waste management plans shall be implemented to collect waste being generated in the project area. This plan includes the collection of solids and liquids. Septic tanks, community toilets, and mobile toilets shall be arranged and costs shall be estimated.

16.10.1.6 Muck Disposal Plan

Muck shall be generated from various construction sites such as excavation at the dam site, tunnel, power house, dike, road, etc. This large quantity of surplus materials shall be transported and dumped in marked dumping areas, and care should be taken so that it does not affect terrestrial and aquatic environments. Costs of the plan shall be estimated.

- Identify sources of muck generation and identify muck disposal areas while considering environmental and social impact.
- Muck disposal areas shall be approved by authorities so that the dumping of muck should not affect the river and any body of water.
- Dumped muck shall be properly leveled and compacted.
- Necessary voids shall be considered during muck quantity calculations.
- Mode of transportation and compaction machinery and equipment shall be decided for calculation of costs.
- Necessary planting shall be done on leveled muck.

16.10.1.7 Fuel Wood Energy Management

Wood is the only fuel available in remote areas like hydro projects. In order to maintain balance, arrangements for LPG, kerosene, solar panels, and other power sources shall be made available, and people should be encouraged to use pressure cookers and solar cookers to conserve energy.

16.10.1.8 Disaster Management Plan

The disaster management plan details worst case scenarios during construction and operations. It may refer to any number of severe nature incidents like earthquakes and dam collapse. This plan is to find the risk of maximum credible failure under maximum flood conditions within the past 100 years. This plan includes prevention measures, mitigation, preparedness, recovery, and rehabilitation plans. It is made for any unforeseen event due to flooding, dam breaks, cloud bursts, or other disasters. Project authorities should adopt the following measures:

- Provision of forecast system.
- Flood-proof communication systems.
- Emergency control center with all possible communication methods such as phones.
- Arrangements of siren and wireless systems to alert people to the situation.
- Arrangements such as boats, rope, medicine, and first aid boxes for emergencies.
- Raised platforms shall be made in strategic locations, level platform shall be above HFL, and adequate numbers shall be close proximity to the health center.

16.10.1.9 Green Belt Development

Green canopies have the capacity to absorb pollution. Green belts shall be developed in project areas and around the periphery of reservoirs in order to decrease pollution and avoid any landslides from the portion of catchment area that are a result of the flow of rain water directly into the reservoir.

- Area to be covered, number and type of trees, and the density of the trees shall be decided in consultation with forestry officials.
- Fast growing trees are preferable.
- Costs associated with each area are to be calculated.

16.10.1.10 Biodiversity Management Plan

The biodiversity management plan shall be used for the protection and conservation of biodiversity within the area.

- Conservation of crucial habitats.
- Conservation of critical and important plant and animal species that are affected.

- Traditional knowledge of locals shall be protected.
- Restriction shall be imposed on laborers so that they do not harm forest and animals.
- Noise shall be kept at minimum levels.
- Blasting operations shall be controlled, as its noise scares animals from their territory.
- Expenditures shall be calculated.

16.10.1.11 Rehabilitation and Resettlement Plan

Displacement is a crucial issue in the construction of hydroelectric projects because it covers large areas. Large land plots from locals need to be acquired. Land acquisition from local populations shall be done systematically, and best practices shall be formulated to avoid future complications.

- Rehabilitation and resettlement (R&R) plans for affected populations shall be developed in accordance with policies of the local authorities.
- Plans shall consider all issues that affect the local population, and land should not be acquired without consent of affected population.
- R&R plans should include compensation, benefit sharing, community development, settlement areas, and active roles in the project for local committees, governments, and power companies.
- Compensation for land shall be calculated based on current market rates.
- Rehabilitation areas shall be chosen in close proximity to the project area so that the benefit of the project can be extended to affected population.
- Employment of at least one member from each affected family shall be provided.
- Provisions for land compensation shall also be included.
- Life-long pension plans shall be considered for old, widowed, and disabled persons.
- Financial assistance and training shall be provided for self employment of local population.

- All basic amenities shall be provided in the rehabilitation area.
- Total expenditures shall be estimated.

16.10.1.12 Fishery Development Plan

Reservoir areas and other bodies of water shall be considered in this plan.

- Hatcheries shall be constructed in consultation with local fishery officials for the conservation of indigenous species.
- Identification of fish stocking zones.
- Estimated costs for the creation of fish stocks.
- Evaluate the feasibility of fish production in new water.
- Search new area for fish production.
- Identification and selection of fish species.
- Design suitable fish farms and methods of fish farming.
- Design civil requirements.
- Plan for minnows.
- Determination of unit and total cost of fish farm development.
- Cash flow analysis, etc.

16.11 Application Prior to Environmental Clearance

The project proponent will identify the project site and conduct the feasibility study. After the feasibility survey is over, the project proponent can apply for environmental clearances with feasibility reports and a draft TOR.

Environmental clearance is required before starting construction work or preparing the land on the identified site.

16.11.1 Procedure for Environmental Clearance

Once the project proponent identifies the location of a project, he is to assess whether the proposed project falls under the purview of environmental clearance as per the guidelines of the statutory authority. If it satisfies the established rules, then procedures for environmental clearance shall be initiated.

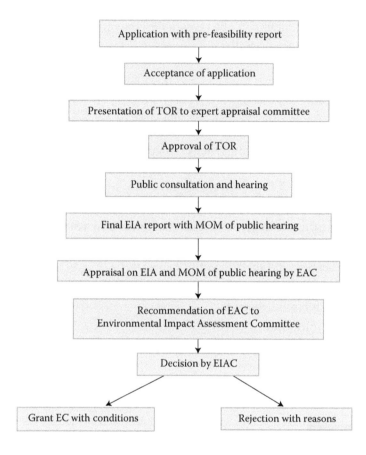

16.11.2 Indicative Application Form (IAF NO 16.1)

An indicative application form follows to give the reader an idea of what the form contains:

Sl No.	Item	Details
I. Basic Information		
1	Name of the Project	
2	Type of Project	
3	Proposed Capacity of the Project	
4	Category of the Project	
5	Location of the Project	
6	Access to Project. Nearest Railway Head/Airport/City/Town	
7	Name of Project Proponent	
8	Registered Office and its address	

9	Name of authorized representative		
	Designation		
	Address for communication		
	Email, Phone number		
10	Name of Project Management Consultant		
	Address		
11	Name of Environmental Consultant		
12	Any interlinked projects		
12	Date of submission of interlinked project (if any)		
Legal and Administrative Framework			
13	Prevention and Control of Air and Water Pollution		
14	Wildlife Protection Act		
15	Forest Conservation Act		
16	National Environmental Appellate Authority Act		
17	Hazardous Waste Management Rule		
18	National Environmental Tribunal Act		
19	Convention on National Trade in Endangered Species of Wild Fauna and Flora		
20	Convention of Biological Diversity		
21	Any litigation pending against project or land where the proposed project is to be built		
22	Name of court		
23	Case information		
24	Forest land involved in the project		
25	Whether any government order/policy is relevant to the site		
26	If project is located in ecologically sensitive or fragile area		

II. Activity

Likely Environmental Effect for the Proposed Physical Changes in the Location (Topography, Land Use, Changes in Water Bodies, etc.)

Sl No.			
1	Land environment Land use Permanent or temporary change in land use, land cover or topography including increase in intensity of land use (with respect to local land use plan)		
2	Land details for various project components		
	Total loss of civil and government land		
3	Agriculture		
	Irrigated, Non-irrigated and Forest type		

4	Soil		
	Pre-construction investigations		
	Soil quality in project and catchment area in monsoon, winter, and summer seasons		
	Geological status		
	Seismic hazards		
5	Water		
	Surface water quality in catchment and project area in summer, winter, and monsoon seasons		
	River water quality		
	Quality of drinking water		
	Quality of ground water		
	Temperature and pH value of water in project area		
	Water availability after and before project work		
6	Air		
	Ambient air quality in dam area during all seasons		
	Ambient air quality in other area during all seasons		
7	Noise		
	Noise level in all seasons in project area		
	Noise level due to working of construction equipments		
	Noise level due to blasting operation		
6	Biodiversity		
	Flora and fauna diversity		
	Major forest products and dependence of local		
	Forest type		
	Trees, Herbes, etc.		
	Rare and endangered species		
	Endemic species		
	Species of special interest		
	Migratory routes of animal		
	Aquatic ecology density or diversity and abundance of aquatic species		
7	Historical, archaeological considerations		
	Archaeological sites or monuments		
	Places of warship		
8	Socioeconomic aspects		
	Demographic profile		
	Rural/Urban		
	Population density		
	Literacy		
	Employment and occupation		
III. Usage of Natural Resources in Work			
1	Land		
2	Water with source		

3	Mineral		
	Aggregates		
	Sand		
	Soil		
4	Forest materials		
5	Consumption of electricity and fuel		
Activities Related to Construction that Harm Human Health or Environment			
1	Use of hazardous substances harmful to human health		
2	Use of hazardous substances harmful to flora and fauna		
3	Waterborne disease		
4	Change of living conditions		
	Endemic health problems		
	Epidemic prevention and control		
IV. Tackling of Production of Solid Wastes During Construction			
1	Soil erosion at construction site		
2	Muck generation		
3	Hazardous wastes (as per Hazardous Waste Management Rules)		
4	Other industrial process wastes		
5	Surplus products		
6	Sewage sludge or other sludge from treatment efforts		
7	Construction waste		
8	Scrap generation		
9	Contaminated soils or other materials		
V. Operational Activity that Harm Health of The Society			
1	Reservoir evaporation loss		
2	Change in water quality and risk of eutrofication		
3	Increased incidents of waterborne diseases		
4	Impact on fish and aquatic life		
5	Public health		
VI. Activities that Lead to Development of Society			
1	Clean and renewable sources of energy		
2	Employment opportunities		
3	Catchment area treatment		
4	Recreation and tourism potential		
5	Additional aquatic habitats		
6	Wildlife and wetland species		
7	Fisheries and aquaculture potential		
8	Economic benefits		
9	Air pollution reduction		
10	Reduction in greenhouse gas emission		

16.12 Case Study: Tehri Hydropower Complex in India: Problems Faced and Lessons Learned

The Tehri Hydro Complex with a capacity of 2000 MW was developed with the intention of supplying much needed power to the northern power grid in India. Two lakes were formed by two dams at heights of 260.5 and 103 m, with both covering almost 44.65 km² that was submerged during operations. Submergence of these areas affected Tehri Town, as well as had full effects on 24 villages and minor effects on 88 other villages, with an additional 13 villages affected due to acquisition of land for project and township construction. Affected people were properly rehabilitated in the new environment.

Hydropower projects generally face obstacles due to their use of vast area. Problems arise for the acquisition of land and the R&R plan. In this project families who had 50% or more land acquired by the project were treated as fully affected, and families who had less than 50% of their land acquired by the project were treated as partially affected. Hydropower is a time consuming project where the degree and nature of problems changes with time. Resettlement of affected people may not be a problem in initial stage, but becomes more complicated as time passes. Affected populations may increase while land available for resettlement simultaneously becomes a scare commodity. All these factors slow the pace of construction resulting in delays in completion.

- *Problem faced*:
  - *Delay in construction of project*: R&R and construction of the project started simultaneously that affected both activities. R&R was delayed due to project design changes that affected funds and new policies were made. Blasting operations were banned due to some agitation which caused delays in construction of the tunnel and the relocation of people was affected. Delays due to all these reasons resulted in frustrations and social disorganization.
  - *Delay in R&R activity*: R&R activity was affected due to multiple changes in the agency handling R&R, land acquisition efforts, non-availability of land, and frequent changes of R&R packages. Due to the non-availability of land, people didn't want to become homeless and jobless, so any proposal for land acquisition was initiated but refused. Ultimately, forest land was converted for resettlement.
- *Lesson learned*: The following lessons were learned during the construction project:
  - Socioeconomic and demographic surveys shall be taken before the project is started.

- Surveys for self employment or any income generation efforts or start-ups shall be conducted.
- Surveys for the identification of facilities affected in areas cut off from the reservoir and in the area above submergence.
- R&R packages shall be created based on surveys done in association with affected groups, authorities and social scientists.
- It is noted that sound R&R packages do not always guarantee success. A detailed resettlement action plan (RAP) shall be formulated to further strengthen R&R efforts which will provide information in the following heads:
  - RAP objectives
  - Parameters of project design
  - Project affected people eligibility
  - Estimate of population
  - Assessment of resettlement site
  - Development of resettlement site
  - Income restoration program
  - Budget
  - Implementing group
  - Participating and monitoring systems

Bibliography

Directorate for internal policy, U.S. Environmental policy.

Environmental impact assessment and strategic environmental assessment towards an Integrated approach, United Nation Environmental Program (UNEP).

Executive summery of EIA and management plan of HEO hydro power project, HEO Hydro Power Ltd.

Guidance on application of the environmental assessment procedure for large transboundry project, European Commission.

Impact of Tehri dam lessons learnt. Water for Welfare Secretariat Alternate Hydro Energy Centre, Indian Institute of Technology, Roorkee.

17

Estimation and Cost Analysis of Hydropower Project

17.1 Introduction

Hydropower is a very mature technology which takes a long lead time for development and construction. It consumes more time for feasibility, planning, design, engineering, construction and commissioning. Hydropower is one of the most reliable energy forms which can help countries promote development.

Developers or investors should have an idea and concept of the costs and benefits of this renewable energy. There should be enough data and information available so that the decision makers/developers/investors or governments are able to assess the status of the energy and they can decide the role of this energy for the development of the countries.

Analysis and cost estimates will provide significant data and information necessary to compare hydropower with other energies for acceptability with useful insight and additional functionality or quality.

17.2 Process of Capital Cost Estimate of Hydro Project

Financial viability is a key parameter, other than the technical viability, of any project. To assess this financial viability, a hydropower project requires an estimate of capital costs, operation and maintenance costs, energy production costs, levelized costs of energy and related tariff. All this information, data and initial assessment will prompt investors/developers to make the decision to go ahead with the project.

The total investment cost of any hydropower project is site specific and it depends on topography, geotechnical and geological characteristics, hydrological features, design choice, accessibility to site and other aspects related to socioeconomics and the environment.

The process of capital cost estimates is a computational process where engineers of different disciplines are involved to develop a specific estimate.

17.2.1 Initial Capital Cost (ICC)

ICC indicates total capital scenario necessary for a hydro plant to succeed from prefeasibility, exploration, investigation, design, engineering, construction, commissioning and operation.

17.2.2 Cost Break Down Structure

Any major hydropower project is large and complex, and the entire project shall be broken down into several capital expenditures (CAPEX) components like infrastructures, site and project development, civil work, electro-mechanical equipment, auxiliary mechanical and electrical equipment, land, audit, accounting charges, etc.

17.2.3 CAPEX Components of Project

CAPEX include all cost items of all relevant hydropower project components, and all CAPEX items are indicated below for computation of capital cost of the project:

   i. Main civil work
- Reservoir development with earthen dyke
- Dam with spillway and gates
- Foundation treatment including consolidation and curtain grouting
- Intake structure, steel trash rack and gate
- Desilting basin along with desilting tunnel
- Low pressure tunnel with adit
- Surge shaft
- Surge gallery/tunnel
- Valve house
- Pressure shaft
- High pressure tunnel/penstock
- Power house
- Tailrace tunnel
- River diversion with diversion tunnel/channel and coffer dams with consolidation and curtain grouting
- Grid connection

  ii. Infrastructure and site development
- Road network for transportation of materials, equipment, etc., with bridges, culverts, etc.

- Haul roads, service roads, etc.
- Temporary and permanent township, offices
- Utility accommodations like guest house, hospital, school, bank, post office, commercial center, police station, fire station, community center, etc.
- Facility building like central workshop, central store and warehouse, laboratory, fueling station, explosive magazine, etc.
- Open space for storing equipment, machinery, etc.
- Development of construction facility areas
- Rail head siding
- Telecommunication
- Water supply
- Construction power with back up DG
- Transmission network

iii. Project development
- Planning, engineering, etc.
- Feasibility study
- Environmental impact assessment
- Licensing
- Fish and wildlife/biodiversity/mitigation measures
- Development for recreation facility
- Heritage and archaeological mitigation
- Water quality mitigation
- Rehabilitation and resettlement

iv. Components of electro-mechanical work
- Turbines
- Main inlet valve
- Generators
- Transformers
- Cabling and control system
- Switch yard
- Grid connections

v. Auxiliary mechanical equipment
- EOT cranes for power house and valve house
- Elevator
- Workshop equipment and test lab

- Fire protection
- Cooling water

vi. Auxiliary electrical equipments
- Transformers
- Cabling and control panel
- DG sets

vii. Other non-equipment/non-construction
- Interest during construction
- Loan cost
- Import charges
- Taxes, etc.

17.2.4 Methods of Estimate

There are many measures of CAPEX estimate and analysis.

17.2.4.1 Method Number 1

The entire project is divided in to three major components:

- Generating plant (components directly necessary for power generation)
- Balance site (all development expenditures, additional cost such as environmental and regulatory requirement, construction of substation and transmission, etc.)
- Financial cost (variable cost of obtaining and repaying the capital used in development and construction of the project)

$$\text{Initial capital cost (ICC)} = \alpha P^b H^c$$

where

P = Capacity in MW

H = Head in Ft

α = Slope Coefficient

b = Capacity Component

c = Head component

17.2.4.2 Method Number 2

Preparation of estimate: Estimation methodology of a hydropower project defines direct costs and indirect costs as the two type of costs:

- Direct cost
  - Civil works including of all structural works
  - Infrastructures and site development work
  - Project development works
  - Electro-mechanical work including all other works related power house
- Indirect cost
  - Cost of land
  - Audit and accounting charges

All components related to each group of activities indicated above shall be taken into consideration for the computation of total project cost.

$$\text{Total project cost} = \text{Direct cost} + \text{Indirect cost}$$

Total investment costs of hydropower project vary with site location, selection of scheme, geotechnical characteristic, design choices, cost of labors and materials. Hydropower projects generally require a large civil workforce. Labor and material costs play an important role in project capital costs.

It is very difficult to get historical trends of hydropower costs due to non-availability of systematic collection data. One of the reasons for it is that a hydropower project is very site specific. It requires extensive time to collect data for the cost breakdown of each project.

For the preparation of an estimate, it is essential to break down the work into various pay items of each proposed component of the project. A detailed bill of quantity (BOQ) of each identified pay item shall be made from drawings reading in conjunction with specification and engineering standards, and the rate for each item shall be deduced based on the first principle of estimation to compute the total cost of the project.

The analysis of each item shall be determined considering the cost of labor, material, handling and transportation of materials, etc. by adding the contractor's overhead and profits (10%–15%) as normal.

Quantitative assessment of material required for any item per unit quantity shall be adopted from a standard coefficient being approved of each type of activity. In default, it shall be derived from data available for construction of similar types of work or from the first principle. The total exercise shall be based on scientific assessment of input of labor, material and machinery required for each item of work.

The basic rate for labor shall be taken from approved daily wages of the country and the rate of material shall be based on current market rate. Sundries shall be considered based on cost index.

17.2.4.3 Example of Rate Analysis for Excavation of Earthwork in All Kind of Soil by Hydraulic Excavator with Necessary Manpower Required for the Work

- Let us assume quantity is 10 m³ for analysis.
- Calculate the average output of excavator per hour analyzing from its cycle time of excavator
  - Excavator's bucket capacity = 0.9 m³
  - Efficiency of excavator = 80%
  - Capacity of excavator = $0.9 \times 0.8 = 0.72$ m³
  - Cycle time includes excavation time + swing + unloading + swing back to position = 20 sec = 0.33
  - 5 min gap shall be considered for rest
  - No. of cycles per hour = 55/0.33 = 167
  - Capacity = $167 \times 0.72 = 120$ m³ per hr
  - Machine hr. required for 10 m³ excavation = $10/(8 \times 120) = 0.011$
  - Cost for 10 m³ excavation = $0.011 \times$ Hire charge of excavator per day 1
- Calculate average output of tripper
  - Cost of tripper = $0.011 \times$ hire charges per day 2
- Add sundries LS = $1.3 \times$ rate 3
- Total cost = 1 + 2 + 3 4
- Add water charges @1% on sl no 4 5
- Total = 1 + 2 + 3 + 4 + 5 6
- Add 15% for overhead and profit on 6 7
- Total cost of the item = 6 + 7 8
- Rate per m³ = total cost in sl 8/10

Rate for all civil items shall be derived, and the cost of civil work shall be computed with respect to the Bill of Quantity (BOQ) prepared for civil work.

17.3 Break Down Cost Percentage

i. Break down cost percentage of civil work shall be approximately 40% of plant cost in green field projects.

ii. *Electro-mechanical equipment*: Cost can be computed based on current market price. Quote shall be obtained from qualified manufacturer as per specification and drawing including transportation, erection and commissioning.

Breakdown cost percentage of electro-mechanical work shall be approximately 30% in green field project.

iii. Cost of engineering, procurement and construction management and owner's cost 7% and 23%, respectively.

Total installed cost of large hydropower project varies from USD $1000/KW to USD $3500/Kw.

Pie diagram of breakdown cost in percentage of a hydropower project is indicated.

iv. *Owner's cost*: Owner's cost includes development cost, prefeasibility and engineering studies, environmental studies, permission, legal fees, insurance cost, property loan, etc.

Contribution of civil work in large hydropower project: Cost of civil work controls the capital cost of large hydropower project. It is mainly site specific which is influenced by site topography, hydrology, geotechnical characteristic, access to site, design, labor and material cost, etc. The cost of civil work in a developing country is less than a developed country due to the availability of large labor pools; additionally, sometimes cost of steel and cement are higher in a developed country.

v. *Electro-mechanical cost*: In a large hydropower plant, turbines of different capacities can be used in combination to have designed capacity due its technical maturity. It is done with the aim to have power generation in case of breakdown of any unit, otherwise generation of power will drop down to zero.

Studies reveal that generation of power depends on plant size and head.

$$\text{Cost (per kw)} = \alpha p^{1-\beta} \cdot H^{\beta 1}$$

where
α is a constant
p is power in kw of turbine
H is head in meter
$\beta, \beta 1$ are coefficient of power and head respectively

vi. *Miscellaneous cost*: The expenditure under this covers the capital cost and maintenance of water supply, sewerage disposal, drainage work, recreational facility, medical, visit of dignitary, etc., and it should be within 2% of civil and electro-mechanical cost.

vii. *Contingencies*: In project activity, there always exists some element of uncertainty in estimation of quantity and cost. Some provisions are kept in the estimate so that some unforeseen activity crops up can be tackled.

- Some unforeseen construction items
- Some unpredictable deviations in geological, geotechnical, topographical and hydrological values

viii. *Operation and maintenance costs*: Operation and maintenance costs of hydropower project are very low. Annual operation and maintenance cost is expressed as a percentage of the investment cost per Kw per year. The average value is 2%–2.5% of investment cost of a large hydropower plant, which includes refurbishment of electrical and mechanical equipment like turbine overhauling, generator rewinding and investment in communication and control system.

A recent study indicated that O&M costs is an average US$45 per Kw per hour for a large hydropower plant.

ix. *Cost of refurbishment and extension*: Life extension where equipment is replaced without any intention of boosting the generation capacity and this repair work may yield 2.5% growth in power.

An upgrade is where capacity and efficiency is increased during refurbishment. Refurbishment of the plant sometimes becomes economic because due to the reduction of O&M cost and higher output with modest investment in repair of turbines and generator, stator winding, etc.

Cost of life extension and upgrades of running a plant is studied, and estimated that life extension cost around 60% of green field electro-mechanical cost.

17.4 Levelized Cost of Electricity (LCOE)

Hydropower is a mature technology and its initial cost is very high but it has the longest lifeline amongst all other energies. Maintenance and operation cost of hydropower is very low. Its investment cost depends on site condition, topography, geo-technical characteristics, hydrology, etc., which contribute to 75% of the cost of the plant. LCOE, discounted life time cost, is divided by discounted life time generation.

Some assumptions are to be considered in obtaining LCOE cost:

- Installed capacity cost
- Capacity factor
- Economic life
- O&M cost

- The cost of capital

$$
\text{LCOE} = \frac{\sum_{i=0}^{n} \left(\frac{I_t + M_t + F_t}{(1+r)t} \right)}{\sum_{i=0}^{n} \left(\frac{E_t + M_t}{(1+r)t} \right)}
$$

where

LCOE = Average life time levelized cost of electricity generation
I_t = Investment expenditures in the year t
M_t = Operation and maintenance expenditures in the year t
F_t = Fuel expenditure in the year t
E_t = Electricity generation in the year t
r = Discount rate
n = Economic life of the system

LCOE is a price of electricity required for a plant where revenue will be equal to cost. An electricity price above LCOE will bring greater return on capital, and prices below it will produce a loss.

LCOE varies from USD $0.02 to $0.19/Kwh for large hydropower projects.

17.5 Determination of Tariff of Hydropower Project

Tariff in hydropower means annual fixed charges of each hydro generation unit. Hydropower is cheaper than any other energy. Its capital investment is very high and no consumable is involved in hydropower and the price of hydropower does not fluctuate with the market. Generating companies generally draw capital in the form of equity, debt, or a mix of both.

i. Key components for determination of tariff are as follows:
- Capital cost
- Debt equity ratio
- Interest on loan
- Depreciation
- Return of equity
- Operation and maintenance cost
- Interest on capital
- Working capital

- Plant availability factor
- Auxiliary energy consumption
- Plant life

Tariff is the summation of fixed cost per unit and variable cost per unit. Fixed cost is derived from fixed cost components like Return on Equity (ROE), interest on loan, interest on working capital, depreciation and O&M.

ii. *How to calculate fixed cost component*: Fixed cost components depend on the capital cost of the project and the tariff calculated will indicate whether power produced by the plant is able to recover fixed expenses and earn a return.

- *Return on equity*: Net income returned as a percentage of shareholder equity. It is a measure of the company's profitability on the invested amount of share equity. The concerned authority indicate percentage of pre-tax ROE for a particular fiscal year and declared percentage shall be considered for the calculation.

  Capital cost and debt/equity ratio is essential for calculating ROE. Debt/equity ratio shall be considered as 70:30 and equity shall be limited to 30% only if it is employed more than 30%. If actual equity is less than 30%, then actual equity shall be considered in the calculation and balance amount shall be considered as a loan example in the guide line of CERC:
  - Capital cost of a hydropower project = USD 3000 million
  - Debt-equity ratio = 70:30
  - Equity = USD $0.3 \times 3{,}000$ = USD 900 million
  - Debt = USD $0.7 \times 3{,}000$ = USD 2,100 million

  ROE = 15% of Equity = USD 0.15×900 = USD 135 million ....(A)

- Interest on loan
  - Interest rate on this loan shall be recovered as a part of tariff

  10% of loan (say) = USD $2{,}100 \times 10/100$ = USD 210 million..........(B)

- Interest on working capital
  - Working capital is cash available for day to day operation after commissioning
  - Working capital = 10% of USD 3,000 million = USD 300 million

$$\text{Interest on working capital} = \frac{10}{100} \times 300 = \text{USD 30 million} \quad \text{..........(C)}$$

- *Depreciation*: Depreciation is calculated on the hydro project based on a plant life of 35 years against capital cost of the project. Rate of depreciation is 5%

$$\text{Depreciation} = \frac{5}{100} \times 3,000 = \text{USD 150 million} \quad \text{.................. (D)}$$

- Operation and maintenance cost
  - Operation and maintenance cost of hydro plant is 2% of capital cost

$$\text{O\&M cost} = \frac{2}{100} \times 3,000 = \text{USD 60 Million} \quad \text{.......................(E)}$$

$$\begin{aligned}
\text{Total Fixed Cost (USD)} &= \text{(A) + (B) + (C) + (D) + (E)} \\
&= 135 + 210 + 30 + 150 + 60 \\
&= \text{USD 585 million}
\end{aligned}$$

iii. Annual energy generation

Gross energy generated in the year

Conversion of MW in to million units

$$1 \text{ MW} = \frac{\begin{array}{c}1 \text{ MW} \times 365 \text{ days} \times 24 \text{ hours} \times \\ \text{Plant availability factor in \%} \times 1,000\end{array}}{1,000,000}$$

Plant availability factor (PAF) is an average of daily capacity for all days and expressed in percentage of installed capacity of the. Plant capacity factor is taken as 90% of plant capacity of 1,000 MW

$$1000 \text{ MW} = \frac{1000 \times 365 \times 24 \times 0.9 \times 1000}{1,000,000} = 7,884 \text{ million units}$$

$$= 7884 \times 10^6 \text{ units}$$

Auxiliary Consumption = 1%

iv. Net annual energy generation

$$= \text{Gross energy} \times (1 - \text{Auxiliary consumption})$$

$$= 7,884 \times 10^6 \left(1 - \frac{1}{100}\right)$$

$$= 7,805.16 \times 10^6 \text{ units}$$

v. Fixed cost per unit $= \dfrac{\text{Total fixed cost}}{\text{Net annual energy generation}}$

$\qquad\qquad\qquad\quad = \dfrac{585 \times 10}{7,805.16} = \text{USD } 0.75 / \text{Kwh}$

vi. Tariff $=$ Fixed cost / unit $+$ Variable cost / unit

$\qquad\quad = \text{USD } (0.75 + 0) / \text{unit}$

$\qquad\quad = \text{USD } 0.75 / \text{KWh}$

Bibliography

Renewable Energy Technologies: Cost Analysis Series, International Renewable Energy Agency (IRENA), 2012.

Renewable Power Generation Cost, International Renewable Energy Agency (IRENA), 2007.

Standards/manuals/guidelines for small hydro development, 1.5 General – Project Cost Estimation, Alternate Hydro Energy Center, Indian Institute of Technology Roorkee, 2012.

Standards/manuals/guidelines for small hydro development, 1.6 General – Economic & Financial Analysis and Tariff Determination, Alternate Hydro Energy Center Indian Institute of Technology Roorkee, 2012.

18

Hydroelectric Power Houses

18.1 Hydroelectric Power House

The power house is the heart of power projects where hydro-energy is converted to electrical energy. Determination of plant capacity and the size of the unit depends on the techno-economic studies for optimum utilization of energy resources.

Maximum utilization of potential energy available at the source leads to the success of the plant in terms of economics, available energy used, and overall efficiencies, which are important considerations in deciding the number, size, and type of generating units and the type and size of power houses.

Before we talk about size and capacity of the power house, we should discuss how the capacity of a plant is rated. Detailed investigation and subsequent techno-economic studies provide various data such as design discharge and design heads by which a plant can be rated.

In hydropower projects, power is derived from water flow while duly utilizing the level difference of the flow, known as the head, and generation of power is a function of the discharge and available head.

$$KW = 9.804.Q.H.\eta$$

where
Q = Quantity of water flowing through turbine in cubic meters per second
H = Head in meter
η = Overall efficiency

Detailed studies determine plant capacity and its operating head, size, and number of units to be installed. Accordingly, size and various dimensions of power houses are deduced. The capacity of plant is expressed in Kilowatts (KW) or Megawatts (MW).

18.2 Classification of Power House

The power house is classified based on various parameters such as load characteristics, capacity of the plant, hydraulic characteristics, and physical features.

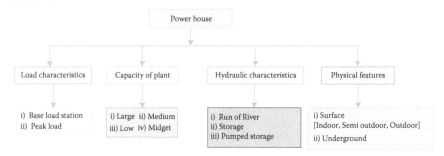

A power station is classified based on physical features such as surface and underground power houses, and are subdivided based on their location as well as proximity to water reservoir and water conducting system.

The power house receives water from the reservoir through water conducting systems, and is known as a detached power station from head works. This type of power house may be constructed on the surface or underground depending on geological conditions of the site.

There is another system where the power house is located in close proximity to the water reservoir and receives water directly from the reservoir. This type of power house is generally built on the surface.

Types of weather protection provided to the main equipment such as generators, generator breakers, and transformer will distinguish between indoor, semi-outdoor, and outdoor types of surface power houses.

18.3 Function of Power House

The power house of a hydropower plant has three functions:

- Water flows to the power house through the conducting system, charging it with potential and kinetic energy. The potential and kinetic energy of water is converted into mechanical energy by the rotation of the turbine.
- Mechanical energy is then transformed into electrical energy by the generator being coupled with the turbine.
- The electrical power is sent to the power distribution system for further use.

18.4 What Does a Power House Need to Accomplish These Activities?

The power house requires the following equipment and facilities for the generation of electricity:

- Hydraulic equipment
  - Turbines
  - Gate valves and relief valves
  - Governor and flow measuring equipment
- Electro-mechanicals equipment
  - Generator
  - Auxiliary transformer
  - Control cubicle and relaying equipment
  - Switching equipment
  - Station battery
  - Cooling water system
  - Drainage and dewatering system
  - Ventilation and fire extinguisher
  - Compressed air system
- Miscellaneous equipment
  - EOT cranes
  - Staff room
  - Facilities

18.5 Layout of Power House and Its Dimensions

Considering the previously cited equipment and facilities, the power house is divided into three areas to accommodate all equipment and facilities: the turbine and generator area, the maintenance and erection bay, and the service area.

18.5.1 Turbine and Generator Area

The main area, which houses the turbine and generator, shall be the central area around which the service and erection areas are located. The generator room is the main feature of the power house and is divided into blocks. Each block should have one generating unit. Unit spacing depends on the discharge diameter of runners. The width of the power house shall be decided by the span of the overhead EOT crane. The height of the generator room shall be kept such a way that the maximum clearance will allow for the dismantling or moving of major items such as parts of the generator and turbines. The location and reach of overhead EOT cranes shall be installed on Gantry Girders to facilitate the movement of turbines and generators during erection and maintenance.

Length of the turbine and generator area shall be adequate enough to house all units with suitable space between them, with the width fixed in such a way that at least 3 m shall be kept between generators and walls of the power house.

18.5.2 Erection Bay

The erection bay is located at the end of turbine and generator area and are preferred to be on the same floor level while the length should be equal to one generator bay. The length should give enough space for the EOT crane. The erection bay is used for the erection of runners, generators, and other components. The space shall be sufficient enough for one runner and rotor plus some additional space.

18.5.3 Service Area

Service areas include offices, control and testing rooms, storage rooms, maintenance shops, auxiliary equipment rooms, and other rooms for special uses. Additionally, some more spaces are necessary such as office rooms, storage, staff rooms, and other facilities.

18.6 Hydraulic Turbine, Heads, and Its Types

Readers must understand different head categories before any discussion on turbines may be had, because the head is the main criteria of turbine selection for power stations.

The head is the main criteria for the selection of turbines for the power house.

- *Gross head*: The gross head is the difference in elevation of the head water level and tail water level when no water is flowing.
- *Effective or Net head*: Effective or net head is the gross head less all losses in the water conducting system including penstock.
- *Rated head*: The head at which the turbine produces the rated output at a specified gate opening.

- *Weighted average head*: This is the net head determined from reservoir operational calculations that will produce the same amount of energy in kilowatt hours between that head and maximum head as it developed between that same head and minimum head.

- *Design head*: The head at which the turbine is designed to give its maximum efficiency.

- *Maximum head*: The gross head resulting from the differences in elevations between the maximum fore bay level without surcharge and the tailrace level without spillway discharge, and with one unit operating at speed-no-load turbine discharge of approximately 5% of the rated flow. Under this condition, hydraulic losses are negligible and may be disregarded.

- *Minimum head*: The net head resulting from the difference in elevation between the minimum fore bay level and the tailrace level minus losses with all turbines operating at full-gate.

- *Critical head*: Critical head is the net head at which the full-gate output of the turbine produces the permissible overload on the generator at the unity power factor (usually 115% of the generator kVA rating). This head will produce the maximum discharge through the turbine.

- *Definition of hydraulic turbine*: A hydraulic turbine is a hydraulic machine that converts potential and kinetic energy to mechanical energy, and it is used to rotate the generator's shaft. A turbine is nothing but a rotor whose purpose is to act as a prime mover providing direct horsepower to the generator.

- *Classification of turbine*: Head is the main criteria for selection of turbines and speed will determine the capacity. The type of turbine depends on the manner in which water causes the turbine runner to rotate.

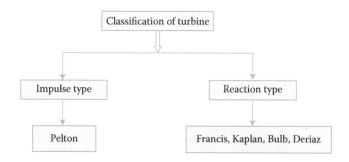

18.6.1 Reaction Type of Turbine

Specific energy is the energy per unit weight of fluids which make up potential energy, pressure energy, and kinetic energy.

In reaction types of turbines there is a pressure difference from inlet to outlet of the runner; at the inlet of runner one component of specific energy is pressure energy. Water flows through the runner where this pressure energy is converted into mechanical energy partly due to the drop in pressure, and this is called the reaction part of the energy conversion. The balance part is converted by impulse forces due to the change in direction of relative velocity vectors. This reaction ratio is expressed as (H1 – H2) hn where þgh1 = p1 is the hydraulic pressure at inlet and þgh2 = p2 is the hydraulic pressure at outlet.

Reaction turbines operate with their runners fully flooded and develop torque because of the reaction of water pressure against the runner blades. Francis and Kaplan reaction type turbines are widely used in the industries.

18.6.1.1 Francis Turbines

Francis Turbines are mostly used for head ranges between 50–650 m. Francis turbines are a reaction type turbine where water enters from the penstock in a radial direction with regards to the shaft. Water flow is guided by guide vanes supported by the stay ring. Water quantity is controlled in the wicket gate and is subsequently distributed equally to the runner area where water starts moving in a counter-clockwise direction and moves to the draft tube in order for the water to leave. The water exits from the turbine axially and is known as a mixed flow turbine. These functions are performed through the turbine's main components including the spiral case, stay ring/ veins, wicket gate, runner, vacuum breaker, aeration device, and draft tube for desired performance and reliability. Hydraulic machines are manufactured by various vendors around the world according to specifications, drawings, and other data supplied by the project authority.

Flow diagram

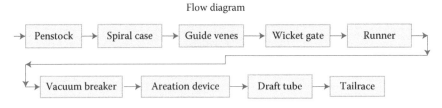

18.6.1.1.1 Components and Their Functions

- *Spiral case or scroll case*: Water enters from the penstock to the spiral case, whose function is to transfer the water to stay vanes and maintain near-uniform velocity around the stay vanes. The spiral case is made of steel because it is the wetted parts of the turbine. The cross-sectional area of this casing decreases uniformly along the circumference to keep the fluid velocity constant along its path towards the guide vane.

- *Guide vanes/rings*: The stay vane area receives water from the spiral case, and the function of stay vanes is to align the flow of water to the wicket gate. Stay ring/vanes are manufactured with steel to enhance performance.

- *Wicket gate*: The wicket gate receives water from guide vanes. It controls the amount of water entering the runners. Water reaches here with high tangential velocity with some angle, and water strikes the surface of the runners. The optimum angle of attack will be at the peak of efficiency. The wicket gate also functions as a closer valve to minimize water leaks through the turbine when it is down.

- *Runner*: The most important section is the runner. In the runner water enters diagonally and exits axially. Its shape is very special when water flows over it; the runner creates low pressure on one side and high pressure on the other side. The lift and impulse force make the runner move, and water moves in a counter-clockwise direction. The runner is also known as the rotor. The runner is manufactured either with carbon or stainless steel, and it is generally manufactured in a single piece. Improved materials are used to enhance performance and reduce cavitations. Its function is to convert potential and kinetic energy into mechanical energy or rotational horsepower that is directly supplied to the shaft. It may be vertical or horizontal, as well as a single or double discharge.

- *Vacuum breaker*: The function of this VB is to admit air in the zone of the runner to reduce rough operation and vibration.

- *Aeration device*: This device allows air into the turbine to increase the dissolved oxygen in the tail water for environmental compliance. Air is blown either by a motorized blower or air compressor to the turbine. Sometimes, modifications are made in the runner to naturally draw air from the atmosphere into the turbine.

- *Draft tube*: The draft tube connects the runner with the tail race. It is a diverging section that gradually slows down the high discharge velocity capturing the kinetic energy and regains the dynamic head. The larger the head deferential across the turbine, the higher the turbine power output. The draft tube should be lined with steel.

There are some components related to reliability of the turbine including the wicket gate mechanism/servomotor, head cover, bottom ring, turbine shaft, guide bearing, and mechanical seal.

- *Wicket gate mechanism*: This mechanism is provided to control the opening and closing of the wicket gate. The mechanism includes an arm, linkage, and shear pin. The servomotor is hydraulically actuated using high pressure oil from the unit generator.

- *Head cover*: It is a pressurized structural member and is used as a water barrier to seal the turbine.
- *Turbine shaft*: Generated torque is transferred by the turbine shaft from the runner to shaft of the generator or rotor.

A Francis turbine may be operated over a range of flows approximately 40%–110% of the rated discharge. Vibration and/or power surges occur when the area of operation is below the 40% rated discharge. The upper limit generally corresponds to the maximum discharge for the generator rating. The approximate head range for operation is from 65%–125% of design head. In general, Francis Turbines within the capacity range of 25 MWs with modern design tools like computational fluid dynamics (CFD) have enabled these types of turbines to achieve efficiencies in the range of 94%–95%.

18.6.1.1.2 Advantages of Francis Turbine

- Work in medium to high head (300–450 m)
- Operate in largest range of head and flow
- High efficiency 94%–95%
- Mixed flow turbine

18.6.1.1.3 Disadvantages

- Efficiency decreases with turbine flow rate below 40%
- Decreased flow increases cavitation risk
- Water hammer effect causes harm to turbine

18.6.1.2 Kaplan Turbine

The Kaplan Turbine is a reaction type and axial flow turbine. It is a low head turbine that works with head of 30 m with specific speed between 750 and 850 rpm.

In this turbine, water enters from penstock to the scroll casing and moves through the scroll casing to guide vanes known as the guide mechanism. This guide mechanism guides the water to the central part of the turbine. The flow of the water in the central part is radial and water drops in a space between the outlet of guide vanes to inlet of runners, known as the whirl section, where the flow of water is changed from radial orientation to axial direction, the disposition of shaft is vertical, and runners are fixed with hubs and water moves through draft tubes to tailrace.

- Perpetual velocity at inlet and outlet is equal
- Velocity of flow at inlet and outlet are equal
- Area of flow at inlet = Area of flow at outlet

The Kaplan Turbine has an outstanding reputation for its high specific flow capacity. As a double-regulated turbine it is most suitable for low heads and large flows as well as for variations in flow and head. Its main utilization concentrates on heads between 1.5 and 30 m.

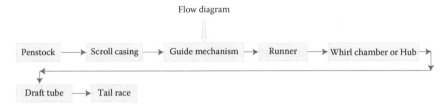

18.6.1.2.1 Main Components and Their Functions

- *Scroll casing*: The cross section of the scroll case is circular and made of concrete head up to 30 m and of steel head above 30 m. Scroll cases receive water from the pen stock and supply to stay vanes/guide vanes.
- *Guide vanes*: Its function is to align the flow of water after receiving water from the scroll case. Water flow is regulated by adjustable guide vanes and guide the water into runner with the appropriate velocity and direction.
- *Runner*: Runners are propeller shaped with adjustable blades. Its function is to convert the potential energy of pressure and waterflow into mechanical energy. The blades, generally in groups of four, are mounted in the runner hub which moves simultaneously with variable water flow. This hub has a blade tilting mechanism which helps to change the tilt of blades automatically depending on the guide vane. These four blades and hub are connected to the shaft.
- *Draft tube*: Draft tubes are the last component of the turbine. It connects the runners to the tailrace where water passes through it and joins the original stream. It helps to utilize runner discharge head if the runner is set above tail water level. It is made of both steel and RCC. It has a diverging section to reduce the velocity to avoid any vortexes and maintain flow uniformity during the journey to the original stream. Generally, impulse turbines do not need any draft tubes because they are set above tail water levels.

18.6.2 Impulse Type of Turbine

Impulse turbine types are used when there is no pressure difference between the inlet and outlet of the runner. The energy is converted into kinetic energy at the inlet of the runner by the impulse force created by the change of velocity vector, and it operates with their runners in the air and converts the water pressure energy into kinetic energy that develops torque.

A Pelton Wheel is the most commonly used impulse turbine. It is used for high head with low flow. The Turgo Turbine, an axial turbine, is used in small units, while a cross-flow turbine is a two stage turbine is also used for small units.

18.6.2.1 Pelton Wheel

The Pelton Wheel is an impulse turbine, and it is the only choice for highest head hydropower projects. It is used in high head from 300–2,000 m with low inflow. The appropriate head range of operation is from 120% to 80% of rated head, with peak efficiency at 90%. It is a turbine in which all potential energy is converted into kinetic energy under atmospheric pressure before it strikes the runner tangentially with 1–6 water jets. It comes in two shaft axis such as horizontal and vertical. The horizontal axis is made up of jets 1 through 3, while the vertical axis is made up of jest 4 through 6. Horizontal jest need a more spacious power house with easier maintenance procedures, while vertical jets can be housed in smaller areas with difficult maintenance procedures.

At the rated power of 423 MW each turbine operates at a head of 1,869 m and a flow rate of 25 m³ per sec with 92% efficiency in the Bieudron Hydropower Project in Switzerland.

Water flows through penstock (a high pressure tunnel made of steel) which is fitted with a nozzle at the end. The nozzle has less cross-sectional area to convert water flow into a jet of high velocity, converting all potential energy into kinetic energy in the form of jets. Demands of the power supply fluctuate as required over the time. A spear head with a wheel at the back is positioned within the nozzle to meet this requirement. When the wheel moves, it moves the spear head. When it moves forward to the discharge end of the nozzle it reduces the space between the nozzle and spearhead and reduces the flow of water. When power demand is low, the spearhead moves forward to the discharge end of the nozzle and reduces the rate of water jet flow. If power demand is increased, it will move backward to increase water jet flow. Therefore, the spearhead produces the high and low rates of water jet flow as required. The Pelton Wheel is synchronized between demand and supply through this mechanism. This system acts as the control mechanism that regulates the flow of water as required. This regulation of water jets has influence over the rotation of the runner, which is connected to the shaft and the shaft is mounted to generator.

The converted and regulated water jet strikes the bucket/vanes/blades mounted with the runner and is fitted with a splitter. When water strikes the bucket it is divided into two parts to avoid any unbalanced thrust on the shaft and water comes out and rotates the runner, thereby producing mechanical energy. The splitter shall be strong enough to sustain the force of high velocity water jets. Finally, water enters the tailrace and the shaft starts rotating duly influencing the shaft mounted generator, which ultimately generates electricity.

18.6.2.2 Component of Pelton Wheel

The Pelton Wheel is made of many components that help the turbine generate electricity.

- *Casing*: It is made of a rigid unit housing all components and guides exit water from the runner. The casing is made of carbon steel BSEN10025 S25JR/ASTM A36.

- *Nozzle*: The purpose of the nozzle is to convert potential energy into kinetic energy in the form of high velocity jets. The cross section to generate jets at the discharge end is very small. The material for nozzles can be stainless steel per ASTMA-743 CA 6NM because it provides better resistance to wear and cavitations.

- *Bucket/Vanes/Blades*: The bucket is a vital component of the turbine. Buckets are spoon shaped and generally cast as a single piece to avoid fatigue failure and is mounted to the runner. If the number of buckets provided in a turbine is inadequate it will result in the loss of water jets. The number of buckets should be spaced in such a way that when a bucket departs from the water jet the next one should immediately follow to become engaged with the high velocity jet, otherwise there will be a loss of water due to wide gaps between the buckets and will cause a sudden drop in efficiency of the turbine. A nucket consists of splitters fitted in the center that divides water into two sides to avoid unbalanced thrust on the shaft. Material for the bucket is chosen per head and stress requirements. A cut is provided at the bottom of the bucket to ensure that the water jet is not interfered with by incoming buckets.

- *Runner*: The runner consists of buckets as indicated above and is connected to the shaft. The runner starts rotating after water jets strike the bucket mounted to it, which induces torque to the connected shaft for the generation of power. Material used for the fabrication of runners is of SS ASTM A-743 CA 6NM or cast Stainless BS3100 Grade 425 C11.

- *Spear*: The function of the spear is to regulate the jet forced out of the nozzle. The spear is attached with a wheel at the back that rotates and pushes the spear forward to the discharge end of the nozzle. Stainless steel internal components are housed in a carbon steel fabricated or cast branch pipe.

- *Deflector*: A deflector arrangement is introduced to direct the water away from the bucket when there is a load rejection to reduce hydraulic torque on the generator.

Some special features provided for improvization of the Pelton wheel

- A hooped runner that minimizes fatigue, stress, vibration, and replacement costs while increasing maintenance intervals.

- Advance ventilation system provides for minimization of losses and maximized efficiencies.
- Optimized manifold is provided for improved jet quality and runner efficiency.
- Water cooled slator uses stainless steel strands to increase reliability while experiencing no leakage, no oxidation, and no blockage.
- Slator core processing system is a maintenance free system that prevents the loosening of core components, thereby increasing reliability.

18.7 Selection of Turbine

A hydraulic turbine is a simple and efficient machine that can have its lifespan controlled. However, the selection of this machine is not simple. Hydraulic turbines shall be specifically selected based on site conditions. Careful observation of flow of stream, operation of reservoirs, and other data are required for the proper selection of turbine, otherwise it will fail to perform at its optimum level.

Proper selection of the turbine depends on many factors such as plant capacity and head, speed and setting of turbine with regards to tail water, weight, and the size of the runner and shaft. All these factors shall be considered carefully to produce power at the lowest cost.

- *Plant capacity*: Plant capacity is determined through detailed studies of the reservoir and water power, and this study indicates at which head the capacity of the plant should be finalized. However, this study cannot be limited to technical specifications and should be extended to commercial aspects as well. This techno-commercial study is required to assess the optimum size of plant, the number of units, and its capacity. This study is further extended to assess limitations like capacity limitation due to size of generator, variations in turbine capacity due to varying heads, and setting of unit with regards to tail water

  Here are some of the techno-commercial considerations involved in the determination of generating capacity to be installed.
  - Optimum utilization of resources.
  - Study of rainfall and stream flow.
  - Study of economic potential focusing on limitation of power producing equipment on a sustainable basis and not on immediate demand.

- Assessment of various ranges of heads and water levels.
- Assessment of efficiency losses.
- Determination of capacity, size, and number of units.
- Operating criteria with expected power requirement including some limiting conditions:
  - Desired capacity at minimum head.
  - Type of operation/part load operation.
  - Character of load demand.
- Information about water conducting system (length and size of LP & HP tunnels, etc.).
- Location of plant and its topography.
- Worldwide and local experience.
- Future provisions.

- *Head*: Head is the main criteria for the selection of turbines, and speed will determine capacity. Types of turbines depend on the manner in which water causes the runner to rotate. In a multi-purpose project where water is released for various reasons, the water level of this project is subjected to fluctuation, resulting in a large variation in water head. The effects of the variation of water head on power producing ability, efficiency, and revenue shall be taken in to account while selecting a suitable turbine. Generally, Pelton Wheels are recommended for high head, Francis Turbines for medium head, Kaplan and Deriaz Turbines for low head, and Bulb Turbines for very low head.

- *Speed*: The capacity of a plant is rated with specific head, and to achieve the lowest cost per unit of power the selected turbine and generator (turbo generator) should run on high speed as often as possible. The specific speed of the turbine is the speed in revolutions per minute at which the turbine will rotate. Low specific speed is associated with high head, and high specific speed is associated with low head. The selection of high speed for a given head will suggest a smaller turbine and generator, and will result in initial cost savings. However, higher speeds require turbines to be placed lower with respect to level of tail water, which will increase costs for excavation and structural construction. The overall economic analysis will come into play during the selection of economic specific speed of turbines.

  The turbine selection process is complex and it is based on technical, environmental, commercial, and especially on location and its application. However, specific speed and head are two vital parameters by which turbines will be selected.

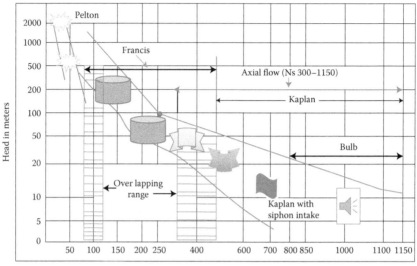

FIGURE 18.1
Guide lines for various type of turbines as function of Ns and Head. (Adopted from AHEC-IITR, "3.1 Electro-Mechanical – Selection of Turbine and Governing System," standard/manual/guideline with support from Ministry of New and Renewable Energy, Roorkee, June 2012.)

Rated head and plant ratings are obtained from power studies.

The specific speed of turbines can be determined by the empirical formula:

$$ns = \frac{n\sqrt{P \times 1.358}}{H^{\frac{5}{4}}}$$

where
ns = Specific speed of turbine in revolutions per minute (r.p.m.)
n = Rated speed of turbine in revolutions per minute
P = Turbine output in kW
H = Rated head in meters

The value of specific speeds obtained can be used in Figure 18.1 to finalize turbine selection for the project.

18.8 Setting of Turbines

The output of a turbine depends on the specified head and the most practical high speed of the turbine, which will ultimately produce the economic rate of power per unit. These three parameters can be controlled by setting the turbine in its most desirable location.

Power generation depends on head and speed, where head is specified and the highest practical speed can be achieved by setting the turbine in the right location. Greater speed requires the turbine to be placed lower with respect to tail water to avoid cavitations, which increases excavation and structural cost.

Cavitations occurs in the runner, so it is desirable to set the runner at a depth below the tail water in order to avoid cavitation pressure and to obtain countering pressure.

The depth at which the runner can be set is calculated by the following empirical equation:

$$Z = (Ha - Hv) - \alpha H$$

where

Z = Depth of centre line of runner below minimum level of tail water
Ha = Atmospheric pressure in meter water column
Hv = Vapor pressure in meters at plant location temperature
H = Head on turbine
α = Coefficient of cavitations

$$\alpha = \frac{(ns)^{1.64}}{50,327}$$

The value of α can also be found per USBR Engineering Monograph Number 20, and respective turbine settings can be finalized.

Generally, cavitation coefficients of Francis Turbines are less than Kaplan Turbines, so the Francis Turbine requires less excavation and the Pelton Wheel can be installed above max tail water level, minimizing excavation costs.

18.8.1 Cavitations

In hydropower projects the water has to pass through different components such as the penstock, scrolling casing, guide vanes, and runners, finally passing through draft tubes to generate power. During this journey two phenomena occur and cause problems; these issues are known as cavitations and water hammer.

Cavitation is the formation of voids in the liquid flowing through the turbine when the static pressure of liquid is lower than the vapor pressure. Voids that are formed due to insufficient internal pressure get filled with vapor, which results in vapor bubbles. The collapse of these vapor bubbles generate high frequency pressure waves that affect the machine. The bubble collapsing near the surface of the runner causes cavitations. Cavitations are a function of high velocity flow, low pressure, and abrupt changes in the direction of flow. The effect of this cavitation causes pitting on the boundary surface of the runner. Pitting is the removal of metal from the surface due to the violent collapse of vapor bubbles formed by cavitations. Cavitation is the cause and pitting is the effect

18.8.1.1 Control of Cavitations

- Set the turbine runner in such a location that pressure and velocity in critical areas do not permit excess cavitation to occur. The runner of the turbine should be placed well below the tail water level which may involve more excavation and structural costs.
- A layer of cavitation resistance material is laid over the turbine by welding it on the base metal. Most of the turbine suffers from cavitations, so a corrective measure is to build up the damaged surface with stainless steel welding rods.
- Some anti-cavitation fins maybe added to blades to minimize cavitations.
- Air bubbles will produce an elastic fluid that cushion the action of pressure pulses, with an amount of 1–2 parts per thousand of air causing a significant reduction in damage.
- A thin hydrogen film on metal surfaces can stop pitting.
- Selection of turbine speed can be used to control cavitations by setting the turbine lower with regards to tail water.
- To control the cavitations, design the shape of the water passage in such a way that the shape and surface will offer minimal opportunity for abrupt changes in the water's flow line.

18.9 Selection Procedure of Turbine

Step 1: Obtain site data for design
- Maximum operable discharge in cubic meters per second.
- Maximum level of water in meters.
- Maximum head in meters.
- Minimum tail water level in meters.
- Normal temperature in degrees centigrade.
- Any economic losses.

Step 2

- Compute rated head from above data.
- Compute design capacity.

  Plant capacity = 9.804 × (Rated discharge in M/sec) × (Rated head in meters) × Plant efficiency. Plant capacity is expressed in Kilowatts.

Step 3

- Select the type of turbine suited to the above information from Figure 18.2 for types of hydraulic structure by plotting discharge versus head, which indicates the type of turbine region. Analyze the combination of head, discharge, and plant capacity to determine the suitable turbine region such as Francis or Kaplan.

Step 4

- Calculate preliminary value of n' (Trial Speed)

$$n' = \frac{n's(hd)^{5/4}}{(Pd)^{1/2}} \tag{18.1}$$

where

n' = Trial rotational speed
$n's$ = Trial specific speed

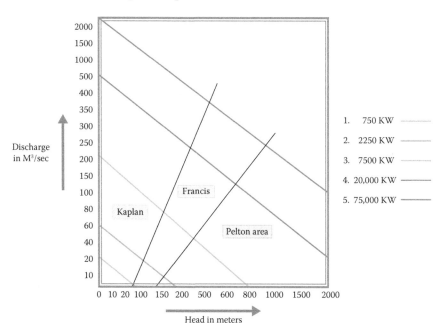

FIGURE 18.2
Discharge, head, plant capacity and related turbine region.

h_{dn} = Designed head
Pd = Turbine full gate capacity at hd

Trial specific speed = $2334/\sqrt{hd}$
Put this value in Equation 18.1 and calculate the trial rotational speed.

Step 5

I. *Rotational and design speed*: Rotational or design speed is selected based on the following considerations:

- A multiple of four poles are preferred, but standard generators are available in multiples of two poles.

- If a head is expected to vary less than 10% from design head, then the next greater speed may be chosen. If a head is expected to vary more than 10%, then the next lower speed may be chosen.

- If the turbine is directly connected to the generator, then turbine speed must be synchronized and calculated as follows.

Rotational speed = n = 120 × Frequency/Number of poles

$$n = 7{,}200/\text{number of poles at 60 Hz}$$

Find out number of poles = $7{,}200/n$ and number of poles shall be chosen.

Calculate synchronized speed with corrected poles = $7{,}200/NP$.

Now, calculate actual or design specific speed by putting synchronized speed

$$ns = \frac{n(Pd)^{1/2}}{(hd)^{5/4}}$$

Refer the preference of turbine with regards to design specific speed.

Step 6

Find the velocity ratio for the selected turbine. Determination of the velocity ratio is required to find out the size of the runner. The actual size is fixed by the manufacturers, and it varies from vendor to vendor for a given power at fixed speed.

Velocity ratios vary between turbine types.

D1 = Entrance diameter of runner
D2 = Minimum opening diameter of runner

D3 = Discharge diameter of runner
ns = Design specific speed
$\emptyset_3$ = Velocity ratio at D3

For Francis Turbines

$\emptyset_3$ = Velocity Ratio at D3 = $0.0211(\text{ns})^{2/3}$

For Propeller Turbines

$\emptyset_3$ = Velocity Ratio at D3 = $0.0233(\text{ns})^{2/3}$

Discharge diameter for both

$$D3 = \frac{84.47 \times \emptyset_3 \times (\text{hd})^{2/3}}{n}$$

using Figure 18.3 Plant sigma $\sigma = (\text{ns})^{1.64}/50{,}327$

$$h_b = h_0 - h_v$$

where
h_b is barometric pressure
h_0 is Atmospheric pressure
h_v is Vapor pressure

Find out value of h_0 from Table 18.1 at given altitude.
Find out value of h_v from Table 18.2 at given temperature.

Turbine Setting Level

$$h_s = h_b - \sigma h_{cr}$$

Where h_{cr} = Normal head water level – mean tail water level.
Now, find out h_s by putting value of h_b, h_{cr} and σ.

18.10 Example

In a storage hydropower scheme the head water level is 1,850 m, the tail water level is 1,690 m, and in-line losses are 0.5 m. The rated discharge of water flow is 50 cubic meters per second. Select the type of turbine to be used in

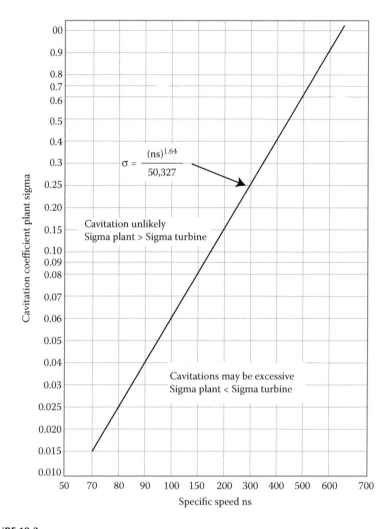

FIGURE 18.3
Relation ship between cavitation coefficient/specific Speed/plant Sigma. (Adopted from AHEC-IITR, "3.1 Electro-Mechanical – Selection of Turbine and Governing System," standard/manual/guideline with support from Ministry of New and Renewable Energy, Roorkee, June 2012.)

the project and the suitable level of the turbine. Temperature is 20 degrees centigrade.

$$\text{Rated head} = 1{,}850 - 1{,}690 - 0.50 = 159.5 \text{ m}$$

$$
\begin{aligned}
\text{Plant capacity} &= 9.804 \times [\text{Rated discharge}] \times [\text{Rated head}] \times \text{Efficiency} \\
&= 9.804 \times 50 \times 159.5 \times 0.9 \,(\text{Assume efficiency } 90\%) \\
&= 70{,}368.2 \text{ Kw}
\end{aligned}
$$

TABLE 18.1

Atmospheric Pressure with Regards to Altitude

Altitude in Meters	H_0, mm of Hg	H_0, M of H_2O
0	760.00	10.351
500	715.99	9.751
1,000	674.07	9.180
1,500	634.16	8.637
2,000	596.18	8.12
2,500	560.07	7.628
3,000	525.75	7.160
3,500	493.15	6.716
4,000	462.31	6.295

Source: Adopted from AHEC-IITR, "3.1 Electro-Mechanical – Selection of Turbine and Governing System," standard/manual/guideline with support from Ministry of New and Renewable Energy, Roorkee, June 2012.

Note: H_0 = Atmospheric pressure for altitude in meter.

TABLE 18.2

Water Vapor Pressure with Regards to Temperature

Temperature(°c)	H_v (meters)
5	0.089
10	0.125
15	0.174
20	0.239
25	0.324

Source: Adopted from AHEC-IITR, "3.1 Electro-Mechanical – Selection of Turbine and Governing System," standard/manual/guideline with support from Ministry of New and Renewable Energy, Roorkee, June 2012.

Note: H_0 = Atmospheric pressure for altitude; H_v = Vapor pressure of water; H_b = H_0 – H_v (Atmospheric pressure – Vapor pressure).

Select from Figure 18.2 where the curve is plotted in turbine discharge versus head and indicates the region of the Francis Turbine.

$$n's = \frac{2334}{\sqrt{hd}} = \frac{2334}{\sqrt{159.5}} = 187.5$$

Selection of trial speed (n′)

$$n' = \frac{n's(hd)^{5/4}}{(Pd)^{1/2}}$$

$$= \frac{187.5(159.5)^{5/4}}{\sqrt{70368.2}} = \frac{187.5 \times 566.91}{265.2} = 400.8$$

Number of poles in generator $= 120 \times AC\,frequency/n'$

$$= 120 \times (60/400.8) = 17.96\,(say\,20)$$

Take the number of poles as even number and divide by 4, so number of poles selected is 20.

Rotational speed (n) = 7,200/Adjusted number of poles = 7,200/20 = 360

Actual specific speed as below

$$ns = \frac{n(Pd)^{1/2}}{(hd)^{5/4}} = \frac{360\sqrt{703,680.2}}{159.5^{5/4}} = 187.11$$

Velocity ratio for Francis turbines $(\varnothing) = 0.0211(ns)^{2/3} = 0.0211 \times (187.11)^{2/3} = 0.211 \times 31.59 = 6.66$

$$D3 = \frac{84.47\varnothing3\sqrt{hd}}{n} = \frac{84.47 \times 6.66\sqrt{159.5}}{360} = 19.13$$

The following parameters to be given to manufacturers:

1. Design head......159 m
2. Turbine output......70,368 KW
3. Rated speed......360
4. Runner diameter......19.13

18.11 Generator and Driving System

A generator is a machine that converts mechanical energy into electrical energy. The mechanical energy that is needed to generate electrical energy is supplied externally by the hydraulic turbine. Generally, generators used in hydraulic power are alternating current synchronous generators that require excitation currents. Excitation currents are provided by an auxiliary generator for the creation of the rotor's magnetic field.

Generator is classified as one of two types: synchronous or induction.

The generator is mounted with the shaft of the generator. It has a moving part that is called the rotor and a fixed part that is called the stator. The rotor is connected to hydraulic turbine blades and the surface of the rotor is covered with electromagnets. The main function of the rotor is to take energy from outside and create a rotation that delivers torque. The wall of the stator is made of copper windings known as armature. When the rotor turns inside, the electrons in the copper windings vibrate and create electric currents.

All generating units shall be synchronized and should maintain exact rotation speed. All electrical equipment is designed to use alternating currents of specific frequencies that depend on the rotational speed of the generating unit, also known as the number of times per second that the rotor magnets travel past the stator's windings. This frequency is expressed in cycles per seconds or Hertz (Hz). In America, alternating current cycles are 60 cycles per second and in Europe they are 50 cycles per second.

18.12 Rating and Various Characteristics

Ratings of generators are designed by performing in-depth studies of economic feasibility, cost benefits, environmental impact, and impacts on operating conditions such as presence of foreign material in air and water to ensure effective operation. Each turbine connected to the generator requires different criteria such as temperature, voltage, mechanical and electrical load, and the ability to operate in altitudes above 1,000 m.

The designed machine should be compatible to local power system arrangements and local geographical conditions, plant arrangements like indoor or outdoor facilities, the number of generating units, and the hydraulic characteristics that are unique to each plant. Any possible future changes in terms of varying heads or turbine speeds shall be considered during the rating of the generator, and the generator shall be capable of continuous output at 115% of rated KV-A at rated voltage, rated power factor, and rated frequency.

Characteristics such as the name of manufacturer, rating, shaft orientation (vertical or horizontal or inclined), type of prime mover, excitation voltage, armature connection, and synchronous of induction machine shall be on the name plate.

Characteristics that shall be on the name plate of the generator includes the following 6 principal ratings

   i. KVA or KW capacity and power factor
  ii. Frequency
 iii. Number of phases
  iv. Voltage between phases
   v. Speed

   i. *KVA or KW capacity and power factor*: The number of generating units, Kilowatt capacity, and speed are determined from hydraulic considerations, turbine designs, and power system availability. Power factors and ratings are determined by requirements such as mechanical, electrical, and hydraulic considerations; anticipated loads and demands; location of plant; suitable transmission lines; and distribution systems. Synchronous generators are rated in terms of KVA at certain power factors. It may be lagging or leading within the machine rating. Operation at lagging power factors lower than what is rated is possible. The rated power factor of the synchronous generator is always neutral or leading. The induction generator always operates at a lagging power factor. The induction generator is rated in terms of KW at rated power factor (PF). The induction generator is normally rated at 0.8PF while synchronous generators have a PF rating of 0.8–1.00.

ii. *Frequency*: Power system frequencies vary. In the United States, it is 60 Hertz and in Europe it is 50 Hertz.

iii. *Number of phases*: Hydroelectric generators have three phases.

iv. *Voltage between phases*: Hydroelectric generators' voltage ratings between phases generator armature at which generator produces rated power between 2,000–37,500 KVA.

v. *Speed*: Synchronous generators operates on synchronous speed which is evaluated by the turbine's rotational speed. This speed is very close to the designed speed which is expressed as:

$$n = 120 \text{ Frequency/Number of poles}$$

The number of poles shall be even and preferably divisible by four. A synchronous generator's diameter varies directly with KVS capacity and inversely with square of synchronous speed. The selection of higher speed reduces the size of machine, so costs can be reduced by selecting suitable speed. Speed range varies from 60 to 900 rpm.

18.13 Components of Hydrogenerator

The basic components of hydraulic generators are as follows:

- Basic supporting frame
- Shafts and couplings
- Rotor assembles (field post, windings, spiders, rims, and accessories)
- Stator
- Bearings and bearing bracket
- Brakes
- Exciter
- Cooler
- Anchor bolts

18.13.1 Shaft, Couplings, and Alignment

Shafts of generators receive torque from the shaft of the turbine. The diameter of the turbine's shaft and the diameter of the generator's shaft shall be matched, and the coupling flange diameter shall be fixed based on the diameter of the shaft at the coupling. The shaft may be horizontal and vertical as required, and may be made of carbon or alloy steel forged with integrally forged coupling flanges.

Shafts may be forged or fabricated as required. The forged shaft shall be continuous without any joint and shall be machined throughout the length of the shaft. A hole shall be drilled through its axis to facilitate inspection and to find any flaws in the shaft. This axial hole may be used as an oil passage on Francis Turbines; it shall otherwise be plugged to avoid any sort of leakage of water through the turbine.

Fabricated shafts are used when a generator shaft receives a large amount of torque that calls for a larger shaft diameter. This type of special shaft is fabricated from steel plates of predetermined thicknesses. The weight of the fabricated shaft is almost half of the weight of forged shaft, and it can resist at least 40% more bending and torque than a forged shaft can.

Steel plates are cut to specification and rolled into shape as required. The seam is welded per approved welding procedures to form a cylinder, and the coupling flange is welded to the fabricated cylinder to complete the shaft.

Many opinions exist regarding the physics of transmission of torque from the turbine to the generator through coupling, with the *Shear Method* and *Friction Method* being the most accepted techniques.

In the *Shear Method*, bolts are tightened properly to prevent the opening PF the coupling due to vertical load, and then the torque from turbine is transferred through the bolts.

In the *Friction Method*, bolts used that have a clearance or are loosely fit in their holes. The bolts are required to be tightened sufficiently to transmit torque.

Alignment of the shaft is an critical activity. Imperfect alignment of the shaft may lead to disaster due to the generation of vibrations, noises, cyclic stresses in stationery parts, and some abnormal stresses in rotary parts. The alignment of the shaft shall be done either in the turbine manufacturer's shop or the generator manufacturer's shop.

18.13.2 Rotor Assembles

Generator rotors consist of many components such as shafts, field poles, amortisseur windings, spiders, rims, and accessories. Synchronous generators are salient poles generators. The rotor consists of a spider where poles are attached to a metal rim and metal rim is attached to the shaft of generator by spokes. These spokes are called spider arms, and the total assembly of arms is called the spider. Field poles, amortisseur, windings, and exciter armature are mounted on the spider. Depending on the speed, the rim spider with bolted poles, the laminated spider with dove tailed bolts, or the solid spider is to be fabricated.

The field poles are comprised of steel. These are pressed and bolted together to withstand rotational and electrical stresses. Field pole windings can be either rectangular wires or flat copper wound poles, and windings are held firmly in position for normal and runaway conditions. The function of the field is to provide generator excitation. The magnetic flux produced by the field converts the rotating mechanical energy into electrical energy. Without magnetic fields turbine power cannot be transmitted to the electrical system, and the quality of electrical power is dependent on the design and construction generator field.

Fabrication processes for rotors depend on the size of the generator. Machined or forged castings are applicable to small generators. For large generators, assembly is fabricated from steel plates or a combination of forging, casting, and welding. Application of casting and forging is limited to high speed and low capacity generators.

Ventilation for small generators may be produced by fan blades that are attached to the rim or by a fan separately mounted on the spider rim assembly.

Separate fan blades or pole mounted fan blades are provided to have uniform distribution of air along the length. Rotor rims have spaces between lamination rings formed by spacers and permits air to flow through rims.

The rotor assembly fulfills numerous functions. The main functions are to transmit torque from the shaft to the poles and also to provide support to the poles. It also transmits combined loads of turbines and generators to the foundation, and it provides ventilation and magnetic paths from pole to pole flux.

18.13.2.1 Stator Frame

The stator frame is fabricated from steel plates, then welded and machined to make a robust grid like structure. Strength and stiffness are the prime requirements of the stator frame, which will be able to withstand unbalanced magnetic pull, short circuit, or faulty synchronizing torques being generated in the core of the stator, as well as to minimize vibration and noise levels.

Problems are faced in cases of extra large frame that have diameter of more than 10 m because of radial expansion and contraction due to thermal changes. A radially free core system is designed to permit the core to expand and contract radially while transmitting torque to the frame.

Sometimes magnetic, thermal, torque, and mass forces become so great that they pose a problem to building a frame. An arrangement should be made to transmit the load directly to the foundation rather than through the frame.

18.13.3 Stator Core

Stator core is fabricated from hot rolled silicon steel in the form of laminations. The steel used may either be grain oriented or non-grain oriented, which possess super magnetic qualities with reduced losses and improved permeability. This core steel is known as core loss type for its max limitation of losses. Core loss varies directly with the machine voltage. The loss increases non-linearly with the increasing of core saturation. Generally, four different steel thicknesses are used in the fabrication of laminations. The laminations are stacked, segmented, and separated by the spacer (made of thicker steel) to make a passage for ventilation throughout the core length. The core is formed by laminations in such a way that armature slots are skewed rather than vertically axial by installing the frame key off vertical, and armature slots made in the core shall be painted with a semiconducting compound. The laminations are stacked and separated by the spacer (made of thicker steel) to make a passage for ventilation throughout the core length.

The laminations are insulated. Varnish is applied to each lamination known as the core plate with the functional role as an insulating layer between each lamination to reduce eddy current.

Stators provide magnetic circuits of the machine, and it concentrates magnetic fields emanating from the generator's rotor to produce induced voltage to the armature. It provides cooling and structural support for armature windings. The windings are constructed and layered with insulated copper wires. The wires are inserted into the core slots and connected together. The assembly is dipped in a pressure vacuum impregnated with resin.

Hydrogenerators are generally not equipped with the permanent installation of instrumentation systems or detection systems for the monitoring of core conditions. Instruments are installed on a temporarily basis such as thermo-couplings attached to the back of the core.

18.13.4 Armature Windings

Winding is a very important part of the hydrogenerator and is known as the heart of the generator. The winding consists of diamond shaped coils that are inserted in slots, one side of which lies on top of the slot and the other part at the bottom of the slot. The coil overlaps in such a way that each slot contains a bottom and top side of the coil and the coil ends form two layers. A double layer winding is one in which there are two separate coil sides in any one slot; a particular coil has one side on the upper half of a slot (i.e., in the top layer nearest to the air gap) and the return coil side lies in the bottom half of another slot (i.e., in bottom layer). The coils in each pole pitch are divided into three groups representing the three faces, within each group consecutive coils are connected in series and corresponding groups under successive poles are interconnected in series or parallel according to the voltage and current rating of the machine. Such a series-parallel connection of coil constitutes one phase of a three phase winding. The windings are form-wound copper coil insulated with high-voltage mica based insulation systems.

18.13.5 Bearings

Hydrogenerators used in hydropower are of different sizes and various configurations such as vertical, horizontal and inclined-shaft generators. Bearing systems are adopted based on the application

Bearing systems are adopted in each configuration of the generator to handle radial and axial thrust in both directions that are imposed on the generator. Selection of bearings depends on cost, speed, speed variation, load and load variation, permissible deflection, available space, and noise resistance.

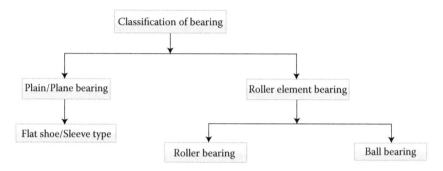

18.13.5.1 Plane Bearing

These type of bearings are used mostly in vertical shaft generators, with recent modifications to this type of bearing called magnetic bearings where part of the load is magnetically supported. In this system, sliding members are separated by the fluid film, which prevents wear and tear during operation. This fluid film produces a pressure in the film through its shearing motion, and the pressure being generated in the film prevents direct contact with the sliding component.

Three possible conditions exist, one being a dry condition when members are not separated by fluid film and no pressure will be developed, which results in rapid wear and tear because pressure or film will not exist at standstill or at very low speed and the bearing will fail as a consequence. The second condition is a mixed friction condition, which is a combination of dry and fluid friction. The third condition is the most ideal one in which fluid friction is separated with separating fluid layers. Bearings in hydromachines use fully developed fluid film lubrication.

The hardness of the sliding member is the sign of capability of bearing that supports the load. The rotary member is made of high quality steel called bearing journal, and stationary parts are called bearings on which anti-friction metal is applied. The harness of anti-friction surface can be imposed by using different alloys. The normal loading of plane bearing per USBR is 35 kgs/cm².

Considerable amounts of shock can be absorbed by plane bearings, and it also acts as a vibration damper because stress is transferred through the film. Thickness of film decreases with the increase of load on the film. When it is overloaded and sliding the component comes in contact with a fixed one, resulting in changes of fluid friction to the plate friction and a large amount of uncontrollable heat is generated from the failed bearing.

18.13.5.2 Rolling Element Bearing

The Rolling Element Bearing is known as an anti-friction bearing where some rolling elements like balls or rollers are used between main parts.

In Rolling Element Bearings friction and wear is very low and it requires less lubricant. This bearing works with elastohydrodynamic mechanisms, where relative friction is facilitated by interposing rolling elements between rolling and moving components. Rolling Element Bearings are preferred at lower speed, and are found in generators having low kilovolt ampere ratings.

18.14 Control System

Major hydropower projects operate at two levels: the Governor Gallery and Centralized Control Room with Digital Control Technology.

Modern control systems control the entire plant from a single location by the help of supervisory control and data acquisition (SCADA) systems. It includes programmable logic controllers (PLC) and distributed computer control systems with graphic display screens. SCADA control systems provide flexibility in control alarming, sequence of event recording, and remote communication.

SCADA systems provide optimization by remotely controlling power generation and transmission through both software and hardware techniques. It is equipped with PC, PLC, RTU, sensors, data communication, and transfer devices. The remote terminal unit is the core of the system, collecting data and transferring it to process equipment with the process equipment transfering the data to the controller unit.

Function of SCADA system:

- Data acquisition
- Data communication
- Data presentation
- Monitoring and controlling

This system is designed to control and protecting hydropower plants, and its control functions are:

- Turbine control
- Generator control and excitation

- Plant control
- Synchronization

18.14.1 Turbine Controls

Speed and load of turbines are controlled by adjusting the flow of water through turbine. The control of water flow is governed by the governor to balance input power and system loads.

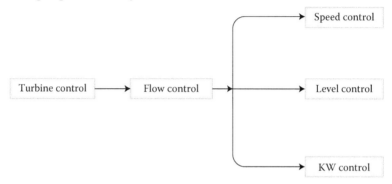

18.14.2 Generator Controls

Excitation controls of the generator are used to regulate generator operations. The main function of the excitation system is to control voltage and synchronization if it is operated isolated, and to control of power factor if it is operated collectively.

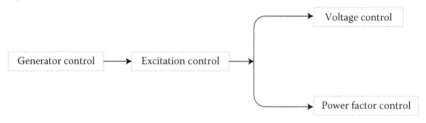

18.14.3 Plant Automatic Control

Plant control means it controls the operation of the plant. Its role is to control the automatic start sequence, automatic normal and emergency shutdown, excitation control, monitoring, and display of the plant condition.

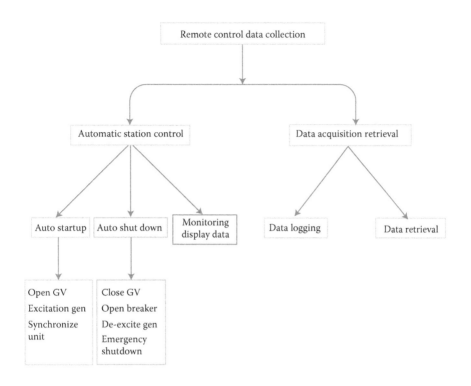

Bibliography

AHEC-IITR, "3.1 Electro-Mechanical – Selection of Turbine and Governing System," standard/manual/guideline with support from Ministry of New and Renewable Energy, Roorkee, June 2012.

AHEC-IITR, "3.4 E&M Works – Selection of Control, Automation, Protection and Monitoring System," standard/manual/guideline with support from Ministry of New and Renewable Energy, Roorkee, July 2012.

Hydro GendesignManual, USBR.

Selecting Hydraulic Reaction Turbines United States Department of the Interior Bureau of Reclamation, Engineering Monograph no20.

Index

A

Abutment, 23, 65, 191–192
Accident, 111, 181, 315–316, 318,
    320–321, 323
Account, 27, 38, 72, 94, 131, 375, 395
Accountable wastage, 168
ACI Method, 240, 246–252
Acrylic, 291
Active crack, 295
Active fault, 186, 354, 357
Activity, 25–26, 68, 81, 89, 105, 111, 113,
    125, 194, 310, 333, 368
Addendum, 94
Additives, 224, 230, 234, 238, 243, 245,
    251–252, 257–258, 297, 308
Adit, 156
Adjustment of bending angle, 180
Administrative Framework, 349, 356, 365
Aeration Device, 388–389
Aggregate abrasion value, 136–137
Aggregate crushing value, 136–137
Aggregate impact value, 136–137
Aggregates, 139, 235–238, 245, 259, 275,
    297, 313
Agitator, 133, 207, 211
Aims and Objectives, 347
Air and Noise Level, 359
Air bubble, 260, 264, 294, 309, 398
Air compressor, 133, 146, 155, 158, 160,
    161, 201–203, 218, 299, 389
Air entrained concrete, 247–248, 250, 252
Alkali-Silica Reaction, 297
Alternative panel, 283
Ambient temperature, 70, 265,
    271–273, 275
ANFO, 145
Angle, 78, 180, 213, 259, 331, 339, 389
Anti-virus, 132
Application Form, 364–367
Arbitration, 84
Area grouting, 67, 196
Armature windings, 409
Armature, 404–407, 409

Artificial channel, 151
Ascending stage, 215–216
Assessment of Foundation, 194–195
Assurance, 303, 344
ASTM, 329, 341
ASTM C Type I, 273
ASTM C, 249, 393
Auxiliary Transformer, 385
Axis of Dam, 65, 67, 135, 191, 198, 213, 284
Axis, 177, 213, 392

B

Baffle Blocks, 62
Bar Bending, 30, 134, 169–174, 178–182, 307
Bar Chart, 103, 113–116
Bar cutting Machine, 179–180
Bar mark, 170
Barrage, 1, 23, 55
Base, 3, 78, 186, 188, 258, 300
Base, Hardener, 300
Batching and Mixing, 255–256
Batching Plant, 31, 108, 131, 133, 137–141,
    256–259, 262, 271, 306–307
Bats, 227
BBS, 169–174, 180–182, 307
Bearing bracket, 406
Bearing, 122, 188, 192, 287, 389, 406,
    410–411
Bedding, 185, 189, 220
Bench, 78, 157, 198–199
Bench Height, 144–146
Bending diameter, 180
Bending moment, 282
Bends, 171–172
Bend Test, 338
Benefit cost ratio, 34, 40–41
BG for SD, 135
Bid system, 91–93
Billing, 85, 131
Bill of Quantities, 84, 87, 90
Bill of Quantity, 33, 375, 376
Biodiversity, 34, 361–362, 366, 373
Bituminous emulsion paint, 284

Blanket Grouting, 68, 196, 200
Blast holes, 141–148
Blast Log, 150
Blaster License, 135
Blasting, 130, 135–136, 141, 147–149, 262, 300, 326–327
Blasting operation, 141–150, 155, 262, 306, 354, 368
Bleeding, 208, 210–211
Block, 62, 135, 152, 195, 198, 208, 214, 260–262, 282, 285, 290, 306–307, 323, 325, 385
Bond, 135, 262, 279–281, 289–290, 294, 300–302
Borrow area, 126, 357
Break up of Activities, 25–26
Breccias, 68, 70, 216
Bridges, 58, 72, 126, 282, 287, 298, 300
Bucket, 58, 60, 62–63, 133, 171, 259, 263–264, 392–393
Bulk density, 136–137
Burden, 143, 145–146, 355
Buried channel, 3, 184, 221–223, 226–227
Butt joint, 176, 294
Butt welding, 176–178

C

Calculation on absolute volume basis, 251
Calibration, 139–140, 307
Canal falls, 18–19, 21–22
Cancellation contractor, 96
Capacity, 10–11, 13, 26, 34–35, 43, 46, 137, 152, 194, 277, 307, 376, 378, 383, 394–395, 399, 402
Capacity factor, 36, 378, 381
Capital cost breakdown, 45
Carbon emission, 1–2, 4–5
Carbon steel, 165–166, 341, 393
Carbonate, 226–227, 297
Carbonation, 290
Carbonic acid, 225
Casing, 213, 390–391, 393, 397
Catchment area treatment plan, 258
Cathodic protection, 298
Cause ways, 126
Cavitation resistance material, 398
Cavitations, 54, 397–398
Cavities, 185–186, 195, 221

Cement, 29–30, 139, 205, 207–209, 211–212, 229–231, 234–236, 239–240, 244–245, 249–253, 269–270, 272–273, 275, 290, 297, 302, 308, 313
Cement composition, 269
Cement register, 313
Cementitious, 195, 200, 204, 243, 246, 275
Centralized control room, 411
Channel bed, 155
Characteristic strength, 230–232, 242–243
Chemical, 166, 222, 225, 233, 292, 294, 297, 331
Classification, 13–14, 56–57, 230–232, 235–236, 298, 330, 331, 384, 387
Clay seam, 3, 67–68, 99, 184, 193, 197, 199, 202–205, 357
Clean Air Act, 356
Clean energy, 2–3, 6, 9, 15
Clean Water Act, 356
Clean weigh dial, 140
Clearance of Foundation, 313
Cleavage, 67, 184, 185, 189, 199, 342
Clinometers, 188
CM, 318
CNC, 329
Coarse filter material, 161
Co-efficient of thermal expansion, 269, 330
Coffer dam, 49, 55–56, 152–162
Cohesion, 77–78, 193, 208–210
Cold, 271, 276
Cold Joint, 259–260, 262, 289–290, 309
Commitment, 86, 109, 131, 303–304, 316
Comparative statement, 93, 98
Competence and Awareness Training, 320
Complete contraction joint, 283
Completion Certificate, 95
Components of Network, 105
Composite Bank Guarantee, 94–95
Compressed air, 281, 299, 385
Compressional ultrasonic velocity, 192
Concrete batching plant, 137
Concrete, 81–82, 89–91, 154, 165, 173, 229–265, 268, 272–276, 277–302, 308–309
Concrete cover, 170

Concrete pour card, 306, 311
Concrete temperature, 211, 247, 262, 265, 270, 274–276, 296
Condolite, 218
Conduit, 60, 66, 152
Consolidation, 211
Consolidation grouting, 67, 77, 99, 133, 154, 160–161, 196, 200, 204–209, 212, 214–215, 217, 223–225, 306, 310–312, 372
Constraints, 21, 68, 102, 110, 112, 128, 152, 200, 343
Construction joint, 70, 99, 259–260, 262, 277–286, 293
Construction management, 125, 127–128
Construction manager, 125, 127–129, 272, 317
Construction of dam, 2, 54, 100, 114–116, 151, 209, 218, 234–235
Construction power, 127, 373
Construction stage, 69, 127, 186, 191, 193
Contour, 16, 20, 131, 135, 186, 198, 220
Contract administration, 84–85
Contract document, 85, 88, 121, 129, 169, 198, 331
Contraction, 70, 272, 285–289, 295–296, 408
Contraction joints, 70, 282–289, 296, 307
Contract management, 83–124
Contract manager, 85, 128–129
Contractual Completion Time, 94–95
Contribution, 165, 234, 261–262, 377
Control gates, 67
Controlled, 5, 8, 66, 148, 169, 202, 232, 262, 297, 309, 397
Controlled concrete, 232, 256
Control of cavitations, 398
Control pillar, 131
Control system, 138, 258, 378, 411–413
Controller unit, 411
Controlling measures, 165, 322–323
Controlling temperature, 275–276
Convener, 128–129
Cooler, 6, 15, 122, 296, 406
Cooling water system, 385
Copper windings, 404
Core concrete, 246, 259–262, 264, 296, 301, 308–309
Core, 58, 65, 76, 159, 162, 191, 218, 271, 275, 408–409

Core drilling, 133, 190, 194, 199, 218
Core of dam, 213, 232, 261
Corrective action, 113, 304
Corrosion, 290, 297–298
Cost, 26–35, 39–40, 42–45, 53, 90, 110–111, 128, 353, 371–376, 379–382
Cost analysis, 34, 44, 371–382
Cost Plus Contract, 88
Cost slope, 111
Coupler, 134, 170, 175–176
Couplings, 175, 407
CPM, 103–105
Cracks, 67–68, 70, 74, 187, 200, 209, 211, 214, 267, 271–272, 281–286, 294–300
Crash time, 111–112
Crawler drilling machine, 160
Crazing, 296
Crest of dam, 66
Critical head, 387
Critical path, 105–108
Crown, 66, 157, 281–282
Crushing plant, 126
CSG, 154, 162
CTE, 270, 272
Curing, 30, 99, 232, 258–265, 272, 274, 293–294, 297, 302, 309
Curtain, 204, 214–215, 223
Curtain grouting, 68, 74, 160–161, 196, 200, 213–216
Curve, 111–113, 263, 403
Cut length, 169–172, 182
Cut sheet, 332–333
Cut-off shaft, 217
Cutting, 30, 134, 169, 334–335, 342
Cyclic tensile test, 176

D

Daily Progress Report, 130
Dam, 3, 22–23, 55–63, 65–82, 89–91, 100, 114–116, 135, 148, 183–196, 258–265, 269, 281, 295, 323–327
Dam based outlet and spillway, 19
Dam site geology, 2, 185, 197, 200
Dam Toe, 18, 22, 23
Data acquisition, 411
Data communication, 411
Dead load, 69–72, 159

Debt-equity ratio, 34, 380
Debt service coverage ratio, 40
De-coiling, 179
Defect liability period, 95
Deflector, 393
Delay detonator, 145–147
Deleterious, 137, 235, 297
Demag, 160–161
Demography survey, 17, 368
Dental treatment, 68, 77, 193, 200, 216–220, 313
Depth, 29, 60, 67, 76, 147–148, 161, 186, 193, 199, 200, 216–217
Depth of hole, 143, 145, 191, 215–216, 224
Deriaz, 395
Descending stage, 215
Designation, 331–332
Design drawings, 332
Designed mix, 232
Design head, 383, 387
Desilting prevention, 18
Detailed estimate, 27–33, 90, 91
Detailed project report, 17, 192
Detonating cord, 142
Detonator, 141, 145–147, 149
Dewatering, 152, 194
Dewatering system, 385
Diameter, 143–144, 146, 156, 168, 170, 175, 216, 340, 401, 407
Diamond, 190, 199, 409
Dip, 3, 185, 186, 194, 196, 199
Direct cost, 27, 375
Direction of the fault, 216, 219
Disappearing stream, 226–227
Disaster management plan, 361
Discontinuities, 195, 218, 225
Discount factors, 34, 38–39
Displaying safety slogan, 317, 319
Disposal area, 126, 360
Disposal of muck, 154, 158, 360
Disruption of concrete placement, 282
Dissipation of energy, 62, 268–269, 296
Dissipation of heat, 268
Dissolved oxygen, 354, 389
Diversion dam, 19, 56
Diversion, 19, 29, 47, 151–155
Diversion tunnel, 49, 156–161, 191
Document control, 334
Documents required, 119, 129

Dormant, 295
Dowel bar, 280–281
Downstream, 4, 23, 47, 50, 54, 58, 62, 67, 78–79, 135, 154–155, 161, 261
Dozer, 133, 158, 160–161, 198
DPR stage, 186, 190
Draft tube, 54, 118, 388–391
Drainage gallery, 67, 70, 213, 216, 263
Drainage hole, 62, 67, 68, 74, 216
Drawing index, 130
Drilling holes, 135, 157, 160–161
Drilling machine, 146, 160, 191, 201, 213
Drying shrinkage, 296, 299
Drying shrinkage crack, 295–296
DSP, 231
Due diligence, 41–46
Dummy activity, 105, 108–109
Dumper/tripper, 133, 160–161
Duration, 38, 95, 105–107, 349
Dyke, 8, 54, 57, 127, 185

E

Earliest Finish, 105–106
Earliest Finish Time (EFT), 105
Earliest start time, 105–107
Earnest money, 94
Earth moving equipment, 133
Earthquake, 10, 70, 72, 76, 78–81
Earthquake force, 70, 76–77, 79, 173, 184, 193, 216, 354
Economic feasibility, 34, 405
Economic specific speed, 395
Economist, 1, 351
Effective cost estimate, 90
Effectiveness of EIA, 353
Efficiency of weld joint, 335
EIA & SEA, 346
EIA Application, 347–348
EIA process, 349–351
Elastic fluid, 398
Elastomeric sealant, 286
Electro chemical chloride migration, 298
Electro chemical reaction, 297–298
Electrode, 176–177
Electro-mechanical, 396, 402
Electro-mechanical work, 25–26, 29, 42–43, 373, 375
Element bearing, 411

Embankment, 55
Enabling works, 25
Energy dissipation, 30, 57–60, 62, 67, 73,
        162, 191, 263
Energy mix, 5–6
Energy of water, 2, 5, 8–9
Environmental and human right, 346
Environmental clearance, 345, 349,
        351, 363
Environmental consultant, 351
Environmental hazards and risks, 316
Environmental impact assessment,
        346–349
Environmental impact, 34, 348,
        353–355, 405
Environmental scientist, 351
Environmental statement, 350
Environment Management Plan, 357
EOT Cranes, 373, 385–386
Epoxy, 298–299
Epoxy coated carbon, 166
Epoxy concrete, 300
Epoxy latex adhesive, 281
Epoxy resin, 298–299, 301–302
Equipment for Concreting, 133
Erection bay, 57, 386
Erection, 30, 88, 154, 323, 325, 329, 333,
        339
Escalation, 32–33
Estimation, 4, 25–26, 89–90, 102, 109,
        375–375
E-Tendering, 97–98
Evaporation, 265, 276, 295
Event, 105–106, 321
Excavator, 155, 158, 160, 161, 176
Excitation control, 412
Exciter, 406, 407
Expansion Joint, 287–288, 293, 341
Experimental procedure, 274
Exploration by drilling, 187, 194
Exploration by pit, 193, 194
Exploration by trench, 194
Exploratory drilling, 190, 194, 199
Explosive cartridge, 143
Explosive licenses, 135
Explosives, 141–145, 148–149
Exposure, 194, 231–233, 240, 242, 246–247,
        249–250, 252, 283
Extension of contractual time, 95

Extension of time, 95, 96
Extension of time limit, 84
External concrete, 246, 260–262, 268, 290,
        296, 306–307
Extra large frame, 408
Extreme load combination, 72, 76, 78

F

Fabrication, 30, 99, 329–344, 393, 408–409
Face of rock, 142–143, 145
Factor of safety, 78
Fault, 3, 67, 70, 76, 79–80, 184–187, 189,
        194, 196, 199, 216–221, 303, 354
Fault breccias, 68, 70, 184, 216, 218
Fault/shear zone, 3, 49, 68, 77, 79, 185,
        192, 197, 199–200, 216–218
Fault zones, 3, 49, 70, 184, 190, 193,
        216–217
Fe, 166
Feasibility report, 25, 190, 363
Feasibility stage, 186–187, 190
Fiber reinforced polymer bar, 166
Fibrous, compressible, 287
Field molded sealants, 291, 294
Field poles, 407–408
Filler material, 155, 159, 299, 337
Final completion certificate, 95
Financial analysis, 34–35, 41, 85
Financing cost, 32–33, 42
Fine filler, 300
Fine filter material, 159
Fineness, 269, 272, 275
Fineness modulus, 246, 249–250, 254
Fiscal incentives, 34, 37
Fishery development, 363
Fissures, 67–68, 74, 184–185, 187, 195–197,
        199–200, 202, 204, 208–212,
        214–215, 221, 224, 226–227, 234,
        298, 342
Fit up, 329, 333, 335, 337
Flame cutting, 334–335, 342
Flat copper wound poles, 408
Float, 106–108
Float or slack, 105
Flocculation, 211
Flood, 22, 50, 54, 56, 66, 68, 79, 81, 96,
        152–153, 161, 197, 355, 361
Floor slab, 280

Fluorescence dyes, 222
Fly ash, 138, 224, 231, 234, 253–255, 265, 270, 272, 275, 297
Fold, 3, 185–186, 189, 194, 196, 199, 357
Foliation, 185
Force mejure, 96
Fore bay, 8, 18, 20, 25, 29, 43, 54, 57, 387
Forest Conservation Act, 356
Forged coupling, 407
Forged shaft, 407
Format, 33, 93, 95, 98, 129–130, 169, 191, 324–327
Formation of cracks, 281–283, 286
Foundation, 77, 138, 148, 194, 197, 204, 218, 300–302
Foundation exploration, 184
Foundation treatment, 200–204, 372
Fractured, 3, 191, 200, 205, 218, 223, 226, 306
Francis turbine, 388–390, 395, 397, 401, 403–404, 407
FRC, 231
Free board, 65, 79, 159
Frequency, 167, 349, 397, 405–406
Friction Method, 407
Friction, 5, 149, 210, 225, 340, 410
Froude number, 60–61
Fuel wood energy, 360
Function of pre award of contract, 85
Fury, 151

G

Galvanization, 298
Galvanized, 165
Gantry Girder, 386
Gas and flame cutting, 334–335
Gate valves, 385
GCC, 94–97, 198
Gelatin, 141–143, 145
Generation of heat, 230, 246, 261, 263, 267, 269–270, 275, 283
Generator, 5, 8, 20, 26, 30, 35, 54, 118–119, 123, 299, 373, 378, 384–387, 390, 392, 394–395, 400, 405–407, 410–412
Generator and driving system, 404–405
Geo-hydrological characteristic of foundation, 192

Geological constraints, 69, 152, 184
Geological exploration, 184–194
Geological map, 186–187, 189
Geological mapping, 67, 187–188, 193–194, 196, 198–199, 220
Geologic structure, 189
Geologist, 3, 67–68, 70, 76, 113, 187–200, 205–207, 209, 216, 218, 227, 262–263
Geophysical survey, 189
Geotechnical condition, 184
Geo technical properties, 50, 192, 378
GGBFS, 275
Global, 13
Gloves, 181, 189, 202, 265, 322
Goggles, 181, 202, 265, 322
Governor, 44, 118–119, 123, 385
Governor Gallery, 411
GPS, 188–189
Grade designation, 252
Grading of aggregates, 236, 247
Granulated Blast furnace slag cement (GGBFS), 275, 297
Gravity dam, 55–56, 59–60, 65–82, 89, 137, 141, 212, 231, 234, 246, 310
Green canopy, 361
Green cut, 262, 281, 309
Grid pattern, 142, 144, 189, 196, 200–202, 204
Grit blasting, 30, 300–301, 338
Gross head, 386–387
Grout ability, 192
Grout mix, 206–211, 225
Grout mixer, 207
Grout pump, 133, 158, 207
Grout, 68, 74, 99, 192, 195–196, 206–211, 225, 298–299, 306
Grouting, 30, 67–68, 135, 193, 195–196, 205–215, 223–224, 290, 294, 298–306, 310, 372
Grouting machine, 160
Guest house, 127, 373
Guide mechanism, 390
Guide Vanes, 388–391, 397
Guide Venes, 388
Gun grade, 291, 294
Gun powder, 145
GW, 7–8, 15, 46

H

Hammer, 188
Hand lens, 188
Hardener, 299–300
Haul roads, 126, 373
Hazard, 191, 321–322
Hazard elimination, 322
Hazard identification, 321, 323
Head, 13, 15, 23, 43, 387, 395–396
Head cover, 389–390
Health, safety & environmental risk analysis, 315–327
Heat of hydration, 232, 246, 269–270, 272
Heavy weight concrete, 230
Height of dam, 65, 215, 217
Heritage, 357–358, 373
High flood, 54, 81, 151, 161, 163
High flood data, 163
Holes, 200, 202, 204, 213, 215
Holing, 335
Homogeneous, 191, 195, 207, 256–258, 282, 300
Honey comb, 310
Hooks, 171–173
Hooped runner, 393
Horizontal channels, 221
Horizontal drilling, 157
Horizontal transition, 280
Housekeeping, 181, 319
How it works, 142
HP-Hold Point, 310
HR, 86, 129–130
HSE, 118, 130, 132, 305, 315–317, 320
HVFC, 231
Hydra, 134, 179, 342
Hydra operators, 181
Hydration process, 208, 211, 233, 261, 263, 265, 268–270, 272, 274–275
Hydraulic characteristics, 384, 405
Hydraulic equipments, 385
Hydraulic head, 13, 43, 73–74, 200
Hydraulic height, 65
Hydraulic jump, 58–62, 66–67, 263
Hydro, 1, 4
Hydro energy, 383
Hydrogen film, 398
Hydrogenerators, 409–410
Hydro-geological map, 186–187

Hydrologic, 196
Hydrologic cycle (HC), 1, 4
Hydrological characteristics, 16, 35, 151
Hydrophilic water stop, 292–293
Hydropower project, 1–11, 42–63, 345–369, 371–382
Hydropower, 2, 4, 6–11, 15, 24, 35, 60, 345, 368, 371, 378–379
Hydrostatic head, 4, 70–71, 183, 215
Hydrostatic pressure, 70, 73–74, 81, 159, 200

I

ICC pressure, 372, 374
Identification and selection of species of fishes, 363
IFC, 198, 339
IFC drawings, 88, 104, 111, 119–120, 130, 169, 180–182, 305, 312–313, 329
Impact of hydropower project, 354
Impervious barrier, 5, 55, 183
Impervious core, 158–159, 161–162
Impounding of reservoir, 354
Improved permeability, 409
Impulse type, 391–392
Indemnity, 97
Indent, 167
Index of drawings, 130
Indian Code of Practice, 331
Indicative inspection and test plan, 310–312
Indicative test plans, 259, 310, 313
Indirect cost, 27, 32, 45, 374–375
Induction training program, 317, 319–320
Infrastructures cost, 375
Injection grouting, 299
Insert plate, 154–155
In-situ test, 192–193, 195
Inspection, 67, 82, 119–120, 167–169, 187, 194, 218, 310, 312, 344, 407
Installation system, 138–139
Insulation, 276
Intake structure, 19–20, 22–23, 54, 57, 372
Intake, 5, 8, 20, 23, 29, 43, 126, 209, 220
Integrity, 185, 218, 227, 277, 285, 301
Integrity check, 81–82, 342

Interest during construction (IDC), 27, 32, 34, 36, 45, 374
Internal friction, 209–210
Internal hydrostatic pressure, 71
Internal rate of return (IRR), 34, 39, 41
Internal review, 304
Interstices, 195, 205–206, 212, 214–215, 234
Invert, 157–158, 281–282
Investigation report, 196
ISA, 330–331
ISJC, 331
ISMC, 330–331
ISMCP, 331
ISO, 304, 330
Isotropic, 195
Item Rate Contract, 87–88

J

Jack Hammer, 158, 160–161, 201
Job procedure, 304–310, 313
Job procedure and ITP, 130
Joints, 184, 277–302, 341, 343
Jumbo drill, 158

K

Kaplan, 15, 388, 390–392, 395, 397, 399
Karst caves, 221
Karsted, 221
Karstic, 221
Karsting, 226
Karsts topography, 225–227
Kentiledge, 148, 150
Kerb, 157–158
K factor, 171
Kick-off, 125–150
Kick off meeting, 85, 128
Knowledge, 3, 119, 125, 196, 209, 215, 319–320, 351, 353
KVA, 405–406

L

Labor laws, 96
Labor license, 86, 130, 135
Laitance, 279, 290, 300
Land use, 354, 356
Lap joint, 176, 294

Lap weld, 177–178
Lapping, 170, 174–175
Latest Finish Time (LF), 105
Latest Start (LS), 105–107
Legal basis for regulation and guidelines, 348
Length of dam, 65
Length of the fault, 216–217, 219
Letter of intent (LOI), 85, 93–95, 128
Lifting equipments, 134
Light weight concrete, 230
Lime stone, 48
Lining, 156–157, 163
Liquefaction, 79, 193
Liquidated damage (LD), 95–96
Load combination, 71–78
Loader, 133, 155, 158, 160–161
Local area development, 27
Longitudinal joint, 281–282
Low pressure, 66, 389, 398
Low pressure grouting, 290
Low pressure tunnel, 5, 8, 34, 49–50, 54–55, 57, 121, 345, 355, 357, 372
Lower ambient temperature, 273, 275
Lugeon, 195, 205, 214, 219, 224, 262
Lump sum contract, 86

M

Machinery, 86, 88, 90–91, 93–94, 102, 112, 125–126, 128–129, 133–134, 158, 160, 258–259, 317, 342, 360, 373, 375
Magnetic flux, 408
Magnetic pull, 408
Main reinforcement, 171, 267, 278–279
Maintenance cost, 28, 31, 39, 345, 371, 378–379, 381
Maintenance workshop, 131
Management system, 84, 102, 304–305, 315
Manpower deployment, 130
Manual, 97, 119, 129, 193, 198, 304–305, 312–313
Mass concrete, 230, 232, 234, 236–237, 246–247, 250, 252–255, 261–263, 265, 267–276, 280, 283–285, 289–290, 295–296, 308–309
Material, 112, 158–159, 167, 169, 257–258, 280, 298–299, 329–333, 341, 393

Material testing, 240
Maximum head, 341, 387, 398
Maximum temperature, 240, 262,
    271–273, 275
MB, 331
MC 200, 331
MDF, 231
Mechanical couplers, 170
Mechanical, 26–27, 30, 117, 122, 168, 331,
    334, 406
Mechanical energy, 1, 5, 8, 20, 57, 384,
    387–389, 391–392, 404, 408
Metal arc, 176
Mill tolerance, 333
Millisecond detonator, 145–147
Minimum head, 387, 395
Minutes of meeting, 86, 92, 131
Mitigation measure, 34, 348–350,
    357–363, 373
Mitigation, 17, 34, 52, 357–358
Mix calculation, 245
Mix design, 108–109, 232, 238–256, 307
Mixed flow turbine, 388, 390
Mobilization, 86, 94, 125, 127–129,
    133–134, 317
Mobilization advance, 86, 94, 129, 135
Mobilization plan, 86, 129, 133–134
Modified commercial bid, 93
Modulus of elasticity, 192, 296, 330
Modulus of rigidity, 330
MOM, 93, 98, 131
Monitoring, 34, 82, 88, 99, 112–113, 262, 317,
    349–350, 357–358, 409, 411–412
Monitoring of temperature of
    concrete, 259
Monthly routine, 141
Movement joint, 287, 292
Movement of slab, 279–281
Muck disposal plan, 360
Multi staged, 133, 195, 203, 205–206
Multiple lines curtain, 214–215
Multipurpose project, 51–54

N

National Environmental Appellate
    Authority, 356, 365
National Environmental Tribunal Act,
    356, 365

National investment, 183
National mapping organization, 186
National Trade in Endangered Species
    of wild Fauna and Flora,
    356, 365
Net head, 386–387
Net present value (NPV), 34, 38–39, 41
Nitro compound, 145
No load test, 122–123
Noise pollution, 357–358
Nominal mix, 231–232, 356
Nominal, 166–167, 240, 246, 252
Non-conformance, 304
Non-destructive test, 301
Non-hydraulic concrete, 230
Normal time, 111
Normal weight concrete, 230
Nozzle, 299–300, 302, 392–393
Number of phases, 406

O

Occupational accidents, 315
Office furniture, 132
ONGC, 186
Operation group, 113, 118, 120
Optimistic completion, 109
Optimized manifold, 394
Organogram, 117, 129, 304
Over burden, 3, 142, 146, 198, 215, 305
Oxyacetylene, 334
Oxygen jet, 335

P

Packers, 160, 195, 207
Panel of size, 259
Participation of public, 348, 350
PayBack method, 40–41
Peak efficiency, 392
Pelton Wheel, 392–393, 395, 397
Penetrability, 208, 211–212
Penstock, 5, 8, 20, 23, 25, 27, 43, 45–46,
    49, 54–55, 121–122, 220–223,
    329–345, 356, 372, 386, 388,
    390–392, 397
Percolation, 194, 212, 225, 290
Percolation test, 187, 212–215, 225, 290, 313
Percussion drilling, 201

Permeability, 68, 183, 189, 194–196, 205,
    212–213, 223, 232–233, 238,
    270–271, 290, 409
Permeability strata, 190
Permissible limit, 33, 135, 183, 208, 211,
    246, 272, 298, 342
Personal protection equipment (PPE),
    181, 189, 319
PERT, 103–105, 109, 113, 357
Pessimistic completion, 109
pH value, 208, 225, 234
Phenomena, 4, 9, 58–59, 79, 220, 246–247,
    268, 281, 286, 289–290, 295–296,
    341, 397
Physical feature of land, 356–357
Physical property, 238, 330–331
PIC, 231
Piezo-metric head, 221
Pink patch, 218
Pipes, 6, 18, 152, 219, 276
Placement temperature, 262, 267,
    273–276
Plan, 56, 65, 86–87, 118, 169, 305, 322,
    359–361
Planning for start-up and
    commissioning, 119
Planning manager, 85, 129
Plans and drawings, 332–333
Plant automatic control, 412
Plant availability, 34, 39
Plant control, 412
Plantation cost, 28, 31
Plastic shrinkage crack, 295
PLC, 258, 411
Political scientist, 350
Political support and commitment, 348
Polysulphide, 291, 294
Polyurethane, 291
Population, 1, 15, 17, 345, 359, 362,
    368–369
Porosity, 189, 195, 208–209, 212, 238, 290
Potable water, 234
Potential energy, 1, 4–6, 50, 62, 383, 387,
    391–393
Pour Card, 259, 312
Pour cards of concrete, 306, 311, 313
Pouring sequence, 277, 279
Power factor (PF), 147–148, 387,
    405–406, 412

Power house, 5, 8, 16–18, 20–23, 25, 29,
    43, 45, 47, 49, 54–55, 57, 66, 118,
    126–127, 345, 356, 360, 372–373,
    383–413
Power of attorney, 129
PPE, 181, 265, 319, 322
PQR, 336
Precaution, 149–150, 152, 167–168,
    193–194, 276, 297, 309
Predecessor, 102, 105
Prediction, 17, 272–274, 346, 349, 351
Prefeasibility study, 16, 41
Preformed sealant, 291
Preliminary cost, 28
Preliminary dimensions, 159
Preliminary estimate, 26–27
Preliminary investigation, 184, 186–190
Present value, 34, 38–40
Pressure energy, 387–388, 391
Pressure, 54, 74, 76, 77, 122, 146, 209
Preventive action, 304
Primary holes, 196, 200, 202–203, 214
Primavera, 132
Primer, 294, 300, 339
Probable completion of the project, 110
Processes, 169, 294, 304, 349
Profile, 58, 65, 194, 198, 292–293
Project closed out report, 47
Project management consultant (PMC),
    26, 84, 239–240, 350–351
Project manager, 85–86, 102–104,
    109–111
Project proponent, 350–351, 363
Project townships, 127
Promisor, 84
Public consultation, 348, 350
Public Health Management Plan,
    359–360
Public health, 355
Public involvement, 348–349
Publicity of tender, 92
Pulling speed, 180
Pumped in stream, 19
Pumped storage scheme, 23–24
Pumped storage, 4, 6, 50–52
Punch list, 117–118, 123
Punching, 335
PVC strip, 99, 284
PVC water stop, 281, 292

Q

QA/QC, 113, 118, 124, 130, 303–313
QA/QC group, 120
QA/QC manual, 206, 312–313
Quality, 128, 136, 303
Quality management, 4, 304–305
Quality manager, 86
Quality manual, 119, 129, 198, 304–305, 312
Quality of work, 88, 97, 110, 120, 125, 182, 294, 303, 310, 344
Quality policy, 304
Quality work, 303
Quarry, 135–136, 357

R

Radiography result, 338
Rated head, 341, 386, 396
Rated power, 14, 406
Rated voltage, 405
Rate of hydration, 211
RCC in cantilever wall, 155
Reaction type, 387–391
Re-alkalization, 298
Recasting of estimate, 33, 91
Reconciliation, 165, 168
Reconnaissance survey, 16, 41, 89–90, 187
Reconnaissance, 186–187
Record control, 304
Rehabilitation, 49, 117, 358, 362–363
Reinforced cement concrete, 30, 55, 229
Reinforcement, 30, 165–182, 229, 246, 261, 268, 290, 298, 301, 307
Renewable energy, 2–5, 9, 15, 345, 371
Reservoir, 1, 3, 6, 8–9, 49, 53–57, 70, 79, 183–196, 220, 226, 354–355, 363, 394
Reservoir and tail water load, 70, 73
Resettlement, 357, 362, 368, 369
Resident Construction Manager, 318
Resistance against over turning, 78
Resource Conservation and Recovery Act, 356
Resources, 36, 93, 101–102, 110–111, 124, 346, 394
Revenue leakage, 84
Revised estimate, 33, 91
Revised financial bid, 98

Revised weight, 168
R factor, 171
Rheology, 209, 269
Ribbed, 165, 292
Risk, 102
Risk assessment, 321–327, 348
River diversion, 151–163, 372
Road network, 17, 126, 372
Rocket science, 125
Rock excavation, 154, 155, 161, 198
Rock fill, 19, 55, 57, 156, 158, 162
Rock fill material, 155, 158
Roller element bearing, 410
Root cause analysis, 318
Rotation, 122, 180, 280, 384, 400, 404–406
Rotor, 387, 389, 404–408
R-Review, 310
RTU, 411
Runners, 386, 389–391, 393
Run of River project, 18, 19, 47–49

S

Safe Drinking Water Act, 356
Safety, 22, 78, 102, 181, 232–233, 317
Safety Bulletin, 320
Safety Manager, 86, 129
Safety Oath, 316–317
Safety performance, 316, 318
Safety Policy, 316, 318
Safety Program, 82
Safety Shoe, 181, 188–189, 202, 322
Safety slogan and Essay competition, 317, 320
Safety workshops, 320
Safe working practice, 181, 315, 317–319
Sale Price of Electricity, 40
Samples Box, 191
SC, 331
SCADA system, 411
SCC, 198, 231
Schedule, 16, 30, 36, 89, 98, 102, 222, 322, 357–358
Schedule of Rate, 33, 84, 88, 90–91, 198
Schwing Stetter, 258
Scope, 1–3, 69, 86, 91, 102, 347, 349, 351
Scoping, 349, 351
SCR, 322
Screening, 159, 349, 351

Scroll, 388, 390–391
SEA, 346
Sealant, 287, 291–294
Seams, 74, 195, 204, 210, 225, 357
Secondary, 68, 70, 200, 207, 214, 221, 268
Section, 55, 58, 66, 153, 215, 269, 281, 312,
    330, 391
Secured Advance, 95
Security Deposit, 94–95
Sedimentation, 211
Sedimentation tank, 20, 25, 54, 57
Segregation, 259–260, 309
Seismic force, 76, 79, 81
Seismic survey, 187
Seismic wave velocity, 192
Seismology of terrain, 357
Selection of speed, 395, 406
Sequence of welding, 177–178
Service areas, 386
Service delivery, 84–85
Service gate, 156
Service road, 127, 373
Setting out, 338
Setting the turbine, 397–398
Shaft, 5, 122, 217, 257, 267, 387
Shape of bar, 170
Shear method, 407
Shear zone, 3, 49, 68, 77, 79, 184, 197,
    216–217, 221
Shearing, 176–177, 334, 410
Shielding gas, 338
Shop drawings, 332–333
Shop floor, 333, 335
Shotcrete, 157, 163, 302
Shotcreting, 156, 163, 301
SHP, 13–14, 16–46
Shrinkage, 208, 231, 236, 244, 283, 299
Shuttering making area, 131
Side slope, 152, 159, 193–194
Silicone, 216–217, 291
Silt Pressure, 70–71, 77, 159
Silting Basin, 34, 67, 191
Single line curtain, 214
Sink holes, 220–221, 226–227
Site mobilization, 125–150
Size, 99, 139, 252
Slake durability index, 192
Slator core, 394
Sleeve, 175–176, 280–281
Sliding Resistance, 78

Slip Test, 176
Sluice gate, 49, 55, 66, 99, 154–155
Sluice, 55–56, 58, 66, 153, 155, 263, 311
Socioeconomic effect, 355
Soil investigation, 130
SOR, 33, 84, 88, 90–91, 198, 305, 310
Soundness of aggregate, 137, 238
Spacing, 67, 142–144, 146, 150, 170, 224,
    264–265, 284, 287, 309
Spacing of holes, 142–143, 146, 224, 299
Spear, 393
Spear head, 392
Specific energy, 387–388
Specific gravity, 136–137, 192, 210, 240,
    245, 252
Specified lift thickness, 261, 263
Speed, 359, 395, 399, 406, 412
Speleothems, 227
Spillway, 19, 43, 54–55, 58, 62, 66, 73, 79,
    261, 263
Spiral case, 388–389
Splicing, 174–178, 292
SPT, 193
Stability, 17, 49, 67, 69, 73, 79, 159, 184, 200,
    232–233, 261, 285, 290, 306–307
Stalactite, 226–227
Standard American beam, 330
Standard deviation, 109–110, 242–243,
    247–248, 253
Static Tensile Test, 176
Station battery, 118, 385
Stator, 378, 404, 406, 408–409
Stay ring, 388–389
Stilling basin, 58, 60, 62, 263
Stirrup making, 179–180
Stone quarry, 135–137, 141–150
Storage bin, 138
Storage for scraps, 131
Storage project, 49–52
Straightening, 334, 342
Strategic Environmental Assessment,
    346–347
Strike, 3, 79, 189, 196, 393
Structural Concrete, 260, 263, 305, 307
Structural cost, 397–398
Structural features, 185
Sub surface mapping, 189–190
Submergence, 368–369
Sulphates, 238, 250, 297
Sun glass, 188

Supplementary estimate, 33
Surface cleaning, 334, 338–339
Surface geological mapping, 187–188
Surface geotechnical mapping, 187
Surge shaft, 5, 8, 20, 49, 54–55, 57, 339–340, 345, 356, 372
Survey equipment, 132
Survey for locations, 135
Surveying, 134–135, 305
Suspension of contract, 96
Sustainable development, 1–2, 9, 347
Swelling, 192, 297
Swelling pressure, 292–293
Switch gear, 44, 118, 356
Switch Yard, 25–26, 28, 30, 118, 122, 373
Switching equipment, 385
Synchronization, 123, 412
Synthetic rubber, 291
System of distribution, 14

T

Tail race channel, 9, 21, 57
Tail race tunnel, 21, 47, 345, 354
Tail water load, 70, 73
Tailrace, 5, 8, 23, 43, 54, 387, 390–391
Target strength, 240, 243
Technical bid, 87, 92–93, 97–98
Technical specification, 92–93, 169, 255, 258, 394
Tectonic features, 186
Tectonic maps, 186
Tehri hydro complex, 368
Tehri hydropower, 368–369
Temperature, 211, 269–274, 402
Temperature gradient, 70, 77, 271–272, 274–275, 284, 296
Temperature log, 221
Tender, 84–86, 87–89, 91–92, 97
Tensile stresses, 232, 267–268, 283, 297
Termination of contract, 84, 96
Terms of Reference, 348, 355–357
Tertiary holes, 200, 207, 214
Tetra calcium alumino ferrite, 269
Thermal behavior, 232, 234, 239, 244, 246, 268, 285
Thermal crack, 272, 296–297
Thermal cracking, 268–269, 272, 275–276
Thermal phenomena, 295–296, 308
Thermister, 274

Thermo-couplings, 409
Thermometer, 274
Thermoplastic, 291
Threaded, 175
Throw, 146, 216, 268
Time Line, 26, 88, 99
Time, 1, 4, 6, 9, 11, 25, 36, 38, 102–103, 109–110, 117, 128, 150, 160, 274
Time-Cost Trade off, 110–111
Tolerances, 125, 307, 333
Tool Box, 318–319
Tool Box Talk, 318–319
Topographic base maps, 189
Topographical Maps, 186, 348
Torque, 175, 201, 388, 390, 393, 404, 407–408
Top width, 153–154, 159
Total station, 132, 193
Tracer testing, 221
Tracer tests, 222
Training wall, 57, 67, 263, 290
Transformer, 26, 30, 39, 118, 373–374, 384
Transit mixer, 31, 154–155, 158, 257
Transverse construction, 282
Transverse construction joint, 282
Transverse contraction joint, 70, 284
Travelling steel form, 158
Treatment of foundation, 184, 197
Trench, 19, 187, 190, 194, 219, 220
Trial Mix, 241, 246, 255
Tricalcium aluminates, 269
Tricalcium silicate, 269
Trogloxenes, 227
Tungsten carbide, 190–191
Tunnel, 3, 18, 34, 47, 49, 54, 57, 152, 157, 163, 191, 213, 281–282, 354, 360, 368, 372
Tunnel or open channel, 20
Turbine, 15, 24, 30, 385, 387–401, 412
Turnkey contract, 88–89
Turning radius, 126
Type of facility, 9, 14
Type of WBS, 100

U

UHSC, 231
Ultrasonic pulse velocity, 301–302
Ultrasonic pulse velocity test, 301–302
Unaccountable, 168–169

Unconfined Compressive Test, 192
Undisturbed, 194
Undisturbed Sample, 191, 193–194
UNEP, 346
United Nation Environmental
    Program, 346
Universal Channel, 330
Uplift Pressure, 22, 70, 74–75, 159, 204
Upstream, 6, 54, 56–57, 70–71, 77, 79, 135,
    156, 183, 199, 213, 218–219

V

V groove(s), 286, 299
Vacuum Breaker, 388–389
Vedose water, 226
Velocity ratio, 400–401, 404
Vendor, 26, 84–85, 87, 167, 400
Venes, 388
Ventilation, 385, 394, 408–409
Ventilation system, 394
Vertical acceleration, 70, 81
Vertical drilling, 192
Vertical hole, 157, 201, 212
Vibratory roller, 160–161
Viscosity, 208–209, 212, 299, 340
Volumetric changes, 267–268, 282–283,
    295–296
Vulnerable, 3, 57, 69, 77, 185, 195, 277,
    322, 360

W

Wagon drill, 133, 155, 158, 201
Wagon drilling machine, 146, 160
Wagon drilling, 146, 160
Washing hole, 99, 202–203
Wastage, 51, 147–148, 165, 168–169
Waste Management Plan, 360
Water, 47–48, 50, 52, 54–55, 57, 60, 62, 66,
    74, 77, 151, 153, 183, 269, 286,
    395, 412
Water-Cement Ratio, 207, 210, 233, 238,
    243–244, 249–250, 253, 255, 265
Water conducting system, 5, 16–17, 22,
    25, 29, 34, 54, 281, 353–354, 384,
    386, 395
Water content, 233–234, 240, 243–244,
    249, 253, 296

Water cooled slator, 394
Water flows, 4, 23–24, 60, 221, 354, 384,
    388, 392
Water Intake Structure, 355
Water percolation test of foundation, 313
Water permeation, 290
Water retaining structure, 56, 68, 71,
    117–120, 122, 197, 286, 298
Water stop, 282, 286, 292–293
Water supply, 21, 24, 52, 68, 127, 373, 377
Wave pressure, 70–71, 77
WBS, 99–100, 113
WBS dictionary, 101
Weeds, 354
Weekly Routine, 140
Weighing hopper, 138, 140
Weighted average head, 387
Weir, 1, 11, 19, 43, 55
Weldability, 336
Welded, 176, 336, 338, 408
Welder qualification test, 377
Welding, 176, 329–330, 332–333, 335–338,
    343, 407–408
Welding machine, 134, 155, 336, 342
Welding method, 176
Welding procedure, 176, 335–338, 343, 407
Wet-Mix process, 163
Wicket Gate, 388–389
Wide flange steel beam, 330
Width of the fault, 216
Wildlife Protection Act, 356, 365
Winding, 378, 404, 408–409
Wire brush, 279
Wire mesh, 301–302
Work breakdown structure, 26, 90, 99–100
Work place safety, 102, 316
Working capital, 34, 37–38, 379–380
Working point, 338
WPS, 336–337, 343
WP-Witness Point, 310, 338
WQPT, 336, 343

Y

Yield stress, 166, 208, 211, 330, 335, 341

Z

Zero Date, 16, 85–86, 93, 95, 128

T - #0594 - 071024 - C0 - 234/156/20 - PB - 9780367670542 - Gloss Lamination